An open letter "TO SELECTED ACADEMICS # 5" Part 2

ISBN-13: 978-1533473844

ISBN-10: 1533473846

http://www.titius-bode-law-explain.co.za/index.html

Author Peet (P.S.J.) Schutte

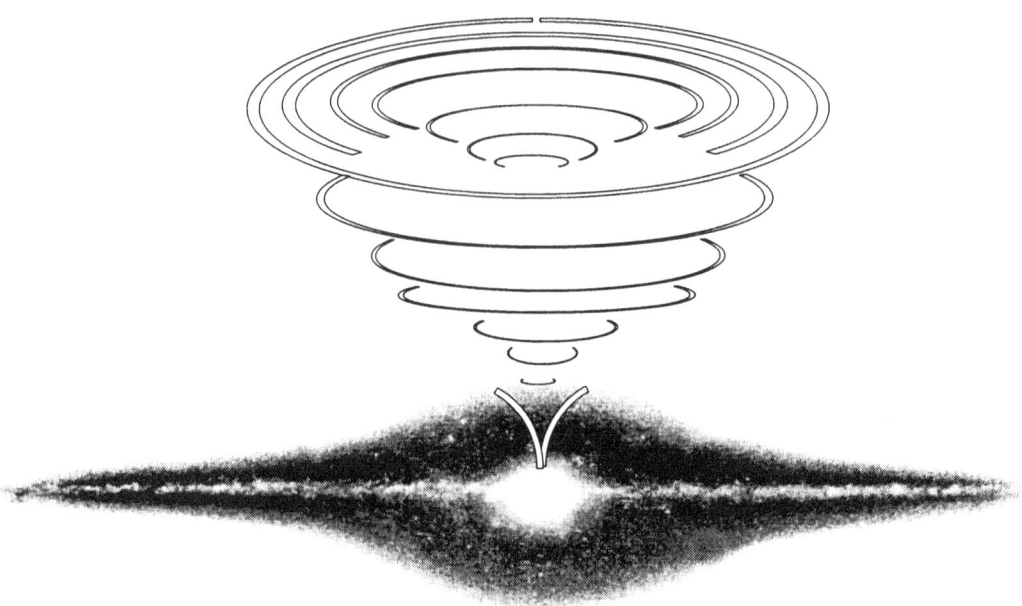

Even in galactica where they are the enormous size they are, they still generate by measure of atoms spinning in future stars that is in a cocoon blanket of heat or liquid time. Still notwithstanding size, the atoms form a unit where the rotation of such a unit generates a singularity governing in the centre with such intensity it forms the spiral we associate with Black Holes.

Let us investigate and try to find a way by using logic how a star applies gravity. Therefore it is not the number of dots that is important. It is not the size of the number of dots occupying the position or the size of the space the dots occupy that is prominent. It is the relation in the dismissing of space and the duplicating of space that becomes important. The less space there is the more the favour will be to reduce the space because of the advantage the dots have in securing space-time that will prevent overheating. On the other hand the more space secured will also prevent overheating and therefore those will opt to duplicate space in order to find space to secure and prevent overheating.

Since the Earth has no singularity demand that is much better developed than the universe sustains, we find on Earth a relevancy of Π to $(\Pi^2+\Pi^2)(\Pi^2\Pi)3$ is adequate. But in bigger units the space-time displacing relating to space duplication presents much more demands on atomic structures occupying space within the star containing through set boundaries. In the presumed to be bigger stars there is much space filled with atoms occupying much space. In the stars more massive but holding lesser space the atoms must also hold lesser space but they also hold more protons by number in the lesser space.

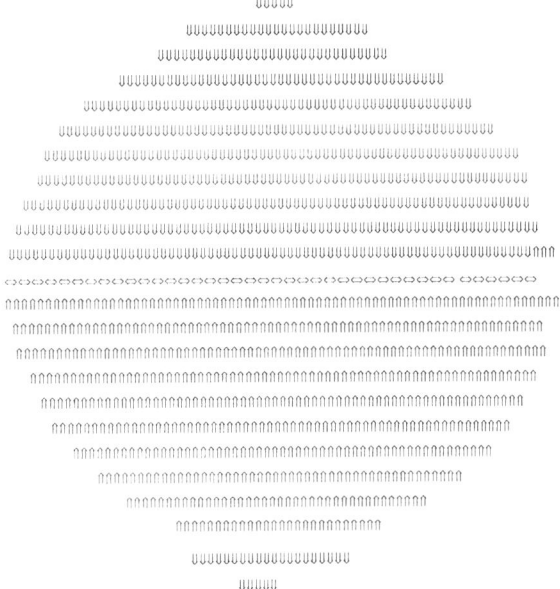

I would suggest we think of stars in the following terms. A star that generates and transmits a lot of light is weak on gravity because their progress started recently. They command a lot of space-time but the demand they have to keep their cooling acceptable is very low. In that they can generate a lot of light but with the demand on cooling low and the gravity in the centre not very developed, those stars cast a lot of light back into outer space. It is just because of the size the stars hold that tell the that the stars are still young and have a weak developed governing singularity. The stars will have very prominent hydrogen and helium layers, with the inner core not very prominent. The control of the star is still very much in the individual atoms and in that the motion the atoms have to produce in order to maintain their individual singularity will only come about through motion. The atom has to make contact with as much space-time through motion as possible since it has a very poor ability in contracting space –time in support of the cooling system.

The entire motion and the entire contraction of every atom culminates as one effort and this produces a single combining effort which is then displaced to the center of the sphere of the star where singularity is normally nurtured as a result of the shape of the sphere

```
                    ⋔⋔⋔⋔⋔⋔⋔⋔⋔⋔⋔⋔⋔
                 ⋔⋔⋔⋔⋔⋔⋔⋔⋔⋔⋔⋔⋔⋔⋔⋔⋔⋔⋔⋔⋔
              ⋔⋔⋔⋔⋔⋔⋔⋔⋔⋔⋔⋔⋔⋔⋔⋔⋔⋔⋔⋔⋔⋔⋔⋔⋔⋔⋔
            ⋔⋔⋔⋔⋔⋔⋔⋔⋔⋔⋔⋔⋔⋔⋔⋔⋔⋔⋔⋔⋔⋔⋔⋔⋔⋔⋔⋔⋔⋔⋔
          ⋔⋔⋔⋔⋔⋔⋔⋔⋔⋔⋔⋔⋔⋔⋔⋔⋔⋔⋔⋔⋔⋔⋔⋔⋔⋔⋔⋔⋔⋔⋔⋔⋔⋔
         ⋔⋔⋔⋔⋔⋔⋔⋔⋔⋔⋔⋔⋔⋔⋔⋔⋔⋔⋔⋔⋔⋔⋔⋔⋔⋔⋔⋔⋔⋔⋔⋔⋔⋔⋔⋔
       ⋔⋔⋔⋔⋔⋔⋔⋔⋔⋔⋔⋔⋔⋔⋔⋔⋔⋔⋔⋔⋔⋔⋔⋔⋔⋔⋔⋔⋔⋔⋔⋔⋔⋔⋔⋔⋔⋔
      ⋔⋔⋔⋔⋔⋔⋔⋔⋔⋔⋔⋔⋔⋔⋔⋔⋔⋔⋔⋔⋔⋔⋔⋔⋔⋔⋔⋔⋔⋔⋔⋔⋔⋔⋔⋔⋔⋔⋔
     ⋔⋔⋔⋔⋔⋔⋔⋔⋔⋔⋔⋔⋔⋔⋔⋔⋔⋔⋔⋔⋔⋔⋔⋔⋔⋔⋔⋔⋔⋔⋔⋔⋔⋔⋔⋔⋔⋔⋔⋔
    ⋔⋔⋔⋔⋔⋔⋔⋔⋔⋔⋔⋔⋔⋔⋔⋔⋔⋔⋔⋔⋔⋔⋔⋔⋔⋔⋔⋔⋔⋔⋔⋔⋔⋔⋔⋔⋔⋔⋔⋔⋔
    ⋔⋔⋔⋔⋔⋔⋔⋔⋔⋔⋔⋔⋔⋔⋔⋔⋔⋔⋔⋔⋔⋔⋔⋔⋔⋔⋔⋔⋔⋔⋔⋔⋔⋔⋔⋔⋔⋔⋔⋔⋔
   ⋔⋔⋔⋔⋔⋔⋔⋔⋔⋔⋔⋔⋔⋔⋔⋔⋔⋔⋔⋔⋔⋔⋔⋔⋔⋔⋔⋔⋔⋔⋔⋔⋔⋔⋔⋔⋔⋔⋔⋔⋔⋔
   ⋔⋔⋔⋔⋔⋔⋔⋔⋔⋔⋔⋔⋔⋔⋔⋔⋔⋔⋔⋔⋔⋔⋔⋔⋔⋔⋔⋔⋔⋔⋔⋔⋔⋔⋔⋔⋔⋔⋔⋔⋔⋔
  ⋔⋔⋔⋔⋔⋔⋔⋔⋔⋔⋔⋔⋔⋔⋔⋔⋔⋔⋔⋔⋔⋔⋔⋔⋔⋔⋔⋔⋔⋔⋔⋔⋔⋔⋔⋔⋔⋔⋔⋔⋔⋔⋔
  ⋔⋔⋔⋔⋔⋔⋔⋔⋔⋔⋔⋔⋔⋔⋔⋔⋔⋔⋔⋔⋔⋔⋔⋔⋔⋔⋔⋔⋔⋔⋔⋔⋔⋔⋔⋔⋔⋔⋔⋔⋔⋔⋔
  ◇◇◇◇◇◇◇◇◇◇◇◇◇◇◇◇◇◇◇◇◇◇◇◇◇◇◇◇◇◇◇◇◇◇◇◇◇◇ ◇◇
  ℧℧℧℧℧℧℧℧℧℧℧℧℧℧℧℧℧℧℧℧℧℧℧℧℧℧℧℧℧℧℧℧℧℧ ℧℧℧℧℧℧℧℧℧℧℧℧℧℧℧℧℧℧℧
  ℧℧℧℧℧℧℧℧℧℧℧℧℧℧℧℧℧℧℧℧℧℧℧℧℧℧℧℧℧℧℧℧℧℧℧℧℧℧℧℧℧℧℧℧℧℧℧℧℧℧℧℧℧
   ℧℧℧℧℧℧℧℧℧℧℧℧℧℧℧℧℧℧℧℧℧℧℧℧℧℧℧℧℧℧℧℧℧℧℧℧℧℧℧℧℧℧℧℧℧℧℧℧℧℧℧℧
    ℧℧℧℧℧℧℧℧℧℧℧℧℧℧℧℧℧℧℧℧℧℧℧℧℧℧℧℧℧℧℧℧℧℧℧℧℧℧℧℧℧℧℧℧℧℧℧℧℧℧
     ℧℧℧℧℧℧℧℧℧℧℧℧℧℧℧℧℧℧℧℧℧℧℧℧℧℧℧℧℧℧℧℧℧℧℧℧℧℧℧℧℧℧℧℧℧℧℧℧
      ℧℧℧℧℧℧℧℧℧℧℧℧℧℧℧℧℧℧℧℧℧℧℧℧℧℧℧℧℧℧℧℧℧℧℧℧℧℧℧℧℧℧℧℧℧℧
       ℧℧℧℧℧℧℧℧℧℧℧℧℧℧℧℧℧℧℧℧℧℧℧℧℧℧℧℧℧℧℧℧℧℧℧℧℧℧℧℧℧℧℧℧
        ℧℧℧℧℧℧℧℧℧℧℧℧℧℧℧℧℧℧℧℧℧℧℧℧℧℧℧℧℧℧℧℧℧℧℧℧℧℧℧℧℧℧
          ℧℧℧℧℧℧℧℧℧℧℧℧℧℧℧℧℧℧℧℧℧℧℧℧℧℧℧℧℧℧℧℧℧℧℧℧℧℧
            ℧℧℧℧℧℧℧℧℧℧℧℧℧℧℧℧℧℧℧℧℧℧℧℧℧℧℧℧℧℧℧℧℧℧
              ℧℧℧℧℧℧℧℧℧℧℧℧℧℧℧℧℧℧℧℧℧℧℧℧℧℧℧℧℧℧
                 ℧℧℧℧℧℧℧℧℧℧℧℧℧℧℧℧℧℧℧℧℧℧℧℧
                    ℧℧℧℧℧℧℧℧℧℧℧℧℧
```

The contracting action is at present the only part of gravity that Newtonian science credit as gravity. There is a lot more to gravity than such simplicity. Every atom in a star is pushing the atom in front by filling the space the atom in front vacated. Every atom in front of every atom behind is pulling the atom behind as the atom behind is urged to fill the space that the atom in front vacated. That is motion, which is the most complex issue one can find in the Universe. Since every atom is driven by singularity and no singularity are able to move it bring about that every singularity must remove and rebuild the space every atom fills or vacate as the atom moves along.

There is a building of an entire Universe going on in every split second and this split second is so fast we cannot name it. By naming it there will be so many time units gone by, by the time we said the name, the Universe might not even be recognisable. We might call it energy but I hate to call it energy because energy is a lot like Holy water. It can come from anywhere and you can use it for everything and in the end it does not even become something durable because its use eventually comes to nothing.
The atom restricts dismissing of space by the containing structure to the atoms relevancy being Π^0 in singularity bringing on Π relating to $(\Pi^2+\Pi^2)(\Pi^2\Pi)3$.
As the layers swap there aligns between duplicating and dismissing the atomic relevancy adapt to comply

Since the star performs as an accumulated atom where innumerable atoms inside the confinement of the star combine to select one centre spot forming singularity that represents the star, I have chosen the to use the same symbols that I found in atoms to describe the relations in space –time to singularity within the space-time of the star. I refer to a star as a cosmic atom in other books.

Early stars still in the envelope of heat within the centre of the Galactica have only space duplication and growth through the cover of such enormous heat. These class stars are not visible but are shrouded in a blanket of heat covered by light. The atoms forming the stars are small and under developed. They remain cool because they contrast with the heat surrounding the star where the star material supports the cool space and does not form part of the liquid heat forming the outer limit. I would like to draw your attention once again to the fact that the Sun at one stage was a cool $18 \times 10^{6\ 0}$ on the inside and a freezing cold at $6500^{\ 0}$ on the outside while all the time outer space was a blistering $10^{34\ 0}$. This was considered the coldest place in the Universe because the Sun was still part of the deep frozen space inside the blanket of heat. Look at any galactica and see in the centre there are stars surrounded by a blanket of heat with stars conversed by heat sitting like a duck frozen in this pond of liquid heat.

With the cosmos the size it is and space so large compared to our smallness we have no chance in finding the centre of the Universe. The Universe started where singularity is and singularity is the sure indicator of the Universe. With all spinning objects holding singularity we then have located singularity in as much as finding the centre of the Universe. The Universe started with a dot forming. That answer arrive from taking mathematics back to a point of being the smallest possible position, far smaller than we may be able to calculate form. The ten dimensions I named the atomic relevancy is also showing the double value of singularity as singularity extends into as well as beyond space. The atomic relevancy is $(\Pi^2+\Pi^2)(\Pi^2 \times \Pi \times 3) = 1836$ that is the mass relation between the electron (3) and the proton. Proton = $(\Pi^2+\Pi^2)$ Neutron $=\Pi^2\Pi$. The atomic relevancy holds the dynamics of singularity control. In the ratio and dimensions we find in the atom, all space-time derives from the atom, whatever the atom is. Our instincts, our logic and our calculating process all indicate that the sphere holds a centre point from where six evenly positioned point's position matter to be. Using The formula $F=G (M_1.m_2)/ r^2$ it indicates to a force pulling objects closer, where each force is coming from each centre point the body in question has. The contraction must commit the two bodies towards a point in each case being spot on in the middle, not withstanding what direction the force is applying, the body will draw to the centre. If the Universe spins around a centre point holding singularity, and singularity confirms the centre of the Universe, then every particle holds the centre of the Universe making the number of universal centres immeasurable many, and every atom and sub atom particle presented outside the atom in smaller bits, are all not pieces of the Universe but they are a Universe surrounded by many Universes. If every atomic particle no matter how small is holding the centre of the Universe, then the gravity is coming about from that point because that is where the gravity applying in the Universe is applying contraction. If the Universe did start from one single point and time, matter and space flowed from that point, then that point must have a relative connecting base because such a point holding singularity must be eternal as space, matter and time link eternal. There therefore must be one point linking the entire Universe when regarding the fact of singularity. Then according to the theory off relativity there has to be one exact point holding time in relevance notwithstanding the fact that time departs from that position and relate differently to all space-time away from such a point.

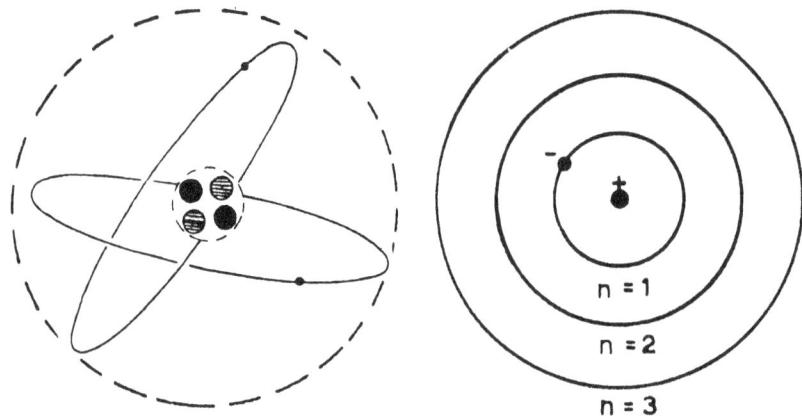

In the final analysis it is the atom that control the Universe because it is the atom that is the Universe.

It then is the atom in the most centre part where space and time meets singularity, that Einstein found a Universe collapsing to a single dimension, and every atom at a point post of the proton where gravity initiates in according with the proton dimensional colas of $(\Pi^2+\Pi^2)(\Pi^2 \times \Pi \times 3) = 1836$

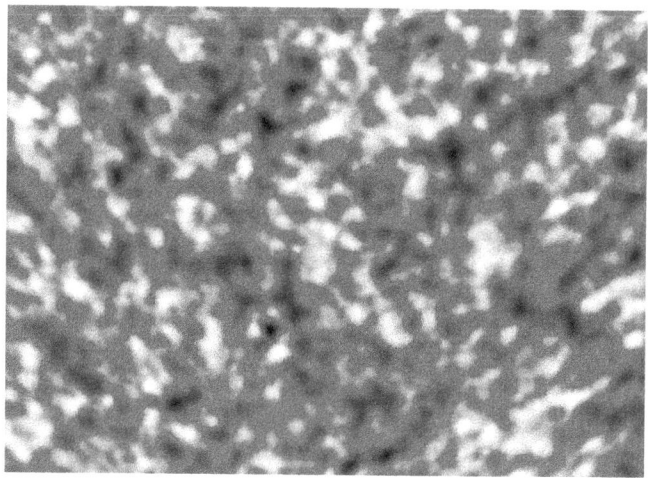

Every person with whom I have discussed the facts concerning creation recollects images in the trend depicted in a presentation as one may find to the above. That would be the most unlikely way Creation came in place. The recalling of pictures representing images about creation must have form, but to mathematics it had no form. From this thought the very opposite arises where Creation came from nothing but such an idea is mathematically simply not possible. The thought of nothing is just what it is, a thought of nothing and although it is in the nature of the human mind, to present nothing as a value in the recalling of something, nothing is a presentation of the figment in the human mind. There can be no number such as nothing and that was (possibly) Newton's biggest error. Nothing represents non-existing and that is just what nothing is, it is non-existing. In order to prove my point I wish to ask the reader to define the shortest line there can theoretically be. If he should answer anything but that the shortest line will be at a point where the beginning and is the very same spot he will be wrong. The shortest line that can ever be anywhere must have a start and finish holding the exact same spot. The line will be humanly impossible to create but we humans are capable of very little.

When the line has a beginning and an end at the very same spot and it wishes to extend the position as to further the possibility it has, which direction should it favour. Humans in the west would naturally think of extending from left to right while in the east humans may want to go from right to left.

Some persons will tend to go up or down, but all of the options are about human preference and not mathematical conclusions. Extending the line in any one direction will favour one direction without a conclusion about not extending in other directions. Such a conclusion has no sound mathematical foundation. The only option about extending will be in all directions equally in order to give a meaningful non-bias flow of mathematical equilibrium

The shortest line in the realm of possibilities must have a start and finish holding one spot and such a line will also be a dot or a circle. Not favouring one direction puts all directions at equilibrium meaning that any form what ever may be can develop from such a spot with the end and the start being the same. This reasoning prompted me to look for singularity in such a spot because if the prime spot from which all came was a spot, then the spot must hold the shortest line but more prominent it will hold the smallest form including the smallest circle. One possibility that the shortest spot can never have is having a starting point on the zero mark. If the mark of zero holds the start it must also hold the end because the end and the beginning has the same position. If the position of zero then is the beginning, the end will also be zero leaving the line without an end as well as without a beginning. The conclusion from this is that no line can start at zero because that will be a mathematical impossibility. A line or spot starting at zero would therefore be shorter than the shortest line possible. A line growing or extending from zero can never leave zero because of the influence of being zero disqualifies any possibility of growth. If the line then had to grow in all directions at the same pace the line must therefore be a circle. The value of the circle is Π, and that is where creation started.

In the centre that holds the line the line is a generated notion that is so thin the line is generated by motion and still the line is not part of the cosmos, while it is supporting the entire cosmos by controlling the entire cosmos. While it is not there, there is no denying that it is there and the control it has over all of the entire cosmos goes beyond question. It is establishing all the dimensions by seven supporting ten.

Every object that spins also generates such a line through the spin. The purpose of the spin places coherency into the Universe to generate the control by command and every atom is the seven. The atom places the seven in relation to the ten and all other atoms in direct linking of the seven points form the ten, which the atom holds as motion or liquid. The line is the diversion of the four by three establishing a parting between infinity and eternity.

Singularity by Motion

Singularity by Time

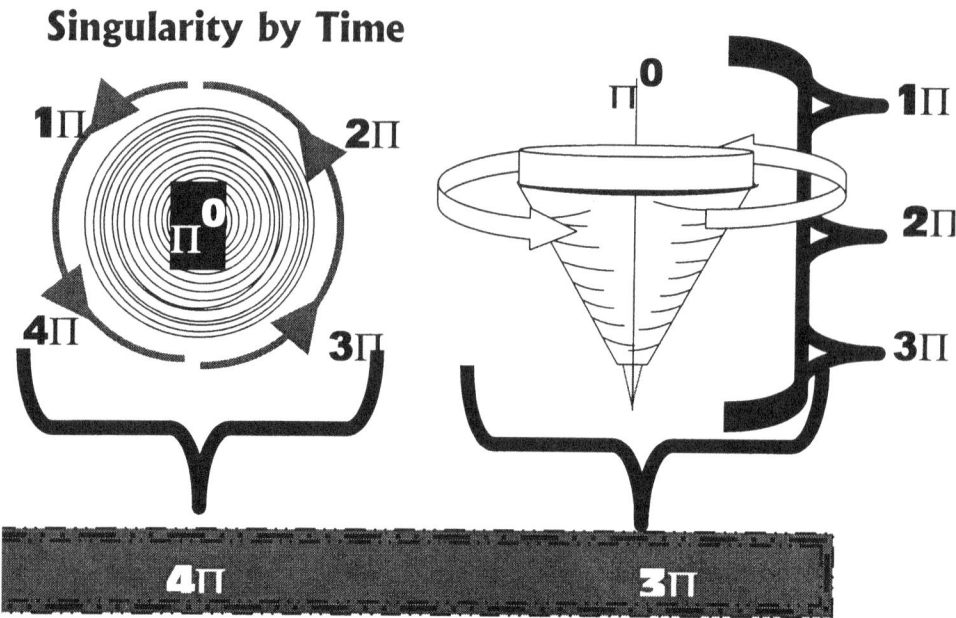

If the alignment is in ninety degrees to each other then Pythagoras has to apply strictly. Should my argument be sound and which it is sound we have to be able to use Pythagoras to determine the value of time in space. When the material rotates or moves the filling of the material is in perspective to the time. However, material can only be in one location in one split time. Since material has to cross over to the other side of the Universe in order to duplicate, which is how material moves, then the material, can be only on one side of the Universe,

Every time matter is generated and moves, it is singularity that is complying with it activating another point in singularity being charged with the motion. The Universe started from allocating singularity charged by heat into positions where such positions contributed to space- time. Every time the spot overheated the spot expanded into four dots and by expanding the spot cooled. In cooling the spot retained heat by which it spawned the dots allocated as time. In overheating objects expand and by expanding objects cool. That is gravity. Gravity is the expanding in relation with the cooling which means it is duplicating material in relation to a generated centre that is contracting the motion by cooling. Every inclination of motion is in fact motion and every movement be it contraction or expansion is moving to the other side of the Universe by bridging singularity because singularity is immovable. Therefore by being immovable, motion has to cross the division singularity applies and by crossing the division the factor that comes in place is $\Pi^2/4$, which results in the Roche limit. But such motion is three and the square of three in addition to the square of four brings about time in space.

As a school going youngster, I was fascinated by astronomy and in particular the cosmology aspect. In a long and strenuous process of self-education I was completely stunned by the behaviour pattern that the comet had in its relation as it orbits the Sun. Please forgive my boyish way of presenting the following but it is important that I bring it across as I saw it as a boy and as a matter of fact still see it today as a middle -aged adult.

Science acknowledges growth as the Hubble constant and then refuses to put the growth in line with the solar system. The growth they reluctantly admit too, they refuse to connect that growth to the solar system in any way. They take a Universal year as a solar year being that of one cycle it takes the Earth to rotate the Sun in the present day. Then they reflect on this as if this was going on since time began, because by doing that, there then is a nice crooked constant that fit mathematicians. Push this double standard applied back to before the Sun took its position and there was not Earth to indicate the year. How small was the year circle at that point in time and space. Take this right down to the:" Big Bang" where "the whole Universe were the size of a man's fist" (To use their words), how far did the circle goes to indicate a year then? The year was immeasurably smaller, shorter and faster than at present. This is logic even the Newtonians must accept. There is no space outside insanity to apply time to the past at the value it is at present and far worse, to use something so extremely insignificant as the Earth to measure it by.

Again I feel that the use of this type of constant just to fit mathematicians to corrupt the truth they in science are using such logic to rubbish the truth. There is just no rational in the time verses events that can explain facts without. Since the time of Newton, science has slowly nibbled at the truth to compensate for the game there is to play. It is as if the one crooked posture corrupts all in general. That is what Kepler's formula is all about? That is what Kepler indicated with his formula $a^3 = T^2 k$. The space of an object (a^3) is equal to the time (T^2), which it is in, in every given instant (k). If the space becomes smaller, the time duration becomes longer every instant of time's progress.

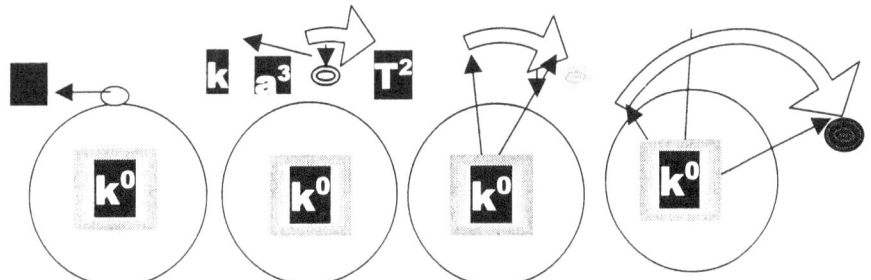

In the accumulating of heat, the object finds motion. The motion brings along structural independence and such independence puts distance between k^0 and a^3. The heat increase will accompany a larger T^2. By increasing heat, the distance between the objects will grow.

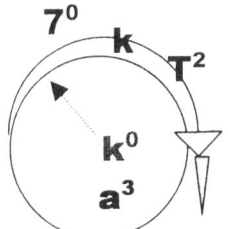

Only by creating a total independent heat centralised in a point holding singularity and feeding k^0 with an independent heat supply can an area a^3 establish a k that will release the independent a^3 from the secure larger k^0. The overall condition is that the escaping a^3 must establish a route following k^0 as k^0 places a diverting 7^0 where that 7^0 then forms part of the object creating heat to secure a release from the established a^3. Providing the heat will bring about a release placing a new object into outer space.

Time has been three since eternity started and time will remain three until eternity ends. When the heat came about eternity spawned the one in infinity that generated a line and then time became the three positions of past present and future all depending on the one line in infinity. In the triangle that time established in conjunction with the law of Pythagoras the three of time goes square that forms nine and when the one marker of infinity is added time by the square in space becomes ten.

The line however forms three and in conjunction with the four positions in space – time (three in eternity and one in infinity that is there eternally) there are four positions relating to the three in the line and from that lying between eternity and infinity is the four eternal position plus the three generated positions which forms the seven in space-time.

However because of dimensional duplication the square of time is ten and five will be on the one side of the Universe and five will be on the other side of the Universe. That then is why the Lagrangian system holds five positions in relation to singularity.

When the four in time spins off one more in infinity as time moves on a fifth spot becomes valid that erects a line by heating and that fifth spot then reverts to the first spot that again parts eternity from infinity. When it has spawned a fifth position that position also goes square and forms by the law of Pythagoras the Lagrangian fifth position.

This puts a huge question mark on the correctness of Newtonian presumptions that currently fondle the idea that rotation has no influence on the cosmos and all gravity goes down to mass where mass has all the influence and control.

All spinning matter has the point where the spin is still there but the radius is to small to measure by any means. That point is standing still in relation to the rest of the spin. In relation to that logic I do not except Newtonian science holding the radius of s spinning object unaccountable in the spin, whether the spin is applying or not.

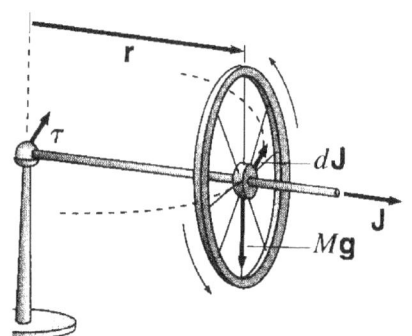

Applying Newton's second law F=ma
One arrive at the formula
$GMm / r^2 = m\,(\omega^2 r)$

By replacing $(\omega^2 r)$ with $2\Pi / T$ we obtain Kepler's third law

This law predicts that $\mathbf{T^2 = a^3}$

What this statement implies is that r does not exist. When anything has a value of zero it is for all purposes non-existent. Only when an object is following s straight line can the radius be non-existent because the radius alters value through time development.

Taking the argument back to Kepler's law,

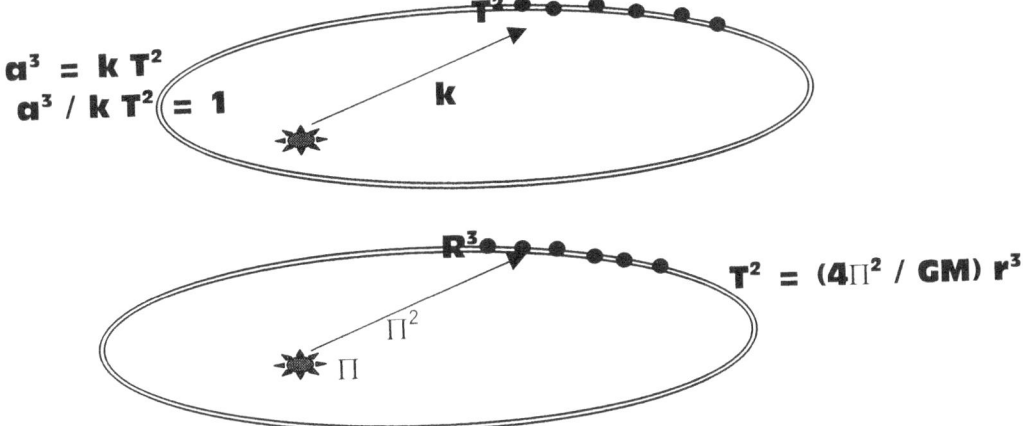

$$a^3 = k\,T^2$$
$$a^3 / k\,T^2 = 1$$

$$T^2 = (4\Pi^2 / GM)\,r^3$$

The spinning or not spinning is not part of the issue because at the point of absolute singularity the object never spins. Therefore spinning or not spinning does not apply to the point of singularity because singularity never spins in any event. In the whole structure with a pivotal centre as the control to the motion of the space the fact of Π is a natural outflow and any adding of Π is totally incorrect. According to Newton the result of spin is zero, however the top will tell a much different story.

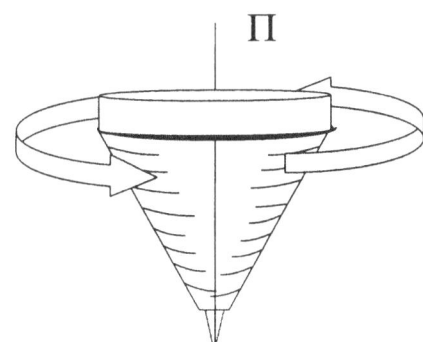

On the surface, at first glance the top is an ordinary piece of dead wood that is machined into a sloping shape. The top is normally fitted with a sharp needlepoint at the bottom and the sharper the point is the better will the spin balance be. It is obvious that the spinning of the top inspired the entire Universe into a reality that is not there while it is in control of the entirety we find as real as life itself

When translating Kepler's mathematical expression into a verbally spoken form of communication such as English we can see what Kepler said also read as $k = a^3/T^2$ where **k** is one point from a centre point that is space a^3 relating to time T^2. From a centre comes space-time

$$k = a^3/T^2$$
$$k^0 = a^3/T^2\,k$$
$$k^{-1} = T^2/a^3$$

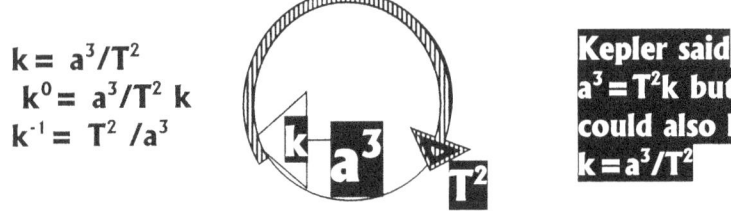

Kepler said $a^3 = T^2 k$ but that could also be $k = a^3/T^2$

Others like Newton and Einstein came much later and coined the phrases but Kepler formulated the concepts. They named Kepler's innovations. That is very clear but only on the condition that Kepler is read correctly and Newton gossip about what Kepler is saying is ignored. What Kepler said in mathematics all the brilliant Mathematicians through so many centuries were unable to read although the coded language was written in mathematics!

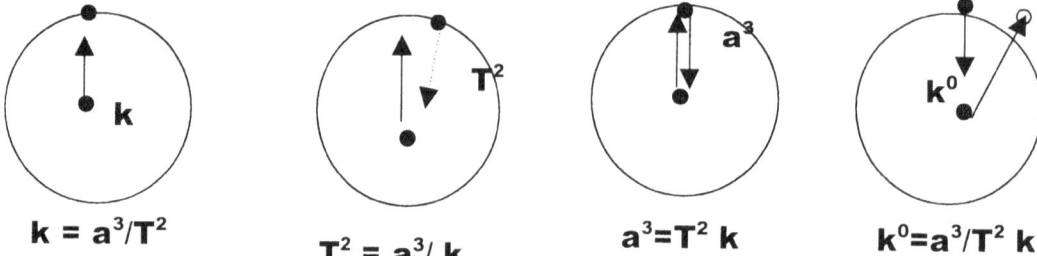

$$k = a^3/T^2$$

$$T^2 = a^3/k$$

$$a^3 = T^2 k$$

$$k^0 = a^3/T^2 k$$

But as one can see I also realised gravity is relations of motion applying in two factors. There is no separation of the two of the factors acting as one but both have different application and values in the unit. It was what gravity was because this action prevented expanding. This is the result of singularity having three parts acting as one but giving three distinctions in application.

Gravity is as much part of dismissing space as it is about making contact with space in time. Since the connection comes about as a circle, the connecting points will relate to Π as the value. Due to the spinning nature of such a point with all surrounding the point will be alternating direction favouring change every second and in that the value to such a point can only be Π because of its constant changing. Using r would specifically oppose another r from every angle because the use of r will bring about a static relation to the previous and following instant and therefore it will cancel the constant spin flow. By reducing the line to its maximum possibility one end with Π being the minimum but that Π is actually Π^0 which can also be k^0 or a^0 or T^0, which all indicate positions in singularity. Only when forming a value past singularity does independent identification come about. When the atom formed that atom applied a relevancy of ten positions where seven positions are included in the atom spinning and three positions are part the exterior of the atom spinning but all the positions relate to singularity but as space flight taught us such relevancies can change when an object is within the space boundaries of a larger structure or roaming free in outer space.

Within the boundaries of the atmosphere where the sphere border touches the space borders the space borders hold six positions and the sphere hold seven points. But at the precise place where the points make contact with the sides one side fall away in favour of the point it connects too leaving five sides relating to seven and where one of the six sides takes control in removing one of the cubical sides by replacing that side with a sphere point position the object then becomes directly controlled by singularity positioned in the centre of the sphere. The object seems then to fall from space and enter the atmosphere becoming a shooting star. What the Coanda effect proves above anything else is that gravity in control of space-time comes about from a centre and such a centre can be created by motion applying to a liquid in relation to a solid. That means there is undisputedly a flow of space-time towards a centre and the centre has to diminish the space-time reaching such a centre to create the flow and therefore the control from such a centre. That's the one pivot of gravity.

Since the Coanda effect shows gravity is control of space-time by motion flowing towards a centre that also prove as it explains the one part of gravity that reduces space by increasing time towards a centre that is established by motion and the lack of space establishes a lack of motion in that centre.

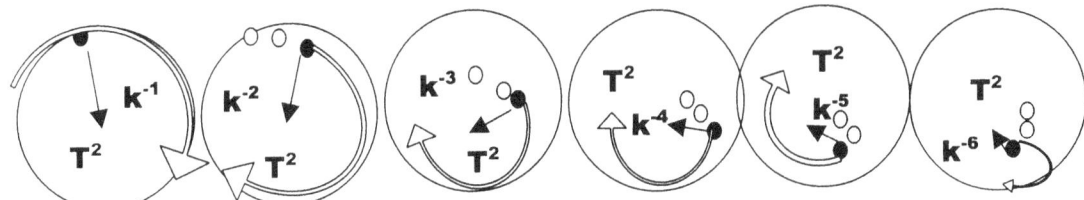

Because the smaller object holds much less space the duplication is in a lesser relation than the main object and because the time factor enforces the duplication period to match therefore something in the applying ratio has to give in to allow the major relevancies to remain in place. Since a^3 has to rematch to apply to the conditions set by the larger object a new relevancy comes about where the new a^3 will bring along a reducing T^2 with the diminished k that the Earth enforces. Since the space that motion reproduce is smaller in relation to the Earth, but the earth enforces the same time value, the relevancy of the time value will deplete by reducing k, but not in a straight line because all factor changes will then only be carried by one factor. I this way the diminishing space produced help the cyclic time factor to decrease with the distance that grows smaller.

When the object is released from the atmosphere of the dominating space, this very same gravity $k^0 = k\,T^2 / a^3$ ratio will still be enforced since it is not the law of the Earth prevailing but it is the law of the Universe applying. Outside the atmospheric borders the Earth no longer have the means to remove one of the cube sides that form the lesser object space and where the cube reinforces position by keeping the rotating object in position floating above the Earth.

The space became too small to allow the time it takes to enter because the distance k decreased faster than the space a^3 could compromise with the time T^2 changing from what is present in outer space comparing that to the time in to atmospheric space. With this information being in hand for a period of four hundred years, one should think that the wise could derive a conclusion. Where the information forms the basis of modern cosmology since the information formulated gravity and not merely produced a name for gravity as our English friend did, it is amazing that such accidents can happen and it is more amazing that no one in Mainstream physics has the slightest idea why this is taking place!

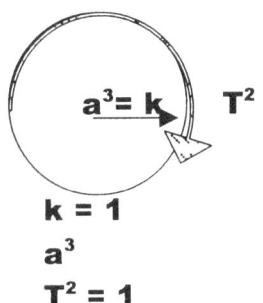

Our most impressive astronautic engineers are assembling a machine that will scramble the ratio Kepler introduced to a level in outer space where the ratio will be more than what the ratio in the Sun is. Surprisingly they are not in the least surprised that not one object in outer space is using an excessive velocity.

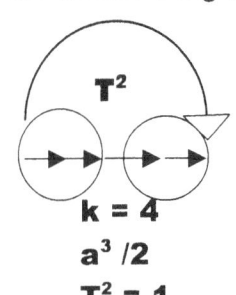

In realistic physics it means double the space will fill in half the time. We know that that is not possible because it can only bring about half the space in double the time or twice the distance in half the time. Space time and distance is a mesh where the lot integrate because Kepler said so. Kepler said the space forming space is the same space forming the distance of the space and that is the same space taking the time to fill the space. If the ratio changes then changes come about the entire ratio. In order to bring about such acceleration much more heat has to be released to gas in order to find such a drive that will sustain such a high velocity. The drive can only be the result of massive quantities of heat being stored around the singularity the atoms generate. The way the cosmic has designed the fight to relieve overheating is by motion. By duplicating the overheating space through motion the duplicating reduces the heat by half

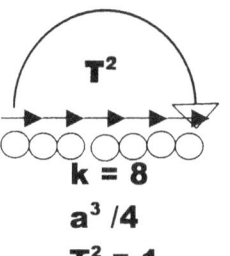

T^2

k = 8
$a^3/4$
$T^2 = 1$

because the heat is spread over half the area that is distributed in double the space. By increasing the relevancy **k** it reduces the area or space by quantifying the number of spaces per time unit in the time from that apply in the atmosphere or outer space.

k = 4 and $a^3/2$ **if** T^2 remains the same but that will not happen and that we know from past experiences. If that happens, we challenger 2004 disaster repeating once

have   the
more.

T^2

k = 16
$a^3/8$
$T^2 = 1$

Increasing space-space by six and centre by twelve.

time displacement by six will decrease the distance the space progresses from a centre. The heat factor of the craft will rise by twelve times as the space decreases by six times.

Increasing space-time displacement by twelve will decrease space by twelve and the distance the space progresses from a centre by twenty-four. The heat factor of the craft will rise by twenty four times as the space decreases by twelve times. Motion of anything in any form is about duplicating the existing into following on images of the same thing. That is connecting space to last a certain period in relation to a specific point holding singularity before the next singularity is enticed or charged to maintain the space-time in motion. Every time (and in this case the referring to time proves to be most accurate) is having another singularity building and breaking down the space it represents for that duration of time. The time duration leaves singularity selected in charge of producing the roving space the extent in which it can duplicate the space it has to duplicate. By reducing the period the particular singularity may lay claim to the space, will inadvertently produce smaller space it is able to reproduce in the shorter period of time.

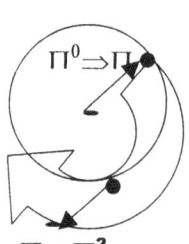

$\Pi^0 \Rightarrow \Pi$

$\Pi \Rightarrow \Pi^2$

There are two ways of looking at this issue. The one is looking at it from the centre that is keeping the rotating object honest or there is the rotating object forming space in relation to the centre and placing the centre in the centre. It will always be one taking prominence to the other and where Kepler introduced the formula it is indicating motion producing gravity which is gravity that is keeping form outside the sphere. Gravity is motion but the motion we see is much different from the gravity we experience while we know it has to be the same with only relevancies changing.

No matter how one looks at the Kepler formula, it signals the same principle. It shows how motion erects the Universe by mathematical equations. It puts singularity, as one in relation to six and that is the Universe decoded.

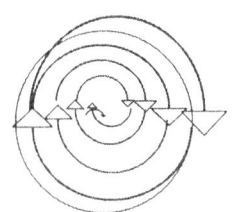

By rotating around a centre that is, standing still such a centre forms a divide that separates the unified unit. **Any point will be opposing itself** within the **rotating of 180°** where it **then changes every aspect** of its **previous flowing** characteristics it had or will once **again have in 360°** from there. While in rotation from the viewpoint of a bystander it all may seem static and never changing. However to the object in spin every next instant in time will be diverting from every aspect it had every second passing, and the direction it held in relation to the direction it held the previous mille-, mille-second as it will totally be incompatible with the direction it holds the very next mille, mille second of rotation. This is why we can use degrees measuring the circle by (6^2) (forming the square relating to matter through singularity) X 10 (square if space) = $360°$ however it is always in motion.

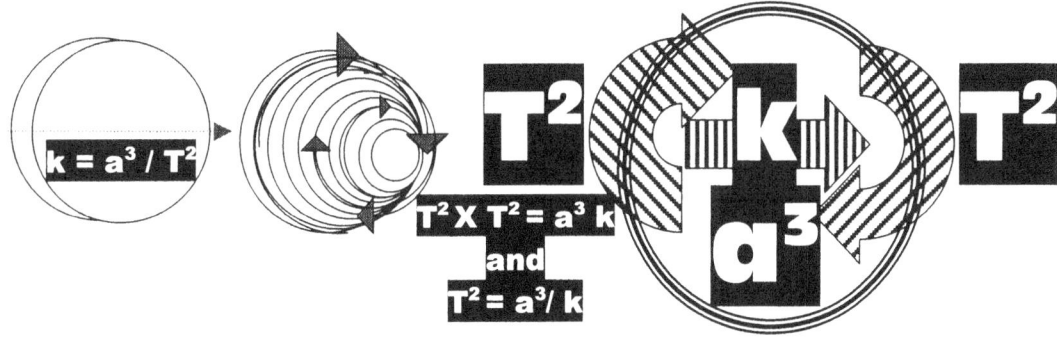

$$k = a^3 / T^2$$
$$T^2 X T^2 = a^3 k$$
and
$$T^2 = a^3 / k$$

The square of motion T2 X T2 forms the square of space a^3 X k . Space a^3 is reducing by the motion of space with the implementing of T^2 having k as the constant. It comes about as the earth spins around the Earth axis. I call this positive space-time displacement

$$k = k^{3-2} = k^1$$
$$a^3 = a^{2+1} = a^3$$
$$T^2 = T^{3-1} = 2$$

$$k = a^3 / T^2$$
$$k = a^{3-2} (T^2)$$
$$k = a^{3-2} = k^1$$
$$k = k^{3-2} = k^1$$

is the same as

If space were zero or nothing as Mainstream science so affectively teaches us, then Kepler's principle formula would need the changes Newton brought about. It is true and stands tested like no other research ever coming either before or after Brae and Kepler's work. By reducing the line to infinity and raising the line again back in the direction of space, the line would erupt as a natural sphere having Π as the natural basic value. That is the value Kepler interpreted. However not realising what he saw he chose to use different symbols.

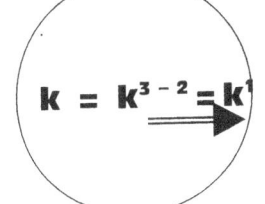

$$k = k^{3-2} = k^1$$

$$a^3 = T^2 k$$
$$a^3 = T^2 k^1$$
$$a^3 = T^{2+1} (k^1)$$
$$a^3 = a^{2+1} = a^3$$

is the same as

$$T^2 = a^3 / k$$
$$T^2 = a^3 / k^1$$
$$T^2 = a^{3-1} = T^2$$
$$T^2 = T^{3-1=2}$$

It is all the same

$k = k^{3-2} = k^1$ is in direct relation to $a^3 = a^{2+1}$ is in direct relation to $a^3 = T^2 = T^{3-1=2}$. With this information staring mainstream science in the face and scream pleading at them to recognise the information they turn around and ask why can man not fly off to other galactica at the speed of light. When the astronaut is departing from space on Earth or filling Earth space it will take the departing astronaut k^2 time to reach k^1 and fill out k^3. At present and in this moment our most impressive astronautic engineers will devise an engine that would cut k^1 by say half. This achievement will come as they increase the power output say for argument sake to double what it is at present. There was no friction of particles destroying the frame of the craft because there are not enough particles in space to do it.

However Newton recognised just the opposite and even allowed a freezing of motion and therefore time.

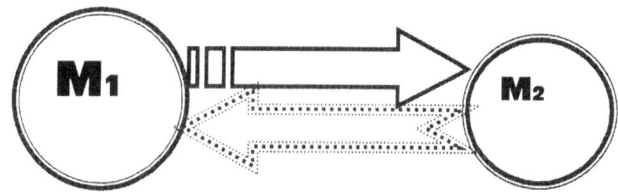

How does one reconcile the behaviour of the top with the foundation of science?

Mass has no influence on gravity in spite of all Newton's unproven claims. The fact that mass in inversely related to the radius as Newton's first formula proved is the proof that mass is not gravity but something after the fact of gravity. $F = \dfrac{r^2}{M_1 M_2}$ The only rue way that mass influence the radius by

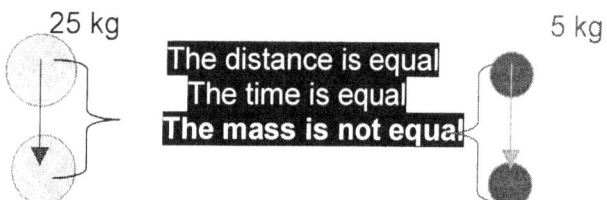

25 kg   5 kg

The distance is equal
The time is equal
The mass is not equal!

diminishing the length is when the mass of both is placed inversely in relation to the radius.
When viewing the findings of Galileo one find that object falling has no mass. To calculate the speed of the object one would require the driving force and since mass must be part of such driving force it has to accelerate the object. At this fact Newtonians threw at me so many answers is differing from north to south where every one was different from the other.

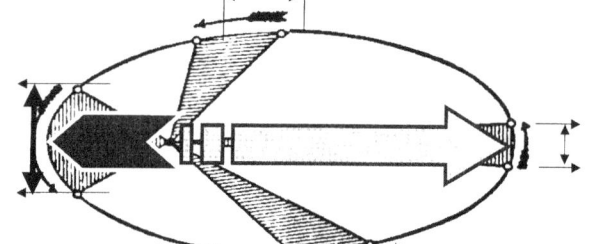

Fact remains if I fall and my mass has any factor in my falling then me being heavier must have a profound affect on the speed of the falling. When I fall I have motion and my having motion eliminated my mass

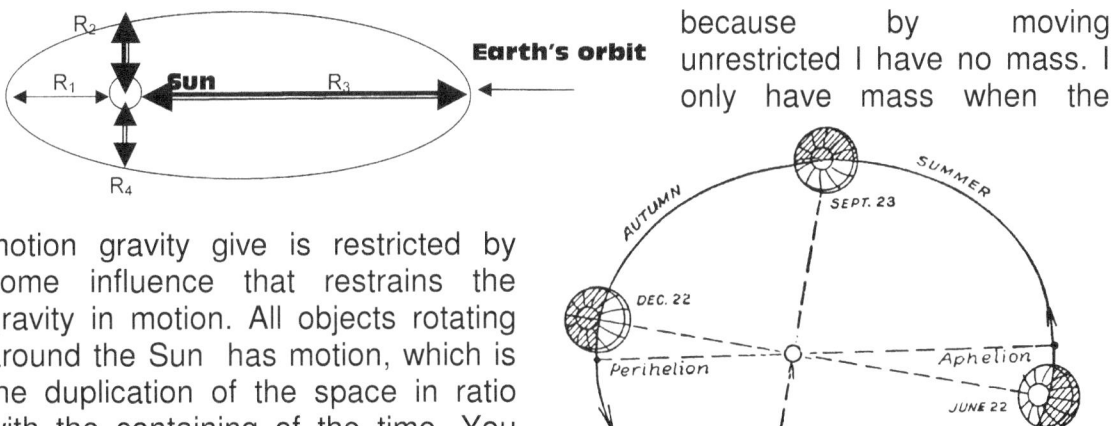

because by moving unrestricted I have no mass. I only have mass when the

motion gravity give is restricted by some influence that restrains the gravity in motion. All objects rotating around the Sun has motion, which is the duplication of the space in ratio with the containing of the time. You Newtonians out there try to be realistic for once in your life in your thinking of cosmic physics without being brainwashed by your education. If the planet mass had the influence of producing the gravity that held the planet in orbit in relation to the centre of the Sun then the planets had to orbit by using the perfect circle.

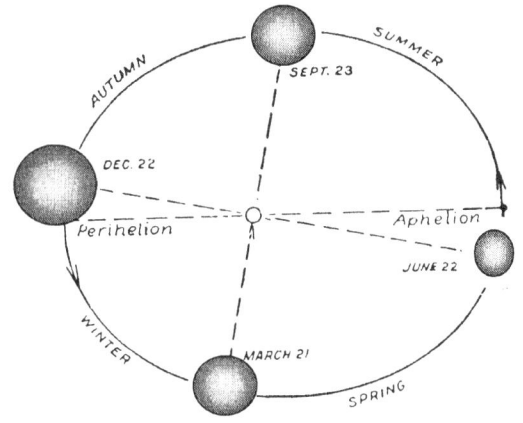

By having a variation radius between planets and the Sun centre it has to mean that either the planet mass show strong variation during the orbit of the year or the Sun shows variation that affect different planets at different time or both must show mass differentiation where the planets become bigger sometimes and other times reduce in size. Since we know that is not the case and we no the orbits do have an eccentric anomaly by the measure of E − e sin E = m it is the M that I dispute. Another aspect of contention about the fact that if mass did play a part in the orbit it is the largest of the lot that should be closer and the smallest being further away. They are as scrambled as coffee with milk and sugar, which again shows mass and size makes no distinction.

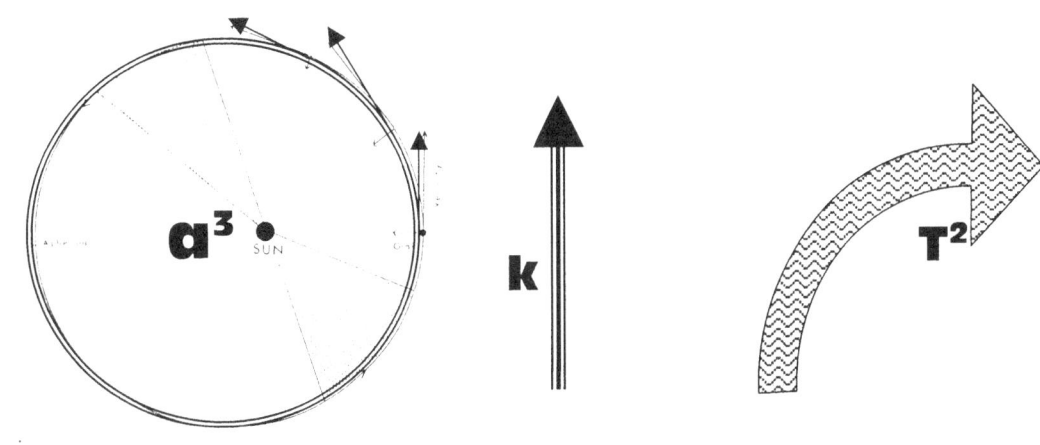

All spinning matter has the point where the spin is still there but the radius is too small to measure by any means. That point in the very and precise centre of all rotating objects is standing still in relation to the rest of the body that is spinning around such a centre. In relation to that logic I do not except Newtonian science holding the radius of a spinning object unaccountable in the spin, whether the spin is applying or not.

Applying Newton's second law $F=ma$

One arrive at the formula
$GMm / r^2 = m (\omega^2 r)$

By replacing $(\omega^2 r)$ **with** $2\Pi / T$ **we obtain Kepler's third law**

This law predicts that $T^2 = a^3$

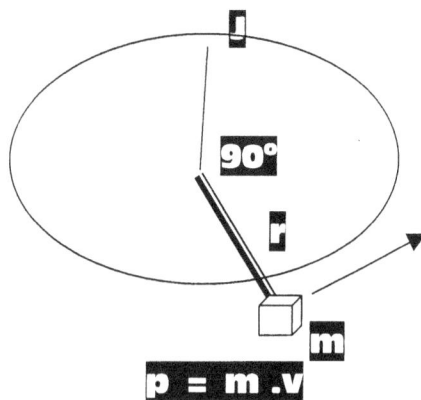

The mass (m) multiplying the speed (v) **forms a new value** J **AND THEREFORE** j **CONTINUOUS TO IMPLY** $J = I \omega$

$= r \times p$ **where** $p = (v = r \times \omega)$

$J = r.m.v = m.r^2 .\omega = I. \omega$ **and becomes interpreted as** $J = I \omega$

This establishes that $r = dJ / dt$

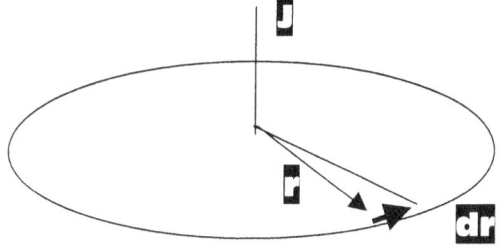

$r = dJ / dt$ **In the case of planets in orbit around the Sun** r **forms a value of zero because** $dJ / dt = 0.$

What this statement implies is that r does not exist. When anything has a value of zero it is for all purposes non-existent. Only when an object is following s straight line can the radius be non-existent because the radius alters value through time development.

To be realistic there is no comparing w wheel spinning on Earth to the planets spinning around the Sun . To have a wheel spinning on earth one require to intervention of life and life is a most alien aspect in the cosmos. The wheel can

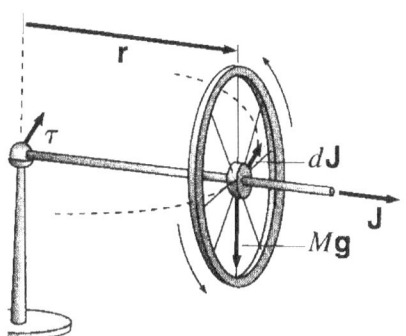

never spin by independent initiative without life supporting such a spin. In that sense it is illogic to compare $a^3 = T^2 k$ with Newton's second law F=ma One arrive at the formula

$GMm / r^2 = m (\omega^2 r)$

This can only be a reality if life provide and actively participate in the support the energy supply that will realise the spinning motion a wheel would have. To work with mass in physics is very earthly bound and that is precisely what life is. But as mass is a very Earthly aspect of physics, we must never spare any intensity in the effort we have to keep mass and the likeliness of life from our minds when considering the cosmos and all aspects about the cosmos.

By replacing $(\omega^2 r)$ with $2\Pi / T$ we obtain Kepler's third law and that is trash because then the third dimension becomes equal to the second dimension and all goes to hell as this formula then would suggest. This law predicts that $T^2 = a^3$

Newton had the revelation of all the above mentioned as an apple fell from a tree apparently very close to him. He was admired as an instant genius and the one the world was waiting for to be born. I do not, for one second, deny or dispute the revelation. What I do encourage is to place the event into its correct context. It was merely, and simply an apple that fell from its branch to its roots. The apple did not pretend to be a meteorite that fell from the heavens. If it were a meteorite, I am sure, with the man's genius, science would be somewhat different at this stage. However, as a young man, being very impressionable, as all young men are, and with the attention this brought about in the world of science, the matter overshadowed the fact.

I am not disputing Newton; I am disputing the relevance of Newton's scientific breakthrough. It was not two objects of cosmic proportions, colliding in a show of the spectacular. It was, after all, only an apple falling from a tree and not that big an event. With this miracle he revealed, Newton found he was competent to improve on the work of Kepler and what Newton saw about what Kepler found was to Newton's mind the proof of total mathematical incompetence. He (Newton) saw a circle and without Π there can be no circle. Further more, since he was the founder of the invert four square principal, the principle also had to be included the make the picture a smart Newtonian picture and with that remove Kepler as such.

$\dfrac{dJ}{dt} = 0$ Newton, and science, made one enormous blunder, from this stance.

They took the radius of a wheel not to have any influence on the wheel. In doing that, they removed the very fact that keeps the universal attachment together.

They put two objects in an attaching relevancy and then announced no relevancy. Doing that is breaking the most fundamental mathematical principle.

$$\frac{dJ}{0} = dt \text{ or } \frac{0}{dt} = dJ \text{ This disputes mathematics.}$$

DJ / dt can have any number from eternity to infinity, only excluding only one possibility; it cannot be 0. By placing the one in division of the other, you bring in relevance. You cannot then say there is no relevance. By doing such, you proclaim that one of the factors is non-existent. In both cases, one of the factors then does not exist. Such a claim is incoherent, because you proclaim that a circle has no radius, or a radius has no circle. When calculating a circle, you multiply either the square of the radius by Π, or the quarter of the diameter at a square by Π.

$$\frac{dJ}{dt} = 0 \text{ constitutes a circle and is also therefore } \Pi \times r^2 = CIRCLE$$

If you remove r it then is $\Pi \times r^2 / r^2 = CIRCLE$.

You cannot then say $r^2/r^2 = 0$ and therefore $\Pi \times 0 = 0$. That is nonsense. $\Pi r^2/r^2$ will always be $\Pi \times 1$, and that is where Kepler placed singularity. By hiding this

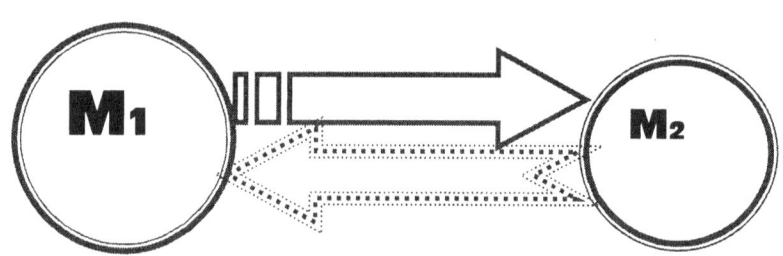

fact Newton went and threw the baby out with the bath water. There is little standing further from the truth than this statement and reality disproves

Newton completely. In the motion every wheel has to have a pivot around which the wheel turns. That is called the axis.

Newton's claim of mass pulling is totally incorrect when compared with reality Instead time can never stand still as time delivers space in ratio to singularity.

 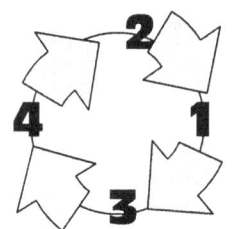 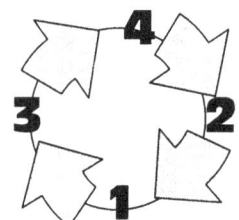

Notwithstanding all the protesting and objection Newtonians may have about the correctness of Newton, the concept is completely fraud in principle.

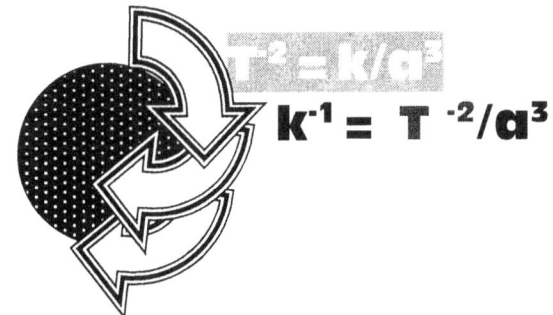

Built into the nature of rotation is the conflict there is between the two opposing sides of the same rotating unit. By crossing the divide the fundamentals in nature changes as every aspect of what was valid change completely to the opposite. On the one side there is a contraction in spin. The thrust draws the spin into the centre by direction of the spin that favours such contraction. This has nothing to do with mass.

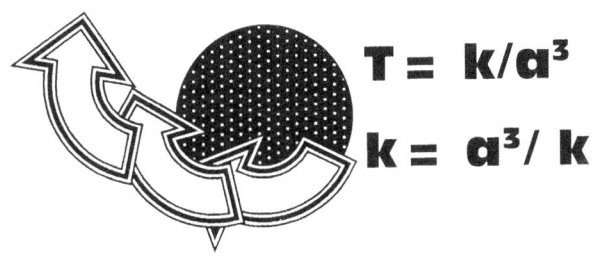

$$T = k/a^3$$
$$k = a^3/k$$

Then by crossing the divide where singularity changes the direction of motion every aspect concerning the ration changes as it actually alternates. That what previously by rotation came down then goes up in the opposing direction. However that is not surprising because every slightest motion involves just such a change in direction and it is the process of interacting changes that manifest in charging motion into singularity.

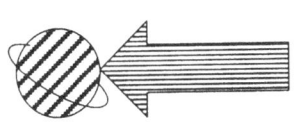

$$k^{-1} = T^{-2}/a^3$$

In the way planets rotate around the Sun this characteristics are also present and that forms the criteria foe comic order or gravity. The same characteristics we find in the rotational spin around singularity as well as a governing centre. On one side there is a directional preference to favour the one side and on the other side of the divide this favouring will swap ends.

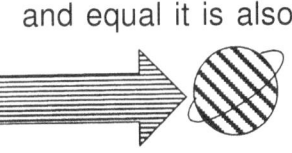

$$k = a^3/k$$

Since singularity is all the same and equal it is also true that that which singularity generates are not equal but depends on the motion that provide the space-time which charges singularity to define space-time. In that the Coanda effect has the role it plays. The motion of the relevant liquid establish the rotation as the motion bonds the liquid to the solid while the solid uses the liquid in motion to extend the space confined by the motion thereof. Since it is two aspects in one unit the defining of the dividing comes about as the two parts perform each its role on either side of the divide. On the one end the expanding party takes privilege position and on the other side of the divide the concentrating partner takes a privilege stance. From where we stand we see a small and a large, but from the singularity it is one side contributing more motion than the other side by performing duplication or contracting. However there is no big or small. It is all based on the contribution in motion that maintains singularity.

In every cycle of every orbit we find four in time forming different allocations in positions and the varying depends on the relation the allocated position has with the centre.

There are relevancies that form location preference to allocated positions just like seasons do. That is a product of time, which holds four positions in one cycle.

By denouncing Kepler and his formula, one must be prepared then to denounce all motion in that manner, and Newton more than most should have realised that. When looking at any rotating object, there has to be a point of no rotation and no rotation means "no rotation", not no existence. No rotation means a factor of 1, not zero. That then is singularity. The eternal Π, the Π that may not have significance but still it is Π of value.

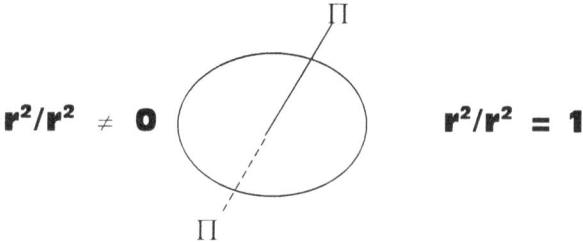

$$r^2/r^2 \neq 0 \qquad r^2/r^2 = 1$$

The relativity remains one, eternally one, but it cannot be zero. Therefore, dJ/dt cannot be zero.

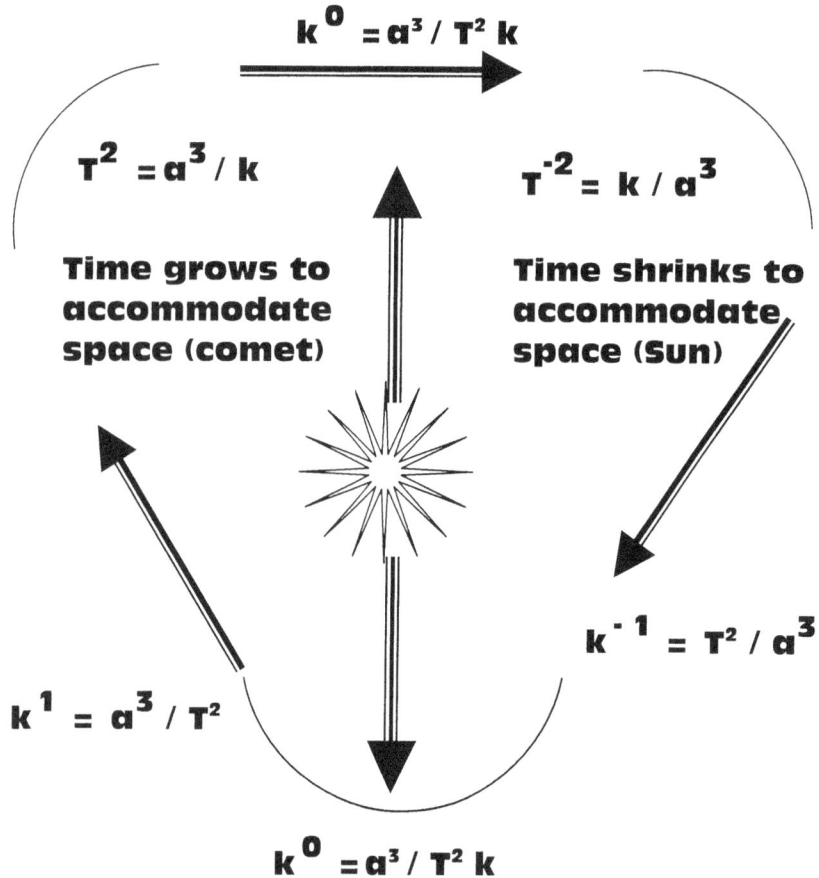

$$k^0 = a^3 / T^2 k$$

$$T^2 = a^3 / k \qquad T^{-2} = k / a^3$$

Time grows to accommodate space (comet)   **Time shrinks to accommodate space (Sun)**

$$k^{-1} = T^2 / a^3$$

$$k^1 = a^3 / T^2$$

$$k^0 = a^3 / T^2 k$$

dJ/dt can be eternal or infinitive or at the worst it can be dJ/dt =1 but dJ/dt $\neq$ 0

Looking at Kepler's statement it seems like a mathematical blunder made by an incompetent not understanding the most basic principles one might have in mathematics. It reads that the space in the third dimension is equal to the calculated motion of the space in the second as well as the first dimensions.

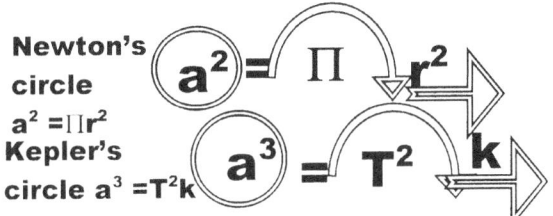

Kepler saw a circle because space is motion provided by singularity from a centre.

A child will know better than that because it is hardly the manner one might use to calculate the surface of the space let alone the cube of the space. Far better is the use of the correct formula to calculate the cube in space as the one used to measure the volumetric displacement accurately as follows:

$a^3 = 4\Pi r^3/3$. Using $a^3 = T^2 k$ is mathematically a fools argument, and yet even when someone perceive it to be wrong, was it accurately changed by Newton?

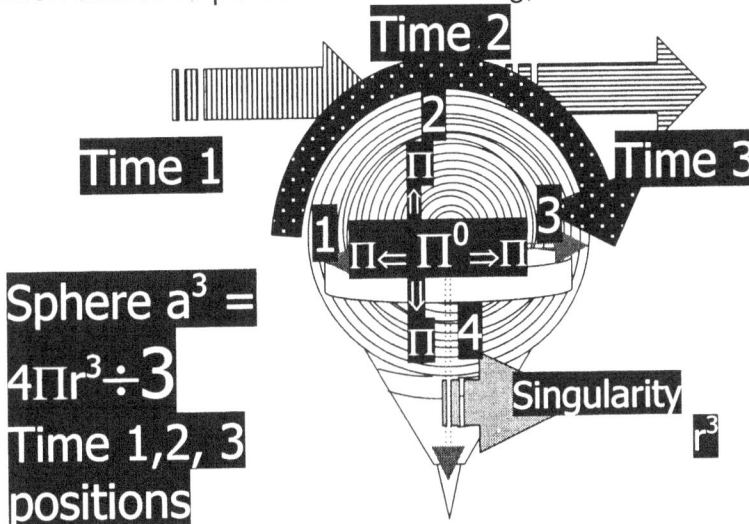

Following the mathematical volumetric formula on can see what is required to generate and duplicate every time the motion allocates space a new position. The centre line holding singularity ($\Pi^0 = r^3$) forms and then charges space by the cube in relation to the outer edges where singularity in Π meets time and where the Coanda principal puts the edge on space. The space in time redistributes the volumetric charged space in relation to three sectors time hold space in. That is the volumetric formula.

Newton said a sphere is $a^3 = 4/3 \Pi r^3$

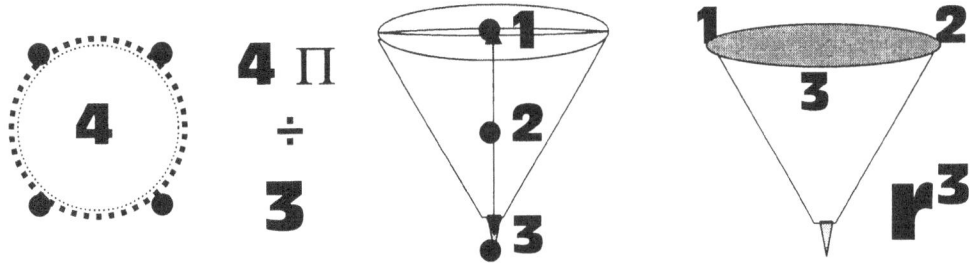

When the top starts spinning the spin generates the time difference that single out the space independence to bring about the space spinning. This is not what Kepler's formula show. Kepler's formula shoes the space turning by rotation as well as lateral motion and by duplicating and contracting the space generates new space as the space carries material through time.

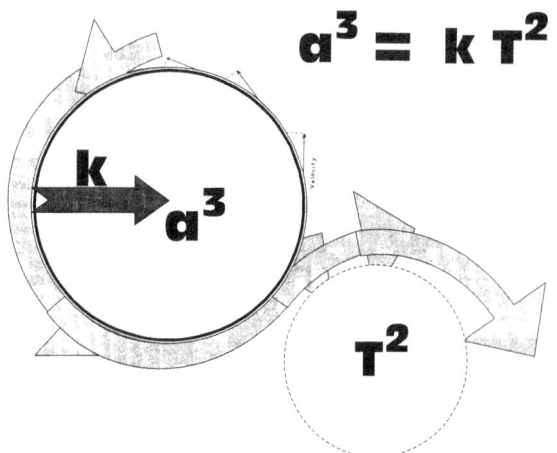

$$a^3 = k \, T^2$$

The space a^3 duplicate the position it had, it has and what it will have T^2k in relation the where it was, where it is and where it will be the very next instant. It does not mathematically reflect on a volumetric space to enable mathematicians to play a game and prove to their compatriots as well their own personal vanity how skilfully they can play a game with numbers and rules. It indicates a cosmic principle on which the four cosmic pillars rests.

Dare I say one has to be more than a mathematician to appreciate this difference? This I say because I am of the opinion that if just one mathematician in four hundred years tried to find out why the formula used by their skills to calculate a volumetric displacement do function in the purpose they use it, that mathematician must then have had the ability to see the difference there is between the Newtonian formula depicting the mathematical purpose and the mathematical language suggesting a cosmic principle.

Newton said a sphere is $a^3 = 4/3 \, \Pi \, r^3$
Kepler said the cosmos told him a cosmic sphere is $a^3 = k \, T^2$.

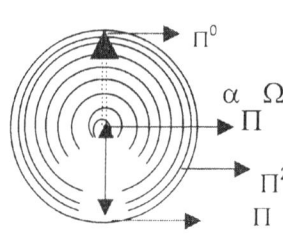

Going down the line that will reduce the radius we finally land on a spot where $r^0 = 1$ leaving only Π as [part of the formula $a^3 = 4/3\Pi r^3$, which then only leaves form as a value with no measure to form. Yet that is not the end because we find in mathematics one more possibility carrying singularity and that is Π^0 which will bring the ultimate value of singularity to the space concept in $a^3 = 4/3\Pi r^3$. At that point a^3 is singularity altogether.

I have tried for so many years to accommodate Newton or parts of Newton into how I see cosmology but with no success. Newtonians leave everything half explained and from that try to formulate perceptions. What is matter? What is in the finest essence that which form material? When in search of an explanation to

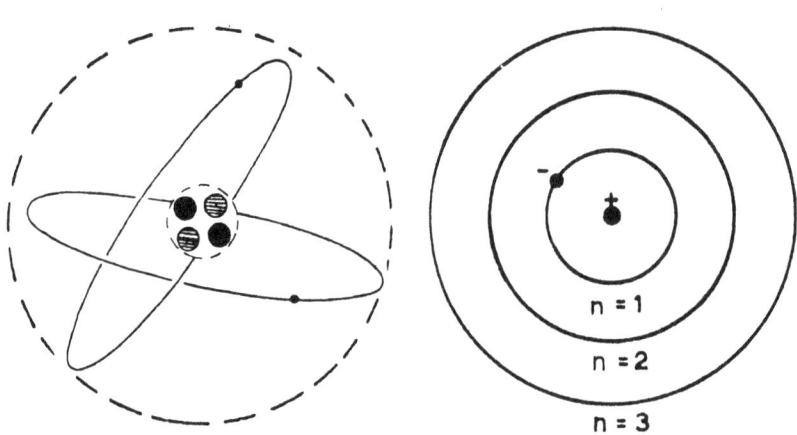

define material I find that there is an ongoing report on the particles forming material, which is not ongoing but is lumps called atoms. Then I suppose you may dissect that into as far as we can see with an electron microscope but that still says nothing on what atoms are. What is that which does the circling electron confine? What is the electron putting into a circle that then becomes a confined unit? What fills the space in the cube of material?

It is well accounted that there is this spectroscopy and the energy bands shifts as the energy levels rise or reduce. It is accounted and it is better well documented that the energy requirements allow the bands to rise or deplete.
The more the energy is in the atom the wider circle the electron has to travel to accommodate the more there is in the atom. In the atom we find a neutron with no mass and an electron with mass. Then down at the bottom there are two protons that expand and that reduce the space they hold.

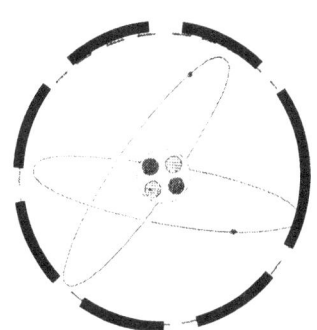

That is so scientific accurate and correct and the only way I have to be more accurate is to remember every useless name some incompetent Muppet gave that enormous discovery that that incompetent Muppet found to be his remarkable creation. The more illustrious the meaningless name sounds the more important it will sound. The most important name will hide the biggest discovery of the lot…that the discoverer has no idea what he discovered and therefore his discovery in the long run has no meaning to science what so eve. Therefore to hide the truth he has to invent a more useless name than the previous useless named other particle has. Never is there a reference to why the electron is spinning in the first place. What is in the purpose of the spin? Why would the electron spin and why would the electron enclose what is in the spin it protects.

What is on the outside and what is on the inside of the electron that is spinning in a wider or smaller orbit. What is the enclosed material that the electron is protecting? What is behind the electron and what is in front of the electron and what is the electron?

That again brings me back to my first question: what is material?

We know that the atom has two positive particles in relation to what seems to be a neutral particle that is between the positive and the negative particles. The positive is a double 2/3 and the neutral is as much negative (2X 1/3) as it is positive 2/3 and they're the explanations tops. To continue the discussion we than have to establish what a positive is and what a negative is

to find a neutral. What will bring about something being negative because being negative is only a position in relation to being positive. What is being positive than? What is positive material and what is negative material because then the neutral has no clear definition in existing.

All the negatives will eat up all the positives and that will leave nothing as neutral. Most senseless is the part that I am the one they frown upon, the one they reject and the one that is uneducated.

In the atom I found a relevancy where the proton has double motion ($T^2 + T^2$) and the neutron has the linear motion of a^3 **X k.** The total positions are equal to every possible position the sphere may offer in relation to a centre singularity

Then when this seven positions are put in relation to time we find that the seven multiplied with the three positions time has the total including singularity expanding and singularity contracting is $(7X3 = 21+0.9991 = 21.9991 / 7) = \Pi$. That concludes that the Universe has gravity at all levels because the atom is a sphere as much as the atom is the Universe.

Present 2

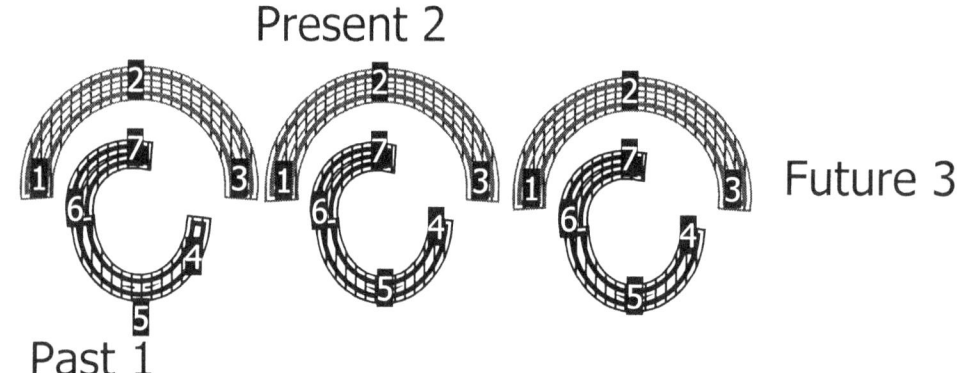

Future 3

Past 1

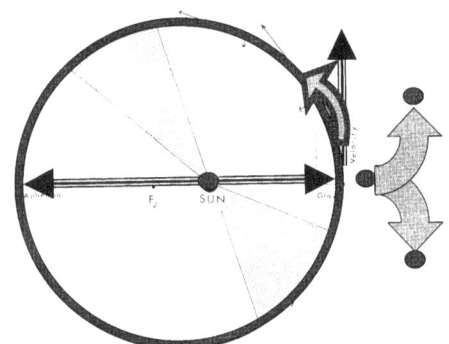

If we look at nature there is no position we can give time to rest at zero because time is the constant flow of all material in ratio to each other. In that the notion of t=0 is incredibly outdated. The fact that there is time is also the fact that there is repositioning of material in relation to each other on an ongoing constant basis. When I see a planet such as the Earth in rotation around its axis the centre line, then we find the rotation is a flow of atoms repositioning in relation to one another. I stop at the atom because I have to stop somewhere and the atom forms the conclusion of the Universe but our conclusion of the atom forms our conclusion of the Universe and not necessarily the atomic conclusion that the Universe comes to define as the final conclusion. In other words there are mostly and many other forms of atoms in the Universe outside our spectrum.

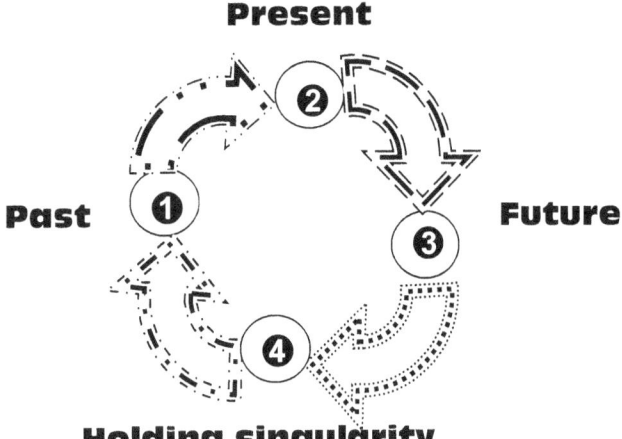

Present

Past   **Future**

Holding singularity

Every time an atom is moving, such moving consists of a series of duplicating and relocating reference positions where every individual atom is constantly proceeding a line where the atom is replacing the location and the position the atom by taking up the place the atom in front had in relation to the centre such moving of the atom. That centre is incapable of moving therefore every time motion is established such motion is the generating of the entire field that is between the two locations of eternity versus infinity.

The atom is not only moving forward it also is moving to the side in accordance with the centre. It is following the leader but the leader is following a line that finds a relation in line with the position the atom in front had and the atom at present then fills. The filling of the position is in accordance with the centre and the centre is the point that claims the dominance of the moving. While from the point the atom has, such moving is straight in line with the allocation the atom in front has. The relation however is a centred one where there is a point that never moves and cannot move.

This part is represented by one side of time that forms the eternal side in time. The rotation is never concluded but always repeat the previous into the future. It is Newton's invert square law where there is four positions always following each other and this position was present when time was eternal. Please take note that time could never stand still and therefore Newton's assumption of t=0 is much misleading. The time found on location in which it rotates and that position holds all four points on one exact spot but that does not remove the location by giving the location a value of zero. However in that there is another aspect where the infinite secures a position and still maintain the point having all sides sharing one point. That is the very inside and to detach infinity from eternity that has to be a moving where that point holding infinity also has to relocate while in truth it cannot relocate.

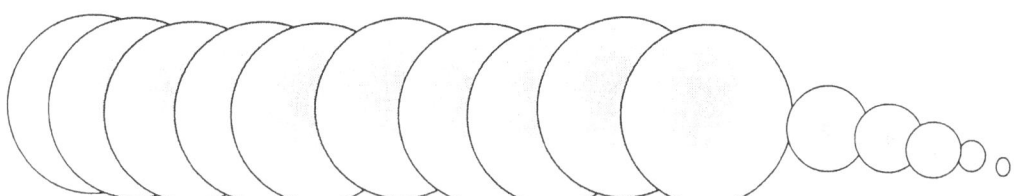

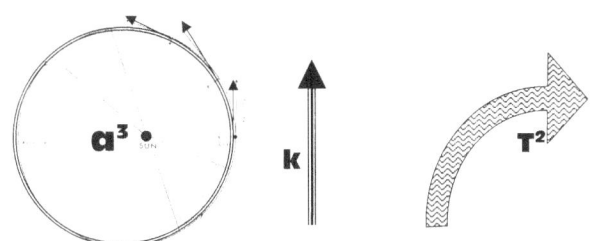

Taking this motion to another level is the repositioning of that which cannot move into a new position where it has move too. This is part of the rotation but also t is the moving of time in a lateral direction. While the lot is rotating the lot is shifting into a new position going from where it came to where it is to where it will be next to where it was every time it is going forward. There are seven points moving in the circle while there is this seven points coming from and going to while it is in the present position. That makes it the seven it is plus three, which is the time aspect ten.

That action is what brings on motion and involves a line as well as an incomplete circle. The space a^3 duplicates what was to the present going to the future in a

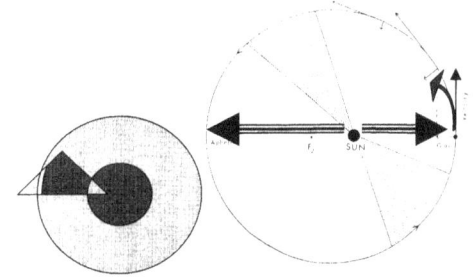

semi circle T^2 as well as a straight- line **k**. The motion of the space a^3 involves the rotation T^2 as well as the straight -line **k**. In order to be in space the space has to be on the move. Time cannot stand still for then there can be no space because the space is the directional movement in a lateral as well as a circular direction simultaneously. **There is no possibility of time standing still at t=0. If there is t=0 there is only 0 and since there never can be 0 there is no possibility of time standing still.** If you find time standing still, chuck away your watch because the watch and not time gave up the ghost. In order to be in space, space has to duplicate by presenting what was in the past, take it through the present and fling that which filled the past and is filling the present into the future. There are both actions serving space where space is going sideways to go forward as it is going forward in a sideways direction. That is motion whether it fits Newton or suits Newtonian perception it is of no consequences because the atom is moving along as the atom is spinning.

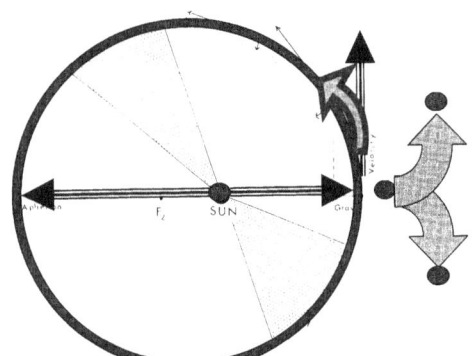

The Newtonian idea that the one direction eliminates the other direction by the sharing of a mutually established centre has no base in reality. Kepler clearly brought to science the fact that space forming include motion of the space but as important is the fact that the motion cannot exclude any one of the factors in the circular or the linear and that is ids the product of both factors that form the flow of space through time.

By supposing that the one factor removes the other factor from the Universe is quite frankly forming an effort to absolutely destroy what Kepler said! Then we get back to the question that I asked earlier as being what material is? What is filling the atom? What is inside the proton, the neutron and the electron? If getting more energy is getting the atom fuller by expanding then the inside must be energy. What is energy because the word or term energy has become an escape goat used whenever Newtonians run out of answers? Einstein was of the opinion that when matter goes past the speed of light it would become pure energy. So what is matter before it goes past the speed of light? Is it then less energy or does it become something other that energy?

That brings about the next question: what is mass. Why would that inside the atom have mass and that outside the atom have a different mass. Why would the proton being so much smaller than the electron be so much more massive? Newton came up with this idea of mass and everyone accepted mass because no one could disprove it and since Newton never was set the task to prove what mass is it was left at that. Why would material have mass?

Taking the argument back to Kepler's law, a^3

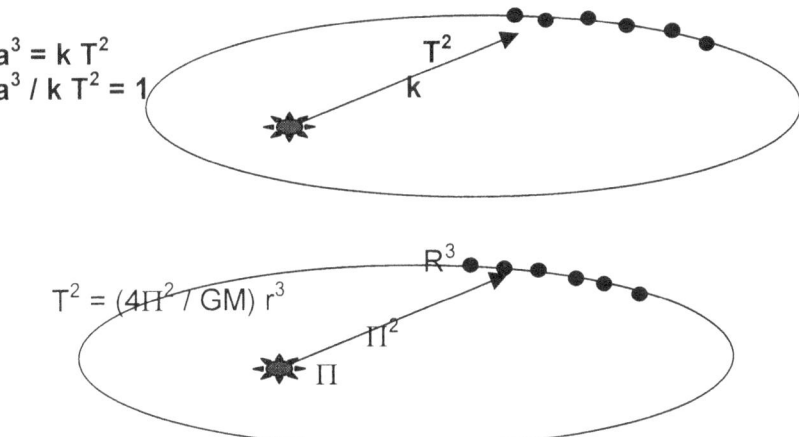

$a^3 = k\,T^2$
$a^3 / k\,T^2 = 1$

T^2
k

$T^2 = (4\Pi^2 / GM)\,r^3$

R^3
Π^2
Π

We all know that all object have mass when the object is on Earth but when the object is in outer space what would give the object mass? If mass was conducting the motion of planets then surely they would be arranged according to mass. The biggest must be to the most inside and the smallest to the further outside. If it was mass that was generating the motion then the most massive must be flying while the smallest must be crawling. That is not the case at all. The distribution of planets relies neither on mass or speed of motion or any distribution in size at all.

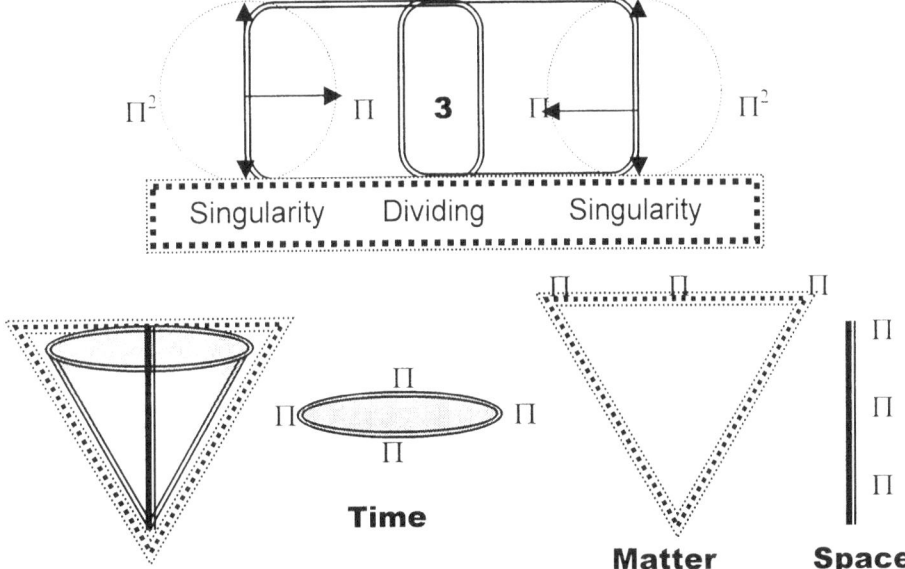

Π^2 Π **3** Π Π^2

Singularity   Dividing   Singularity

Π Π Π

Π
Π Π Π
Π Π

Time

Matter   **Space**

There is no correlation that would even suggest that mass plays a part in the orbits of planets or any cosmic structures. The dynamics we contribute to the realisation of mass should be far better defined if there is to be any clarity on the matter. According to Kepler space needs motion to produce space $a^3 = k\,T^2$. Let us see how true that is. How does that which fills the inside of a particle move from one point to another point during time? The rear must back the front by pushing the front forward while the front is pulling the rear by vacating the front. If the front does not move the rear will smash the font out of line and in some cases destroy the front from behind. Think of a car crash where the front becomes blocked by other objects while the rear is coming to fill the front from behind.

That which fills the front has to vacate the location it holds in an agreeable direction within a suitable time while that which is filling the relocated position of the front requires the filling from behind as to find the required flow so that the motion can go about spontaneously. If that which is coming from behind is retarding in motion as to fill the vacating position in front the unity in the movement will tear into parts. In order to accomplish motion, the motion has to accomplish the sphere. The motion not only fills the vacating space by rotating in a "follow my leader" process but also establish from one point in infinity a line of points running along a line that is never there. The line establishes four rotating positions parting infinity from eternity. In that I do not find grounds to blame mass for anything. I do find that the matter behind the matter in front is only retarded allocated filling of a position in time. The following is in a retarded stance to that which it flows and that makes that which follows behind that which is in front being ahead.

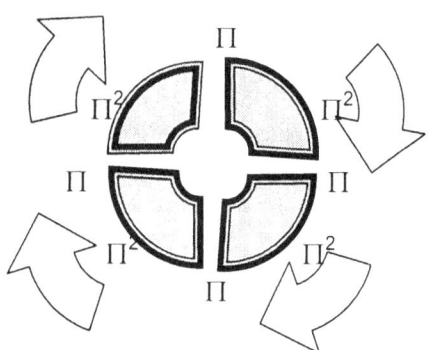

The one is always behind the other which is in front and that is material because being behind that which is in front gives both that which is in front as well as that which is behind an independence from one another. The proceeding to follow and the following that proceeds placed both in separate compartments during time and that puts the one in motion while the other is filling the motion. Is till do not find mass. What I do find is the relevance that Galileo found. All objects notwithstanding size or mass fills the position the one in front left vacated to be filled by the one coming behind.

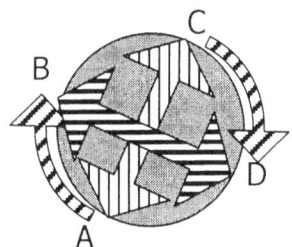

There is more to motion than just that because the rotating plays follow my leader while the lateral plays follow my leader and with that there still are two identifiable substances in relation to the space and the motion that time represents. The lot is duplicating and by duplicating every aspect is relocating what it represents in terms of space –time to a new location.

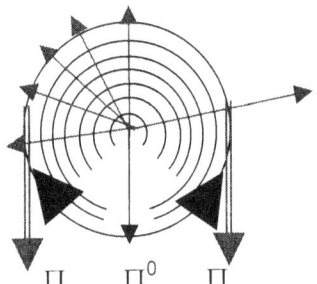

We have the rotation there is where the one point serves as a guide to the previous point and follows the next point into the future while honouring singularity. That is one part of the story. There is more and that is time in the lateral or eternity running without ever stopping.

However that which cannot move does move not only by repositioning the lateral but also by relocating the rotated into ea new relevancy. By shifting point A to point B that shifting also involves moving point K to where point L was before. However that centre is not movable and on both flanks the lot shifts. That means what ever is relocating a position is breaking down all there is and shift the lot to where it is going to be by generating what there was in the previous location onto where it will be in the next location.

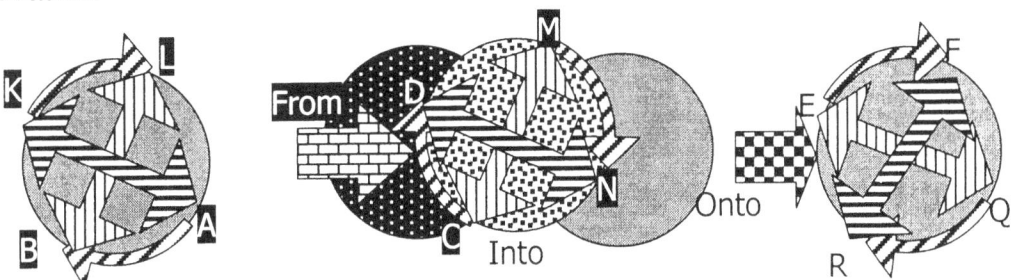

Then this lot in rotation also has to relocate that which rotates along a single file. The point in singularity cannot move and by it being unable to move it has to break down all relevancies there was and shift all that to new positions in relation to an entire new Universe. In that there is no mass. What there is we can describe as controlled duplication where the duplication is represented by the expanding while the control is represents by the contracting. It is just as Kepler said it will be when it is $a^3 = k\ T^2$ and on the one side there is $k = a^3\ /\ T^2$ as singularity generates space-time in the space generated by time finding a new location during the time such generating takes while $k^{-1} = T^2\ /a^3$ the motion produce a new space in the allocated position it has to be at that particular time. Still in all the dissecting I find no evidence of mass doing anything and even less generating gravity.

It is a pity Newtonians never get more specific as to where one may find mass and how mass go about in producing gravity. Being as explicit as I am by going into time between eternity and infinity I still find no mass. I do find eternity forming the three positions on the side that has no end and those three positions do correlate to the most accurate detail in relation to that which has no start. In between that which has no end and that which has no start space-time is generated through the motion that activates space-time.

Newton made the error Newtonians still do after so many years. He took the fact of life as a cosmic reality. He took the motion that life can achieve as standard cosmic occurrences. Newtonians go much further than Newton did by claiming life comes at a dime a gross throughout the Universe while there is no evidence of that. To swing a weight on a string from the hand has as much cosmology in it as trying to pump a tire with air and then compare that result to a star…and yes Newtonians do just that! When considering the motion applying then first see that being tested fits freely in cosmic reality. A rock cannot roll up a hill and a brick cannot be while a cloud can't cycle a bicycle.

That which Newton supposedly discovered being gravity is that which Newton denounced as zero. When Newton discarded the motion part by putting the value at zero he threw away that which produces gravity on both sides of the border of motion.

By replacing ($\omega^2 r$) with $2\Pi / T$ we obtain Kepler's third law
This law predicts that $\mathbf{T^2 = a^3}$

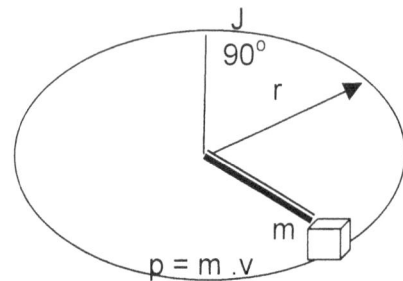

This establishes that r = dJ / dt

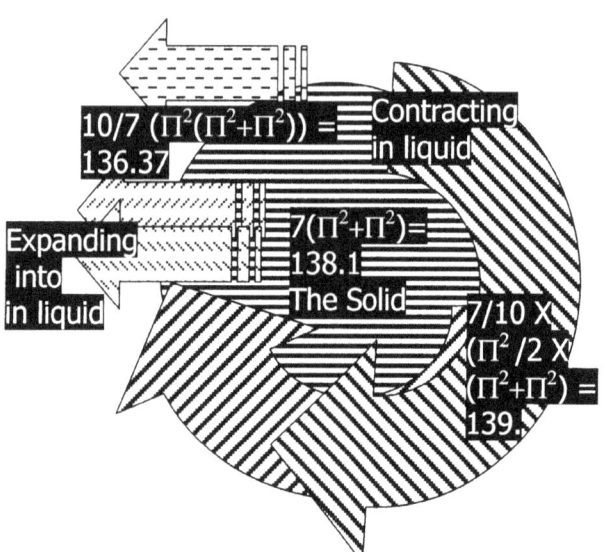

r = dJ / dt In the case of planets in orbit around the sun r forms a value of zero because dJ / dt = 0.

Where the motion of time forming the liquid interacts with space forming the solid we find the two parts motion offer on either side of the divide. The motion is the same but crossing the divide that motion then falls into the other side of the Universe where all changes to become the opposite of what was.

Because of this we find that by nature and not by mass there are elements that favour duplicating more than contracting notwithstanding the number of protons or the resistance the number of protons may show to the blocking of motion. Then there are other elements that favour contracting much more than the duplicating side and again mass plays no part. Mass have a place in Earthbound Newtonian physics where one my substitute the correlating motion with the restricting of the motion because there gravity is the tendency to move whereas mass is the restricting of the motion turning the motion into a tendency to move.

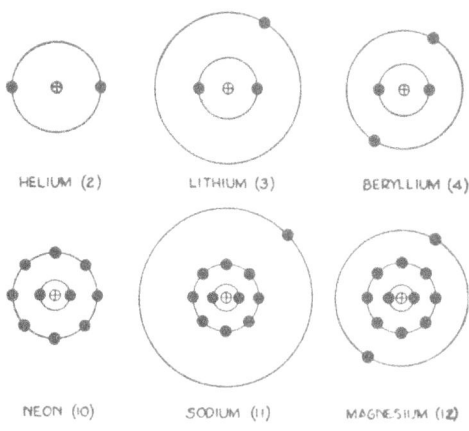

That however is not related to cosmology because as long as there is unrestricted motion there is no mass and then all the planets are equal as they circle about the centre of the Sun . Mass is the frustration particles experience when unable to contribute to free gravity - motion.

Getting back to the fictitious mass and the question about what material is. What is material and why is material what it is?

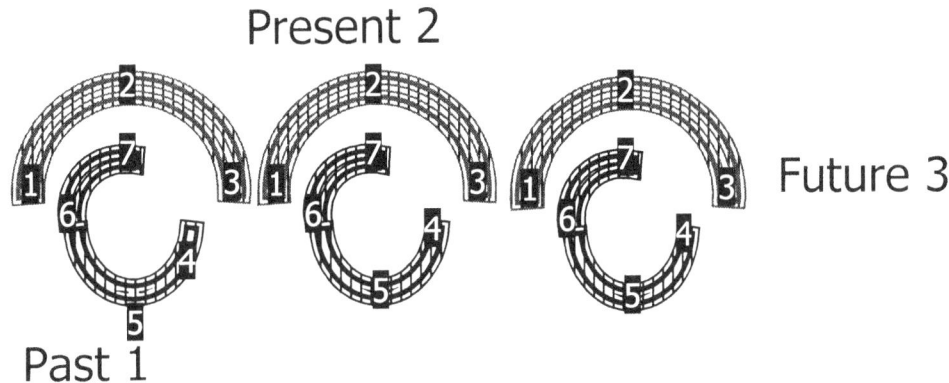

Present 2
Past 1
Future 3

The flow holds seven points that forms the atom and the atom was able to retard time so dutifully and preserved heat so faithfully an entire Universe with immeasurable Universe coming from an array of immeasurable possible Universe can now serve as multitude stages of universe developing eras.

Yet in all that, it never deflected one measure from the original retarding by rotating principle it had in the culmination of starts we preserve to put a Universe in.

Expanding

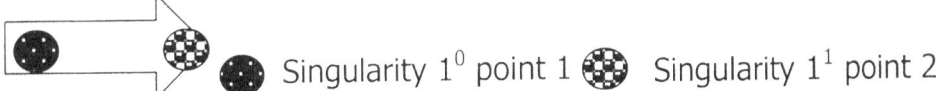

Singularity 1^0 point 1   Singularity 1^1 point 2

There was a point in eternity that held infinity and combined singularity. Both are still in our presence and both are wall established. We still gauge the one in the centre of all spinning material and where one the point is not there is a line that is not and in the line the line holds seven points all holding the precise same position. Then light or heat came about and since light or heat expands because it takes more space than what was taken before, there is more of the same that was before.

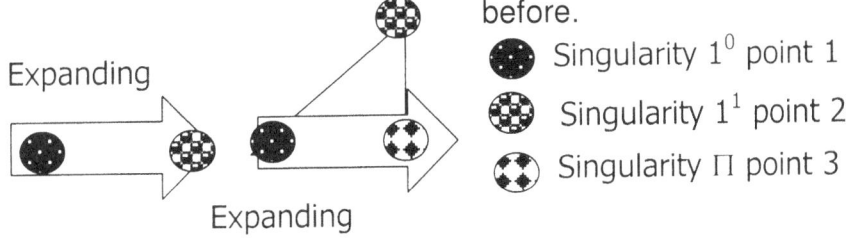

Expanding

Expanding

Singularity 1^0 point 1

Singularity 1^1 point 2

Singularity Π point 3

Since the expanding brought what was not in between what is not eternity parted

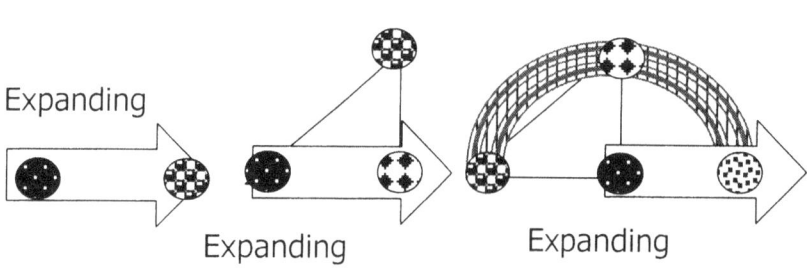

from infinity. But time is motion and since the one became parted from the next the next placed the previous on the other side of the Universe while the following did the expanding. At such a point the law of Pythagoras started to develop the cosmos by putting in the triangle in relation to sides.

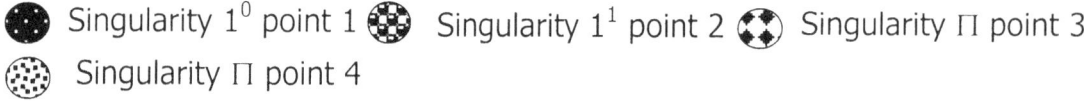

The flow of time continued as new positions established while point once established did not vanish because once anything is part of the cosmos it stays part of the cosmos, as there is no other place to go but remain in the cosmos. The points continued as time moved on.

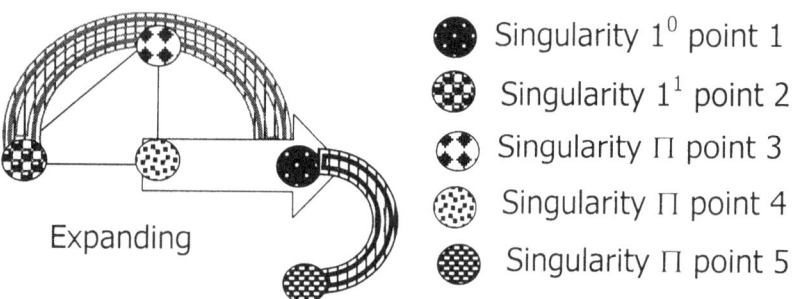

Then after the four points expanded eternity produced an additional point at a location where infinity parted company with eternity. At such a point, point five came in place and this was where the cosmos became different from what was previously applying.

The cooling set in where the expanding brought more but also the expanding distributed more of what expanded and since that which expanded was covered by more, the more made the expanding less and by cooling the expanding retracted. It was not the progress that retracted but the direction the progress had that retracted.

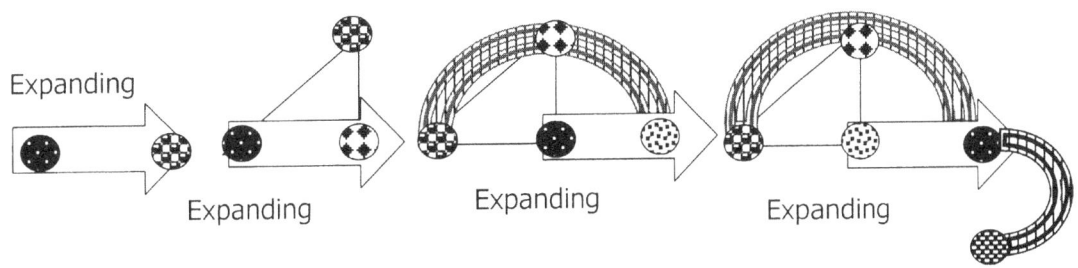

At this point eternity parted from infinity. Infinity is there for all to witness. It is not a hypotheses but a reality. Eternity is there where everyone is seeing it as long as man had a mind. That too is not new and that too is a reality and not a hypothesis. It is a reality within every human and is as concrete as the blood running through the observing person's veins. It is as real as the Universe itself because it is the Universe itself. Any one arguing this reasoning has no mind to understand the smallest concept any human can form. Then came the rest.

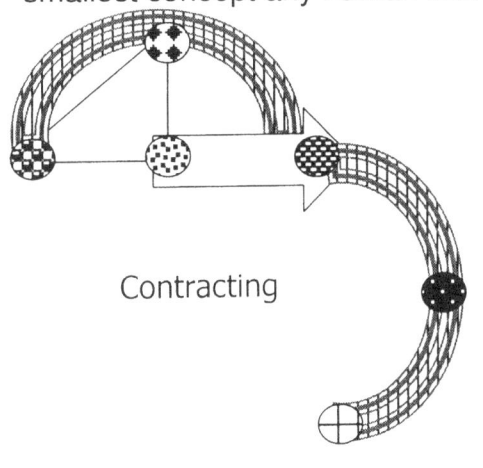

Contracting

Singularity 1^0 point 1

Singularity 1^1 point 2

Singularity Π point 3

Singularity Π point 4

Singularity Π point 5

Singularity Π point 6

By mathematical implication and the influence of the law of Pythagoras material found a limit at point six.

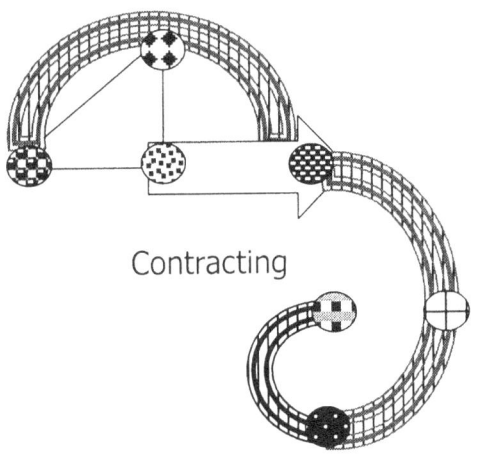

Contracting

Singularity 1^0 point 1

Singularity 1^1 point 2

Singularity Π point 3

Singularity Π point 4

Singularity Π point 5

Singularity Π point 6

Singularity Π point 7

That which was inside the seven connecting points

serving singularity is contained heat by spin. That which is outside the seven points is expandable heat that expanded without control. That inside was controlled by motion that those outside provided. Any one with doubt that it is controlled heat look at photo's of Nagasaki and Hiroshima and the Bikini island later and see what a tiny bit looks like when the control is released. It is heat that is retarded time and the time is heat dragged on by the lagging behind of heat in a different era of time.

That is material and every seven points holding singularity confirm the backlog of heat dragging behind time. Still I see no evidence of mass and if mass was not then, then mass cannot be now except for in the head of Newton and in the imagination of Newtonians suffering from mental programming. Material is heat that is responding according to a time delay where the spin reduced the time and in that controlled the expansion as it expanded the reducing. That is the essence of any atom.

The material is there because the motion retarded the time to preserve the heat that bonded a unit by a dividing of one unifying point reserved by singularity in order to maintain singularity.

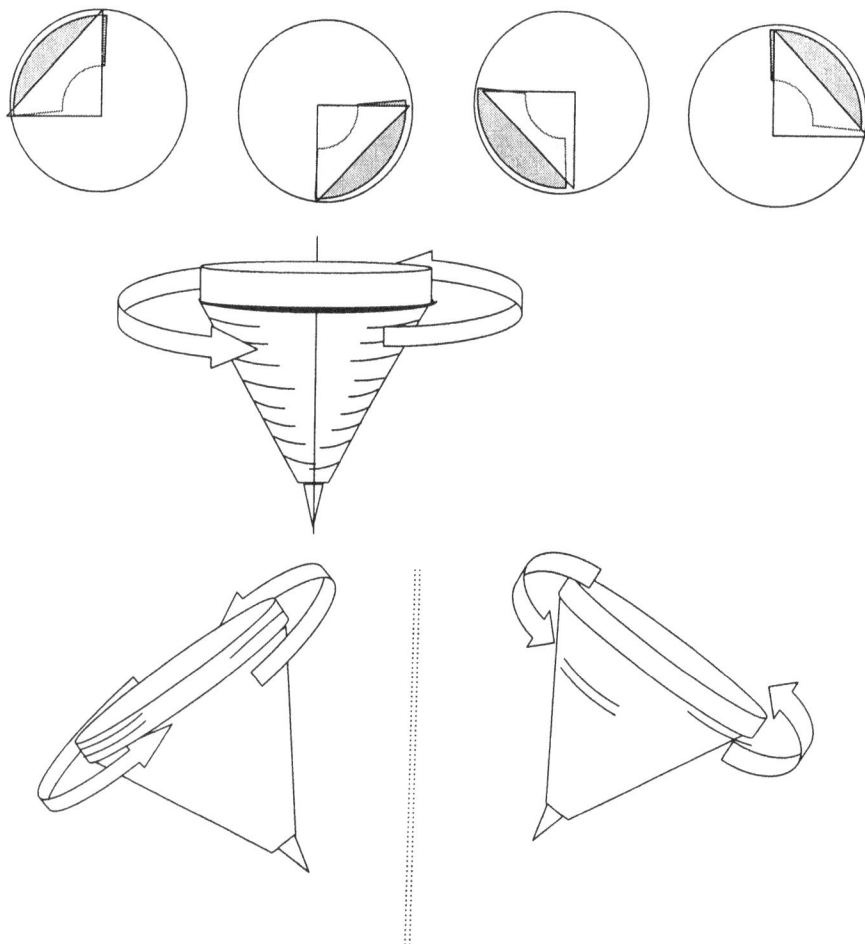

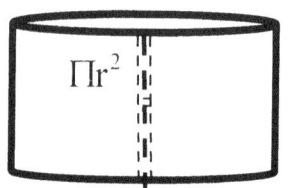

Πr^2

What is it the Newtonians fail to see? If an electron is orbiting around an atom, the inside of the atom must be a circle. If the atom was not a circle, it then had to be a cube. The electron cannot rotate around a cube; therefore, the inside of the atom is a circle.

The radius r runs from the circle outwards, from a circle centre point towards Π, the value of the circle. In the centre of the circle, there is a point where the radius starts. It runs outwards from that point in all directions towards the circle Π. Technically, there then has to be a point where r is infinite and not zero, an absolute infinite. However, the circle therefore remains Π. The circle does not disappear; it remains there for all to see. It is only the radius that almost disappears into the infinite, but it does never become zero!

In a circle, there is a radius that initiates the circle. The calculation of such a circle is $\Pi \times r^2$.

$$\frac{\Pi r^2}{r^2} = \Pi$$

If one removes the radius from the circle, the circle remains, only holding the value of Π. By removing the value of r, Π becomes singularity with no place to be. Singularity is the place where there is no space to be in place. However, Π remains because once r receives the slightest of space Π will find space. Then the circle will grow to Πr^2 and r would determine the space. Without space, there is no r but there is a circle with the value of Π. Singularity is in every single rotating object, be it the proton or the combining effort of all particles in the Universe. That is what light and the photon is. It is concentrated heat that the Sun (or any other generator of electricity) connects heat to singularity where the heat receives either temporary connection to singularity or a small piece of individual singularity.

All spinning matter has the point where the spin is still there but the radius is to small to measure by any means. That point is standing still in relation to the rest of the spin. In relation to that logic I do not except Newtonian science holding the radius of s spinning object unaccountable in the spin, whether the spin is applying or not.

What this statement implies is that r does not exist. When anything has a value of zero it is for all purposes non-existent. Only when an object is following s straight line can the radius be non-existent because the radius alters value through time development.

The spinning or not spinning is not part of the issue because at the point of absolute singularity the object never spins. Therefore spinning or not spinning does not apply to the point of singularity because singularity never spins in any event.

$\Pi \times r^2$ = CIRCLE

If you remove r it then is $\Pi \times r^2 / r^2$ = CIRCLE.

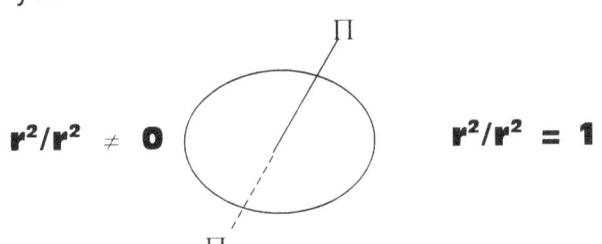

$r^2/r^2 \neq 0$ $r^2/r^2 = 1$

You cannot then say $r^2/r^2 = 0$ and therefore $\Pi \times 0 = 0$. That is nonsense. $\Pi r^2/r^2$ will always be $\Pi \times 1$, and that is the eternal circle.

That then is singularity. The eternal Π, the Π that may not have significance but still it is a Π of value. The relativity remains one, eternally one, but it cannot be zero. Therefore, dJ/dt cannot be zero.

    dJ/dt can become eternal or infinitive or at the worst it can become one
    dJ/dt = 1

When explaining this to any child, they can immediately see that. Explain this to any Newtonian High Priest and he may have you removed forcefully from campus. I cannot find one Newtonian, of any significance being large or small to accept that. By not having a wheel rotate, the wheel becomes the factor of one, and the rotation becomes zero. The wheel does not disappear. In the cosmos, everything is rotating because nothing ever stands still. Therefore the mean equilibrium, the common factor there is to share, has to be one, eternity, the eternal Π, because all rotating objects has Π in singularity, and sharing singularity, gives every object in space a relation with all other objects in space. After trying for many years to bring them the candle, I concluded that Newtonians are incapable of realizing that mathematical principle as reality.

If Newton had said that $dJ / dt = 1$ then that is exactly what Kepler said when he said that in the centre of space–time singularity is allocated a position of control $k^0 = a^3 / T^2k$. Kepler also said the motion brings about the filling of singularity $k^0 = 1^0 = 1$ when he said that the space is filled by the matter in the motion through the time period. Motion establishes space in time and cannot be zero because THAT is what gravity is. It is the motion of space-time and that can't be zero. If gravity were equal to zero the entire Universe would stop existing.

The centre and that which connects the centre is of such importance that every Universe holding singularity at $k^0 = 1^0$ pivots around it. With every one of the four points taking form to the value of Π at a measure of $\Pi /2$ each brought about the Roche value of $\Pi^2 /4$ in relation to the developing centre. One has to remember that the star of today takes on the characteristics of the form of that era. The barrier there is relates to this precise limit being the end of the Universe at $\Pi^2 /4$. If there was no bearing of the centre on the why would that be a factor in the formula we use to calculate the size and why would it have any role to play.

In the same manner the ring cannot remove, because the spokes will then still imply where the ring must be. The only way to cheat yourself out of the situation is to remove the wheel and spokes altogether, and you are left with what you say there is: NOTHING. But that does not apply in cosmology. The object rotates the centre structure and therefore there has to be a radius holding the circling orbit in relation to the centre structure.

Removing the radius from a circle does not be removed the circle, because the circle is there, securing the ring If the Universe started from a point of singularity, then there was initial spin at a pace where the spin did not apply and that spin included the entire Universe, still in non-existence. That is singularity. That is the only singularity there can be. The spin was going on for eternity because the spin does not apply, it has a value of infinity and infinity was running along the line of eternity.

By receiving the command, singularity received a value outside eternity as Π^0 received edges. Granted the fact that the edges were so small there still was no r to present a circle.

Having edges where Π^0 duplicates to present the edges, singularity lost the value of Π^0 to the value of Π^1 with the same value singularity had being Π^1 to the one side and Π^1 to the other side, the cosmos received the eternal value of the first dimension outside eternity. It was the square of Π^1 being Π^{1+1}.

That was the first dimension outside singularity Π^0 where singularity has a value of Π^1 in the form of $\Pi^{1+1=2}$. The first claim to space had a value of Π^2. This applied to both sides of the claim to space outside singularity, and the double proton became the dominant factor on matter.

That, which formed the Universe was the growth of time expanding while overheating. It is the expanding by the measure of Π^2 that has much significance as where the expanding went Π and the expanding was limited to the value of Π^3.

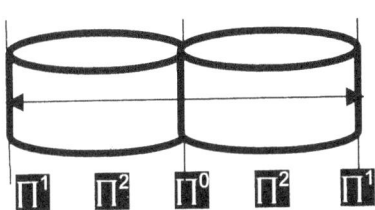

As singularity burst out into matter forming space as much as occupying space inside singularity, the protons started flying around, spinning around singularity, as each individual proton occupies matter in space. For every space there has to be spin and every spin is defined by a relevant. It truly makes me feel bitter thinking about the many times I tried to explain the facts to the Brainy Bunch with no luck. You know you are correct, but that person holding the establishment secured for Newton just push your argument aside, because he has the authority to investigate and lacks the interest to initiate change

In the action of the inseparable drawing closer and moving closer gravity finds the dual value of linear and circular gravity. There is no separation of the two factors acting as one but both have different application and values in the unit. This is the result of singularity having three parts acting as one but giving three distinctions in application.

How many dots was there is a question no person can answer because everything was un-dividable solid and yet it did group together to form every atom located in the 3D.

In the circle T^2k which consists of the atmosphere the space surrounding the rotating object will also extend by k as the concentration of the spinning motion draw or drag on past T^2 extending the influence of T^2 by the value of k. Very clear

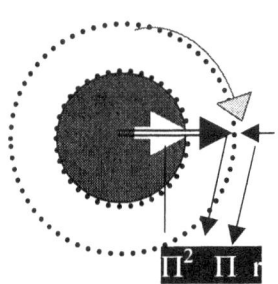

evidence about this one can see in the Coanda effect. This extending of T^2 to accommodate **k** we refer to as the atmosphere, but physics apply to this extending in the normal fashion. The soil of the structure represents the solid proton being $\Pi^2 + \Pi^2$. From the spinning motion T^2 does not stop at the end of the solid structure but the influence of **k** extends and this then becomes the atmosphere. The influence of T^2 stops at the end of the solid structure but the influence of **k** extending plays a most dominant role in the cosmos, although not yet recognised and that factor is most crucial to a better understanding of the implications of laws governing the cosmos.

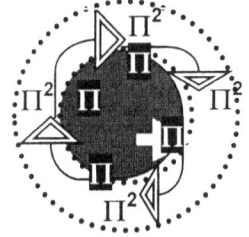

With the circle being T^2 **k** the T^2 will reflect the circle in the square with **k** forming the extending of T^2. This is an extending of the six **k** forming in alliance with the centre **k**. This produces that any extension of 6 forming material one further extending goes into space and relates to a seventh

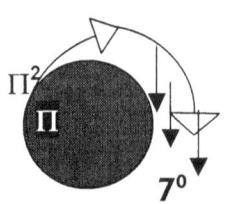

dimension. The extending of **k** will not end immediately but will carry to the surrounding space the circle influence through rotation. The influence immediately above the circle will have the biggest influence and reduce gradually as the value of **k** reduces in the leverage that the space has on **k** and a gradual but definite change from Π to r will affect the extending of **k** progressively more. The decline of **k** will follow the same contour of the circle at 7^0. Every one of the dimensions indicates an individual significance as I shall show later and the increase into space runs by 7^0.

Individual singularity and governing singularity and group singularity enhancing the gravity every time singularity find an accumulation. With looking at Kepler's in a mathematical sense it is clear that from singularity comes space by three duplicating space in time by three. $k^0 = a^3 / (T^2k)$. Very clearly the dimensions produced space and produced more space by applying time and gravity as

Any point will be opposing itself within the **rotating of 180°** where it **then change every aspect** of its **previous flowing** characteristics it had or **will once again have** in **360°** from there. While in rotation from the view point of a bystander it all may seem static and never changing but to the object in spin every next instant in time will be diverting from every aspect it had every second passing, and the

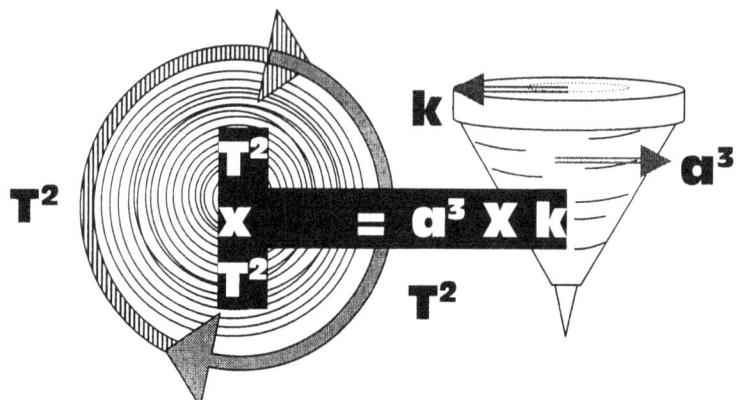

direction it held in relation to the direction it held the previous mille, mille second will totally be incompatible with the direction it holds the very next mille, mille second of rotation. This is why we can use degrees measuring the circle by (6^2) (forming the square relating to matter through singularity) X 10 (square if space) = 360^0 however it is always in motion. That proves no point can be static or constant, though it may seem that way to outsiders. Although matter is matter, matter can also be anti-matter and moreover form its own anti-matter at the same time. This degeneration of structure is very likely to occur with overheating. Revaluing Π to Π^2 will bring about a new contact point where Π meets **r** forming another relation in Π^2 **Time is** the **changes in relation** where Π **contacts a different r** not withstanding the many r points there may form because **every r constitutes a different value** to the universe through other ratios and relevancies brought about **by heat and light. Time is the duration it takes Π to rotate between any two given points of r** and therefore must always amount to **a square (T^2)** moving from point to point through the **cube of space (a^3)** in that **duration of time (k)**. With that it proves **Kepler's a^3 (space) $=T^2$ k (time in the instant of motion)** but motion must continue through a specific value in space where the space-time is maintaining relevant equilibriums throughout singularity connecting.

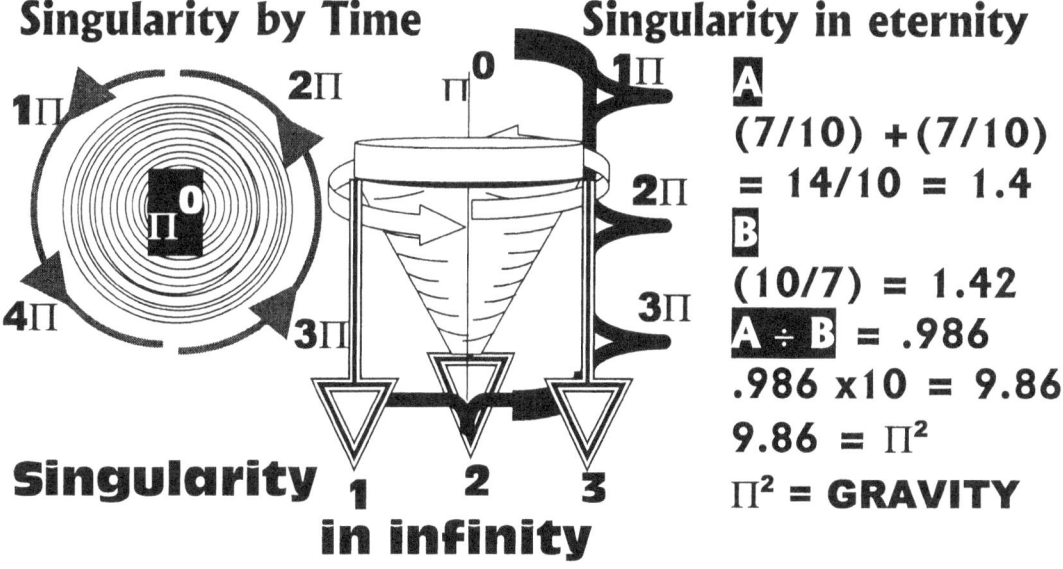

With the dimensional change from space in the cube to space in the sphere a relation of 5 to 7 comes about depicting gravity. The principle of 5 sides in space relating to 7 in the sphere holding matter forms the basis of the Titius Bode and the Lagrangian principles.

Newton, and science, made one enormous blunder, from this stance. They took the radius of a wheel not to have any influence on the wheel. In doing that, they removed the very fact that keeps the universal attachment together. They still insist that rotation results in nothing.

The state that is the time component called outer space, is coming about from the fact that outer space is the Titius Bode law because the Titius Bode law is evidence of how the Universe was compacted by motion. The time zone called outer space is the motion called gravity and is the neutron factor in the Universe.

$r^2/r^2 \neq 0$ $r^2/r^2 = 1$

$$\frac{dJ}{dt} = 0 \quad \frac{dJ}{0} = dt \quad \text{or} \quad \frac{0}{dt} = dJ$$

$\Pi \times r^2 =$ CIRCLE If you remove r it then is $\Pi \times r^2 / r^2 =$ CIRCLE.

You cannot then say $r^2/r^2 = 0$ and therefore $\Pi \times 0 = 0$. That is nonsense. $\Pi r^2/r^2$ will always be $\Pi \times 1$, and that is the eternal circle. When looking at any rotating object, there has to be a point in the infinite middle where the one side rotates in one way and the other rotates in the other direction opposing the opposing direction. That point in infinity is the point of no rotation and no rotation means "no rotation", not no existence. No rotation means a factor of 1, not zero.

That then is singularity. The eternal Π, the Π that may not have significance but still it is a Π of value. The relativity remains one, eternally one, but it cannot be zero. Therefore, dJ/dt cannot be zero.

dJ/dt can become eternal or infinitive or at the worst it can become one dJ/dt = 1

When explaining this to any child, they can immediately see that. Explain this to any Newtonian High Priest and he may have you removed forcefully from campus. I cannot find one Newtonian, of any significance being large or small to accept that.

The comet rotates the Sun , and the Sun by itself has a point of singularity where Π remains without r. The comet, holding the orbit, also has a point of singularity, but since there is space separating the two objects, they cannot share a mean point of singularity, the very point of existing. Since singularity means just that, being single, there cannot be two. The comet and the Sun have a mean point of singularity but the space they occupy divides their common singularity. That is why they orbit in an oval path, a path where the one structure holds on to more space from its point of singularity towards the space it claims. Since they do not claim equal space, BY THE DENSITY they hold, the space will not be in proportion. Singularity is a mathematical reality. Einstein may be the first to name it and Galileo (unwittingly) may have been the first to define it as Kepler was the first to formulate singularity, but in mathematical terms singularity is the most basic principle. It is singularity that attaches the top to the orbiting comet. At this point I wish to establish a fact that seems lost in all other grandeurs of cosmology. A straight line cannot begin at zero or nil it can only start at infinity/ Such a statement will hardly seem appropriate but the relevancy of this fact has no limits.

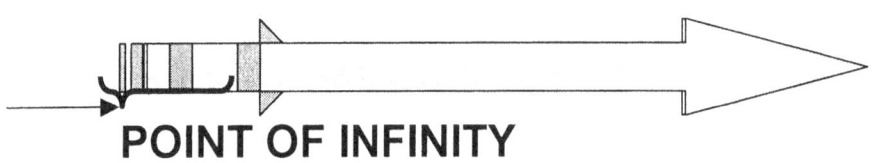

POINT OF INFINITY

If the line started at zero there was no line to start because zero multiplied by whatever results in zero as the answer. That must also be the cosmic starting point. Einstein introduced such a point and named that point singularity.

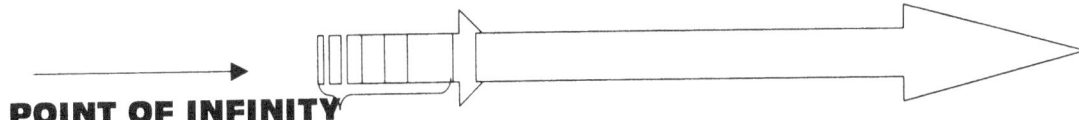

POINT OF INFINITY

The Universe does not change because there is not one single item that is in the Universe that can change. What the top evokes is what was established at Moment-Alfa when time was interrupted by space for the very first time. The spin or expanding that was introduces still present the very same principles as the spin or expanding did when it did for the very first time. The line of time is eternal and was interrupted by space with infinity bringing an end to eternity in time. However in infinity the line was interrupted extensively but briefly while the line continued intensely small but eternally long.

Einstein introduced matter time and space and I can see where Einstein was heading with the three concepts forming one Universe but Einstein got his wires slightly crossed, because for one, what Einstein saw as space is time and what Einstein saw as matter is time in general as it is a connecting that singularity has with the flow of time throughout the universe. In that there are two strands of time that formed with one massive time delay that compacted the heat in the time delay and that time in delay compacted in nice units, which now forms the part that are the matter Einstein referred too. In that there is overall just time relating to time in time. However the only existing Universe is the Universe, which is not part of the Universe. The rest is creation by generating motion and it is not created as such in measure of time flow so very long ago and back in the most distant part. It is created by motion of time delay through time in time. By moving from one point to another point the flow of time goes square while the points duplicate. The duplication is a product of overheating in one specific spot where that spot exaggerate the space by expanding. However in duplicating there is cooling and cooling is reducing of heat. Heat is what there is becoming more of the same thing and therefore cooling is what there is taking away some of that.

The motion brought about the square of the value but in the square initially was only singularity.

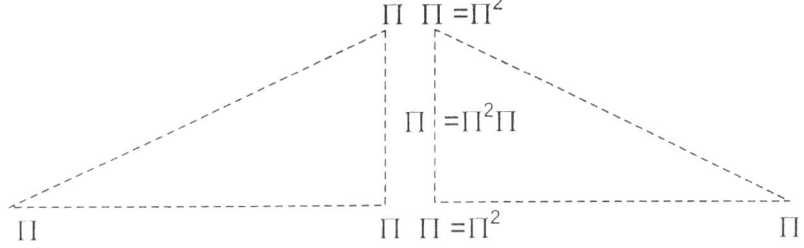

As we already determined on a previous occasion we now accept that expanding comes about from heat and only heat brings about expanding. By heating whatever one may find in the universe expand.

By expanding the space becomes more and the space in doubling cut the heat by half. It is an immaculate and genius way of controlling heat.

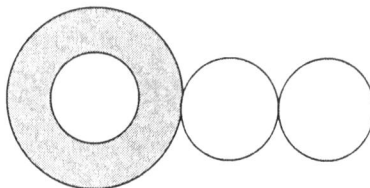

As it is still the case, the contraction results in duplication and by distributing the overheating space over a bigger area, the cooling contracts half (in the beginning but at present the proportionate) the heat back to secure the original singularity while the other position secures a new point that activated singularity. The distribution results in the contraction. When 1^1 expanded from 1^0 there was a linear motion established. The motion took what had no start away from what had no end. In that the expanding had a direction as lateral that connected to the cross of the lateral that connoted what remained of the eternal to the lateral.

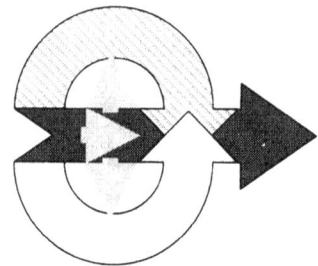

The lateral motion connected 1^0 to 1^1 but in reflection as a mathematical response there was too a connection with 1^1 the upper casing to 1^1 in the lower casing because there was an immediate contraction to the expansion.

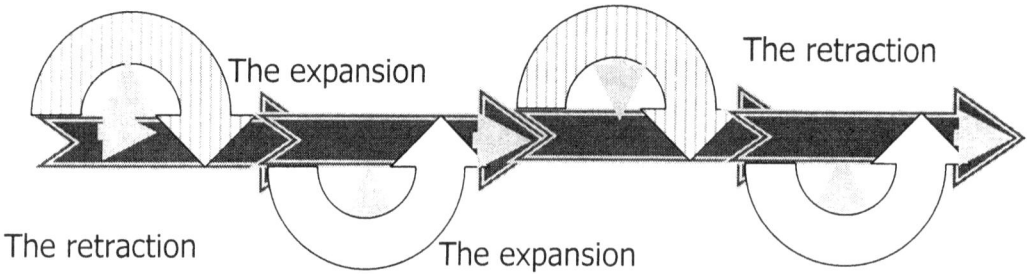

The expansion

The retraction

The retraction

The expansion

The flow of time from the past through the present onto the future had three dimensions resulting from the oblong time in the perfect eternity strayed from by going imperfect. This brought to the future where half confirmed the past, half conformed the future and believe it or not but half converted the future.

Therefore with one the Universe was in eternity and the value of the one was confining everything in singularity. By duplicating two relevant points forming three positions in singularity came into being a form in the universe. The number arriving in the Universe was two, but I am somewhat reluctant to say that what ever formed at this stage, was already part of the Universe. That is still miles off.

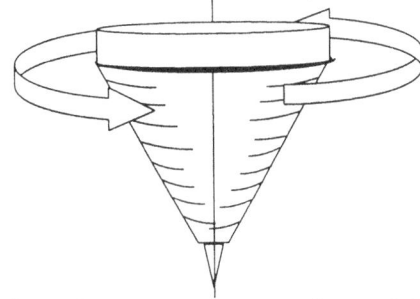

$T^2 = a^3 /k$ and $T^{-2} = k / a^3$. In this period of development the time associated with eternity much more prevalent than it did with a break on continuity in infinity.

What happened back then is precisely what we are able to gauge from the behaviour we find the top shows because the Universe doesn't ever change.
This brings us back to the spinning top I presented at the beginning.

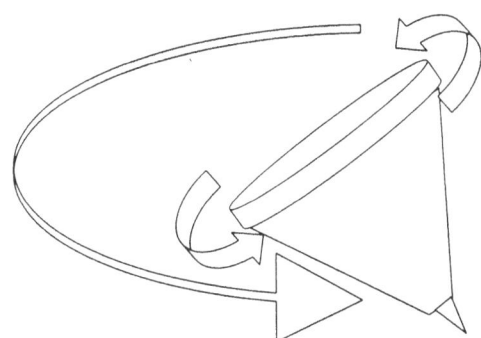 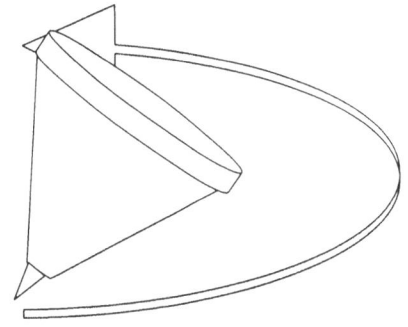

I have asked as many persons as I do not care to remember why the top sinning will remain spinning around one point while turning. The answer I receive from the most educated to the schoolboy is always about momentum. That is a very simple answer and to say the least a little too simplistic by further analysis. Why would the spinning top go off centre when spinning higher than a specific velocity and lowering the velocity it would stabilize and run square to the Earth only after that it will go oblong and then fall.

I could go on about different positions bringing across different momentum of thrust but I do not wish to insult your intelligence because I am aware that you are familiar with all the law. When the top is spinning it is spinning about its own axis and when it is not spinning it still remains spinning about the Earth's axis, therefore when it is spinning it is also spinning about the Earth's axis.

Therefore the limitations applying can only result as an influence coming from the Earth's axis. The second question now comes screaming across and that is in what manner could the Earths axis ever affect a spinning top since the spin and the spinning top is a gross mismatch to what ever standard the Earth may introduce. It is clear that spinning objects do influence each other in contrast to Newtonian opinion.

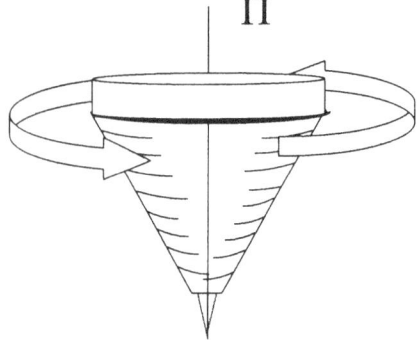

Every round object has a point establishing a very centre, a middle dividing one side from the other. That division determines the space from one side away from the other side. At one point there must be a point that does not fall on either side of the divide. Such a point will still be a circle, because from that side the circle divides into two sectors.

In every spinning object there is a point of infinity, a point that does not turn because it holds the dividing spin. From that point running in all directions the spin is opposing the other side. All spinning activity starts at that point diverting outwards and from that point the spin is either clockwise or anti clockwise in all directions. As I pointed out no line can start at zero because then there is no line and no rotating point can start at zero because then there is no rotation.

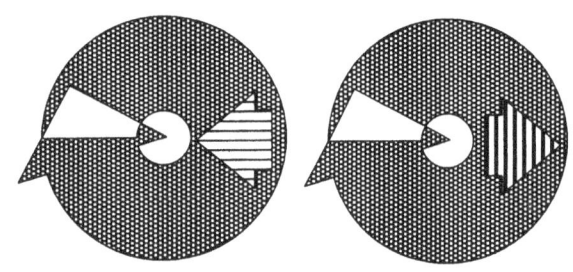

I have indicated that motion creates space $T^2 \times T^2 = a^3 \times k$ and space finds limits of space in motion $k = a^3 / T^2$ as well as $k^{-1} = T^2 / a^3$ where the motion confirms the space while the space conforms the motion. The rotation of the motion in both relevancies completed the space that formed by the motion in linear as well as rotary.

Saying that 1^0 moved to 1^1 sounds senseless but in that motion is so much potential locked in that no other movement ever came close to such giant step. It happened when the first relevancy came about and everything that ever could be in the Universe and would be was locked in one spot becoming one dot. To our mathematical genius of numbers and measure the movement had no meaning because it still remained what it was before in $k = 1$. To us with small minds that reality left little cancelation with k being either 1^0 or 1^1, which remains at 1, but the moving of the space gave the space a reason to be and the space gave something that could be moved a distance from 1^0 all the way to 1^1. The emerging Universe in its entirety was in prominence.

By reducing the one line the other line can never reach zero because then there were no such a line to begin with. That makes a straight line also inevitably always a potential square and that makes the straight line half the value of the square being 180°. At a later point I shall continue with this argument, but for the mean while I wish to come back to the circle. This same principal applies to the cube and that means everything there is and ever will be is either a square being part of a cube or a circle. With the straight line forming half the value of a square $360^0 / 2 = 180^0$ in as much as being one line and reserving one line in infinity to eternity. The straight line is just half the value of a square. In that manner the triangle is also half a square and therefore holds the same dimensional value as the straight line being also 180^0

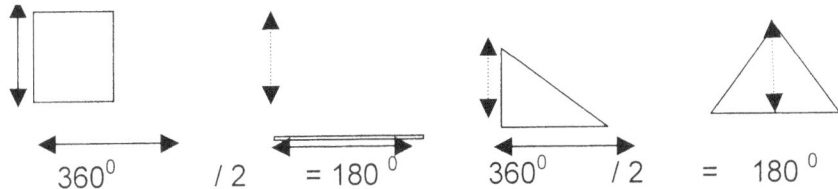

The circle is a square holding a round shape, as the straight line is a square holding one side to infinity. Calculating a circle involves two aspects where the one is either the radius or the diameter that is double the radius. The other is the factor Π

$\Pi \times D^2 / 4 =$ circle and $\Pi \times r^2 =$ circle

The point of singularity cannot be in space at large because space is not there and secondly what ever is there spin to slowly to have a connection with singularity directly.

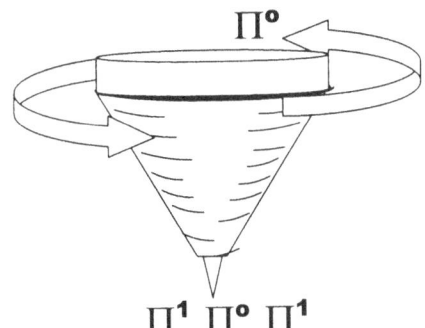

$$\Pi^1 \ \Pi^0 \ \Pi^1$$

With everything in a cube or a circle or a potential of the two, brings about the implication of eternity in a form of singularity or the point of creation. Removing the radius of a circle does not remove the circle, because the circle is there, securing the ring. If the line (or imaginary line if you wish) holding the value of Π^0 = 1 there has to be a point where the circle is no longer in infinity but claims existing outside the imaginary. At that point the radius may be slightly more than infinity, but to all calculating purposes it still remains as infinity. The spin was going on for eternity because the spin does not apply, it has a value of zero and zero is another expression for eternity.

Having edges where Π^0 duplicate to present the edges singularity lost the value of Π^0 to the value of Π^1 with the same value singularity was being Π^1 to the one side and Π^1 to the other side, Π^0 must be the point splitting singularity into two parts of eternity, the eternal value of the first dimension outside eternity. It was the square of Π^1 being Π^{1+1}. That was the first dimension outside singularity Π^0 where singularity has a value of Π^1 in the form of $\Pi^{1+1=2}$. The first claim to space had a value of Π^2. This applied to both sides of the claim to space outside singularity, and the double proton became the dominant factor on matter.

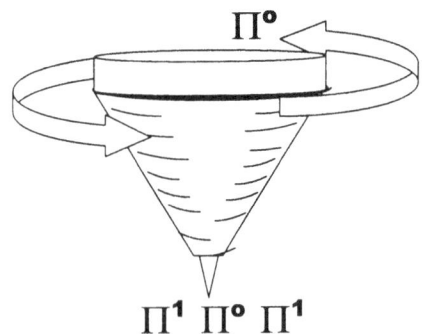

$$\Pi^1 \ \Pi^0 \ \Pi^1$$

Right at the start before space and time became developed the motion produced space in the principle of the Coanda effect. By receiving space, singularity received a value outside eternity as Π^0 received edges. Granted the fact that the edges were so small there still was no r to present a circle. The manner that the top use to evoke singularity, which enables the top to maintain independent motion could be, traced right back to the very first line that came about as Singularity initiated spin

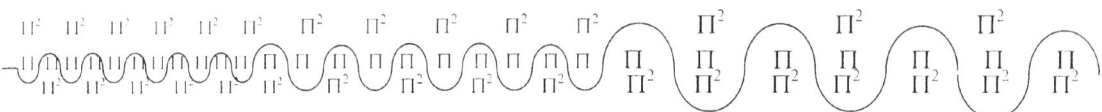

In the beginning there was no space in which to move so therefore the only way to move straight was to move in a circle. The movement k producing the line was the same as the motion T^2 that produced the circle and the space a^3 achieved was the compliment that two factors combined and that formed the space developed.

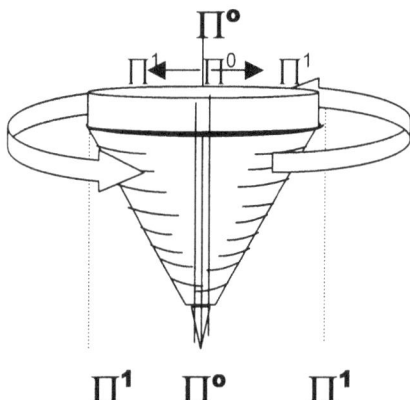

$$\Pi^1 \qquad \Pi^0 \qquad \Pi^1$$

Taken from the point of rotation the two sides are in opposition to each other in every aspect that they may contain and with all that they hold. The motion is the extending of singularity and singularity reacts on the motion by establishing a proton value.

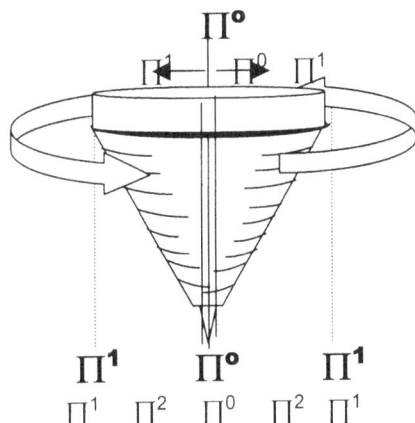

With Π^0 little more than a figment of the imagination there is actually to values of Π^1 facing each other in a relation combining Π^1 to hold the value of $\Pi^{1+1=2}$ $=\Pi^2$ and with two sides being the very same but opposing each other there will therefore also be Π^2 to every side that holds Π^1.

At last I can come to the one part that I disagree with Newtonians, and what I regard as Newton's second biggest infamous or famous blunder. Science, made one enormous blunder, from this stance. They took the radius of a wheel not to have any influence on the wheel. In doing that, they removed the very fact that keeps the universal attachment together.

Singularity controls the Universe by establishing a Universe but that is done in a specific manner.

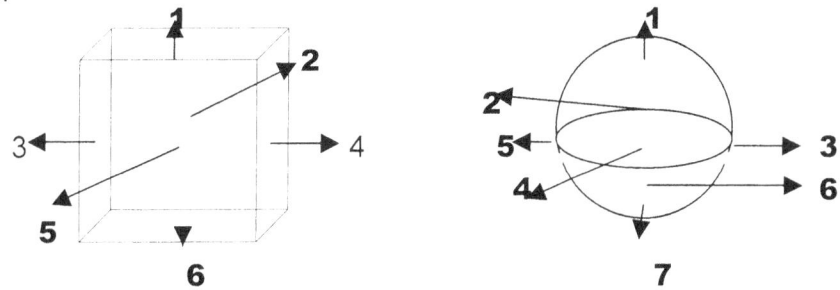

In the very centre of the sphere the form of the sphere dictates that the shape will relinquish space as the line run from the outside towards the very centre. With this natural state of affairs the sphere are naturally inclined to dismiss all space that it can form in the form as the sphere holds space inside and the form will finally be without dimension. All that I attribute to the line shrinking by reducing actually takes pace in every sphere as the diameter reduces to the centre. In the centre where the radius line goes single the form relinquish the three dimensional form it has inside. Being without dimension in the very centre means that at a point in the extreme centre of all spheres there are a point that holds singularity because this point with no space has a mathematical position although it is invisible since there is no sides to such a point to give that point any dimensions. The shape of the sphere is calculated by using the formula $4\Pi\,(r^3)\,/\,3$.

By reducing r to a point where r is r^0 singularity steps in because only the form remains as Π. Going even further we find that there then comes a point where Π goes singular Π^0. At that point absolute singularity is present but so is absolute gravity present at that point. When holding the strength of the shape of the sphere in mind as well as taking into account that all cosmos objects of importance is in the form of planets or stars and they are all in the form of a sphere, we therefore may contemplate that it is where gravity originate. We now only have to find the reason why gravity will hold a base in a space less ness as Einstein predicted. It is clear to be seen that gravity is in the centre of the sphere controlling from the centre everything that is outside the space less centre. We can reason with confidence that gravity is the strongest where space is the least. We can further reason that it is gravity that is holding the sphere in true form and since the sphere allow gravity the best working opportunity, gravity can form the sphere in as strong a shape and form as the sphere seems to have.

From every point on the surface of the sphere is where that point connects with the other side of the surface of the sphere by a line that runs through the space less ness of such a centre of the sphere. Such a line also connect by an angle of 180^0 as well as 90^0 to six other lines running from top to bottom, right to left, and back to front, where all join and cross in the centre of the sphere. There are therefore six lines crossing and connecting by a centre from any given point on the surface of the sphere. Such points connects in total six surface points on each side of the sphere while they all support one another through the space less centre. In that absolute space less ness in the centre holding singularity we find gravity supporting and controlling all space within the sphere as well as space connected to the sphere. That is where gravity control and guide the space, which falls in the parameters as well as under the influence of the form of the sphere. In the gravity centre space goes singular meaning space becomes space less or flat.

It is from the layout that the sphere uses as natural form that we are able to locate singularity. In the case of the sphere the material naturally reduces by measure of the radius becoming smaller to a point where the radius is r^0. At that point the line that will form the radius has gone single dimensional r^0 and that is equal to 1^0, which is singularity.

Also it is true that the entire form that is the sphere is controlled from a centre within the sphere. That centre holds the sphere in form and shape. Therefore the strong form is dictated from that space fewer centres where there is no space and no form left. The natural inclining is in the form of the sphere. It is part of the roundness that the overall shape of the sphere represents and this structural strength is carrying down to the very centre.

Because the circle is forever reducing that reducing which is inherently part of the form of the sphere becomes a tool in distorting of space in the sphere and is eventually removing all forms of space from within the centre of the sphere. The very centre ends up as having no space because of the reducing that continuous down to become the space less inner centre. The all roundness is the ingredient that forms the backbone of the absolute strength that the sphere has and that is the component that the sphere is so famous for. The form the sphere has allows the sphere to have a control that is coming from the centre deep inside the sphere where the space vanishes and being without space seems to keep the entire structure rigged. From the centre the sphere shape shows strength that the shape as tough as it is. How does it work in its most basic analyses?

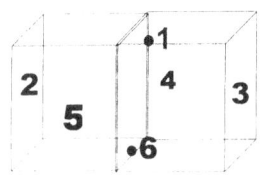

There is one more point in the sphere in the centre forming an addition in the sphere. That point holds gravity secure.

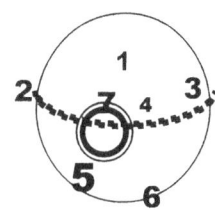

The cube has sides and the sides form a rather weak and flat surface that connects four corners. The flat surface produces a rather indifferent contact point with no special features on the surface. The corners connect to other sets of corners and those corners form a weak structure without any direct support coming from the other five sides. Without material to fill the body of the cube the cube has no direct connecting between any of the sides other than corners connecting at the edges of the sides.

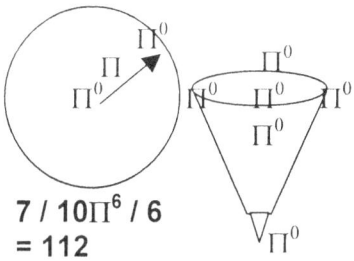

$7 / 10\Pi^6 / 6$
$= 112$

Taking the vantage from the point the sphere is holding from the centre out into space there are ten points connecting to the centre. In that are the dimensions of singularity connecting to space where five connects to space in the second dimension of singularity, and five connects in the third dimension of singularity. On the other hand, the cube does show a very different characteristic, which involves only six sides (at least) connected.

The spinning of Π^0 around the centre Π^0 establishes Π and Π is what produces the form gravity has. Still it is the relation or relevancy there is between the centre Π^0 and the spinning Π^0 that gives status to the form that Πrepresents. In out Universe we are accustomed to and are familiar to the rules we want to place seven points holding singularity to the centre holding singularity in a relation of $7/10\ \Pi^6 / 6 = 112$. In that Universe everything less that a duplication ability to the value of 112 protons fit but only atoms to a maximum of 112 protons fit.

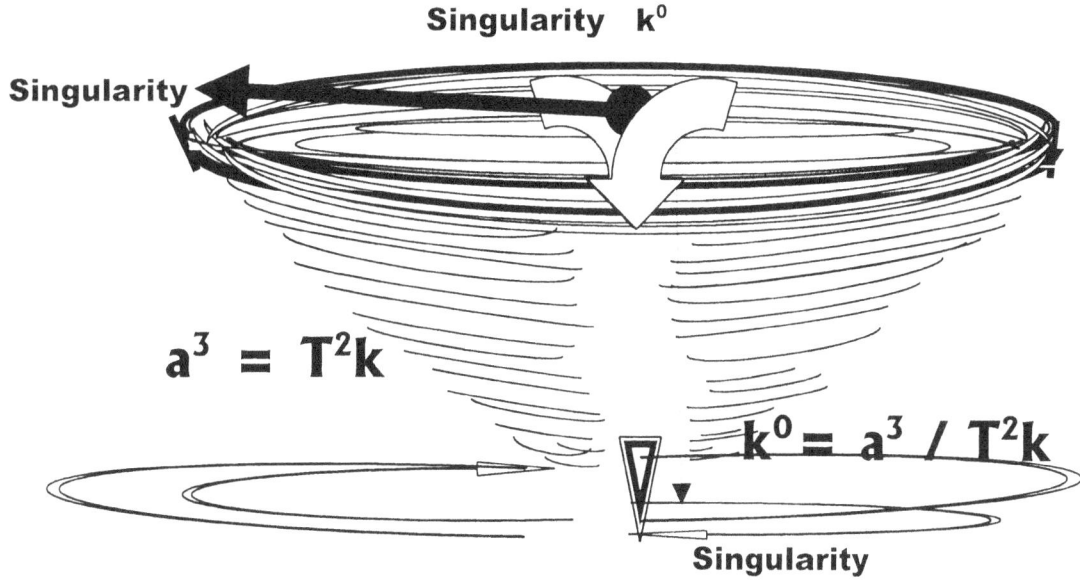

Singularity k^0

Singularity

$$a^3 = T^2k$$

$$k^0 = a^3 / T^2k$$

Singularity

If gravity is motion T^2 the process may sound deception ally simple but it is very complicated to rein act. By motion T^2 space comes about forming a^3. But the motion T^2 will mean a crossing to the other side of the universe since singularity divide the Universe into sectors.

 Material produces the dismissing or the concentration of space by applying the motion. Surrounding all elements are a layer we call the atmosphere and even Pluto and the moon must have the atmosphere because the have gravity. In this the relevancy of ten to seven forms this layer and it results in forming a circle because of the combining of the motion duplicates the singularity factor Π forming from that gravity as Π^2

THE PROCESS PARTED USING THE ROCHE PRINCIPLE

By establishing motion and creating motion singularity quadruples to 4Π in rotation. But since the rotation is motion duplicating the space established as four times the value of singularity the motion divide the space coming about by halving such space by the dimension, which is putting a square root over the quadrupling of space. In this comes about the direction gravity takes the universe. The expansion is always double the square root but the square root is neutralising the expansion and that brings about that the neutralising of the expansion creates a contraction that seems dominant to us but it is not. The contraction is doubling the expansion by halving the effort of the expansion.

Singularity is a mathematical reality. Einstein may be the first to name it and Galileo (unwittingly) may have been the first to define it as Kepler was the first to formulate singularity, but in mathematical terms singularity is the most basic principle.

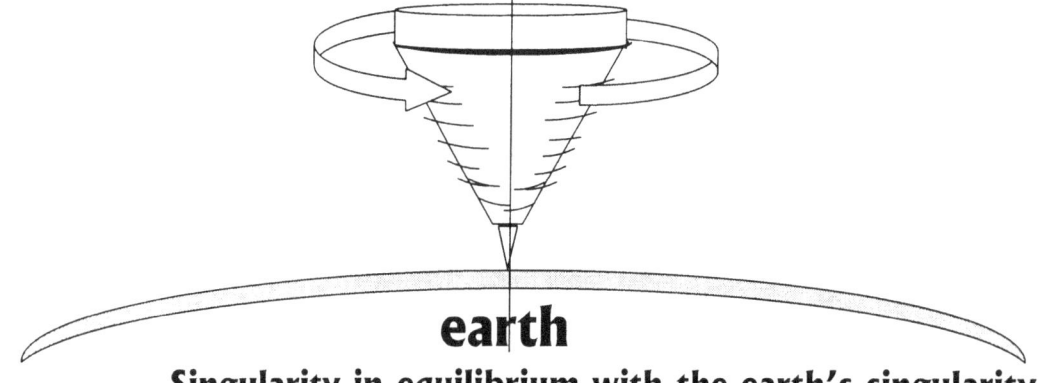

earth

Singularity in equilibrium with the earth's singularity

Singularity applies gravity by charging motion and the motion (not the pulling) of space-time is gravity. It is motion that moves the gravity that moves the Earth and it is also motion the moves the top that forms gravity. Gravity is not the pulling of but the motion or tendency to move to the centre of the Universe, which is the next domineering, point holding singularity in a control or a governing mode. The motion or tendency to move is that which forms gravity and mass is the occupying of space and therefore restricting the motion of gravity.

The greatest minds in the entire world are missing the smallest line in the entire Universe.

By rotational motion, the top creates a line confirming singularity running down the line and by generating the line the line charges gravity. The gravity is what drives the top as the top and as long as the top spins. There is an influence generated by the spin of the top that keeps the top upright while the top is spinning.

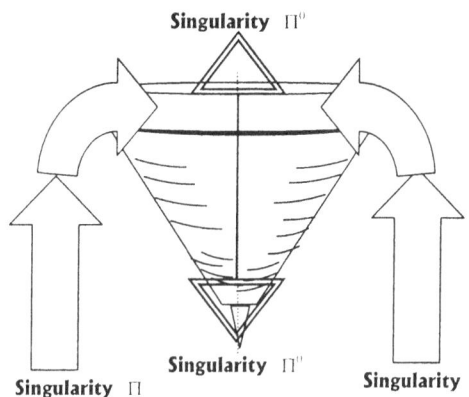

The line is generated but the line is far from magic. The line is where the centre of the Universe is which the Universe is then that what the top filled by particles from the line to the edge of the sphere. The particles in motion generate motion by electing a centre from the centre of every particle in the spinning top. Such an elected centre becomes the centre of the Universe as far as the top relates to a Universe because all the atoms in motion elect the centre of the Universe.

In this, it is clear why the Titius Bode ([10 + 10 + 1 + .991] / 7) and the Lagrangian 5 \\ 1 systems part their ways when applying the different processes they hold. With all the differentiating, the observer must also consider the dual message that light uses in travelling through the vastness of universal space. The thought of nothing is just what it is, a thought of nothing and although it is in the human mind common nature to present nothing as a value in the recalling of something, nothing is a presentation of the figment in the human mind. There can be no number such as nothing and that was (possibly) Newton's biggest error. Nothing represents non-existing and that is just what nothing is, it is non-existing.

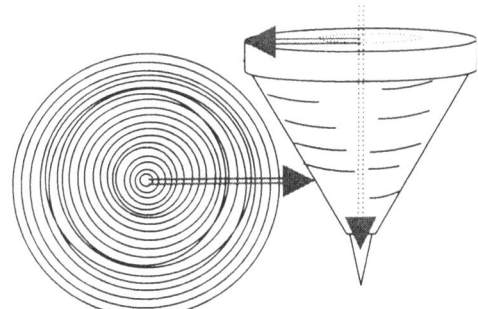

The centre may or may not spin and the fact that it does or does not spin is all the same because that centre part never spins in any case. Therefore the boundaries set by the spinning motion does not depend on the spinning motion of the object but has to stand related to another bogy bringing about a larger spin influence. Granted the fact that the influence the Earth has on the top may be that of gravity but if that is the case then surely the Sun has also influence on the Earth and other rotating objects through gravity. It needs more investigation because it may bring about evidence we are not aware of.

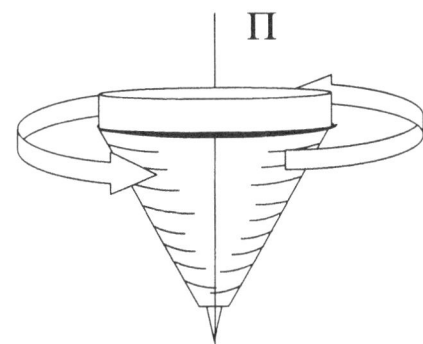

This observation places a much bigger question mark on the statement of Newton where he proclaims no influence on two rotating cosmic structures.

By rotation such rotation is a duplication of what singularity retracts. The rotation involves the three factors in time coming from the past through the present and onto the future, which holds three positions excluding the one position allocated to singularity. The four I named the eternal motion because singularity being without motion was contracting as well as expanding at the same time it was not moving. That placed singularity in eternal motion without ever moving being singularity.

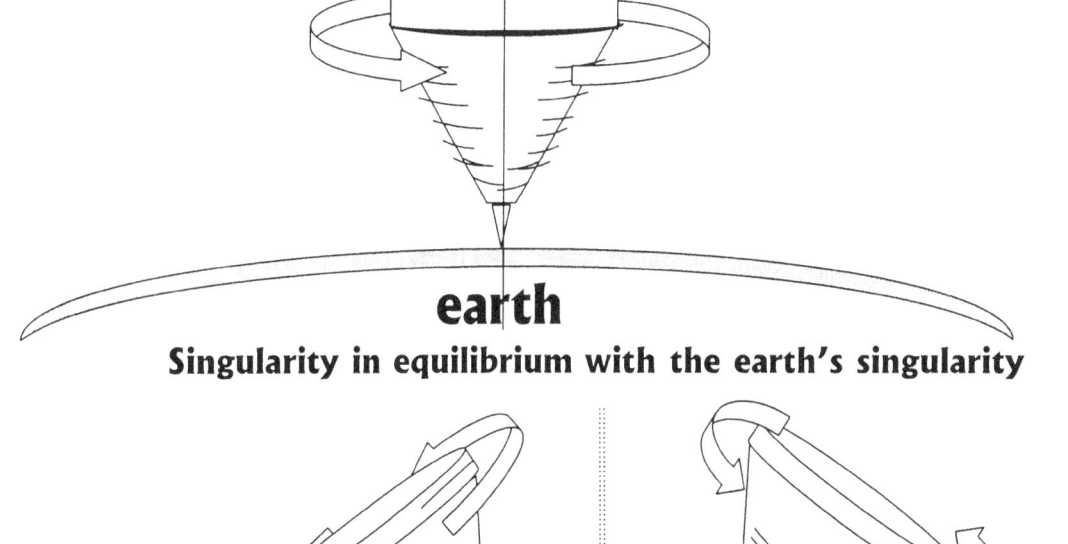

Singularity in equilibrium with the earth's singularity

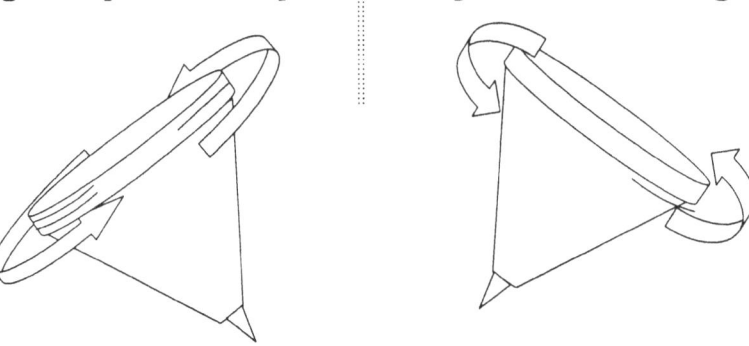

Singularity of the top exceeding the earth's singularity

The earth's singularity dominating and exceeding the singularity top

Understanding all the following is connected intimately and all conditionally to the fact of accepting that all individual particles in the universe use motion and therefore spin.

Every quarter provide a distinct value that indicates the progress of the flow of time from the one point Π to the next point Π.

Any changers occurring in Π will lead to a an unequal triangle providing two different values to r and will alternate the link between r and Π^2 bringing about different form (Π) and

time (Π^2). When singularity forming the lines of the triangle is not in equilibrium the triangle will destroy the matching of half circle.

In considering the spinning motion in the fraction of time in the detailed instant every aspect of rotation will turn in every instant of change in time. Although the points had the same characteristics only one instant before, they oppose the characteristics it had just before and just after the very instant in which they are and to which they relate by similar points also in rotation. The fact of the graph proves my point in quarterly opposing dimensions and values. As the rotating direction moves inwards, the rings will become smaller and smaller. Move the rotating line progressively to the middle by reducing the length the line have from the edge to the middle. At one point all further reducing ends.

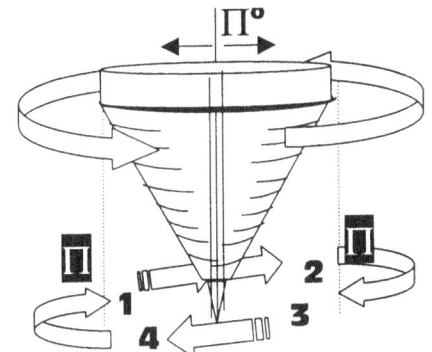

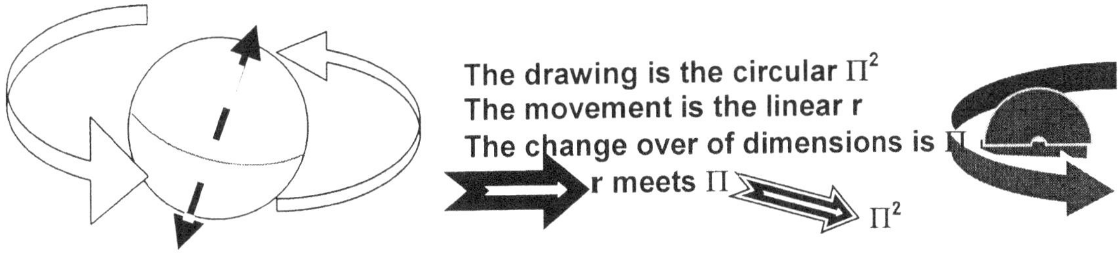

The drawing is the circular Π^2
The movement is the linear r
The change over of dimensions is Π
r meets Π Π^2

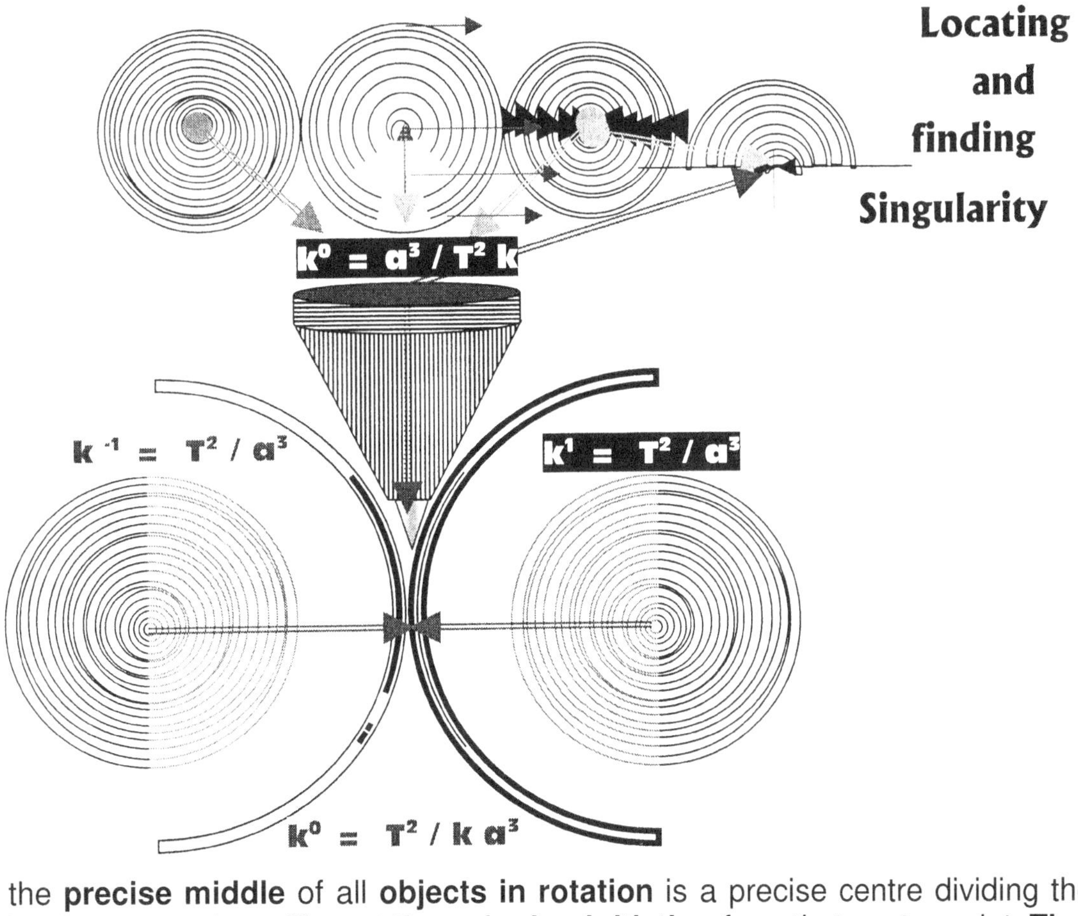

Locating and finding Singularity

$$k^0 = a^3 / T^2 k$$

$$k^{-1} = T^2 / a^3$$

$$k^1 = T^2 / a^3$$

$$k^0 = T^2 / k\, a^3$$

In the **precise middle** of all **objects in rotation** is a precise centre dividing the object in sectors that will **start the spinning initiation** from that centre point. **That**

point albeit hypothetical, is also as much a reality none the less and is where that point **must be standing still** because every line **running from that point** in **opposing directions** are also **in opposing directional spin to each other. That point** is completely hypothetical, is also as much a reality none the less and is placed where that point **must be standing still** because every line **running from that point** in **opposing directions** are also **in opposing directional spin the other or opposing side.**

In considering the spinning motion in the fraction of time in the detailed instant every aspect of rotation will turn in every instant of change in time. Although the points had the same characteristics only seconds before, they oppose the characteristics it had just before and just after the very second in which they are and to which they relate by similar points also in rotation. The fact of the graph proves my point in quarterly opposing dimensions and values. Due to the spinning nature of such a point with all surrounding the point will be alternating direction favouring change every second and in that the value to such a point can only be Π because of its constant changing. Using r would specifically oppose another r from every angle because the use of r will bring about a static relation to the previous and following instant and therefore it will cancel the constant spin flow. There must come a point where the ring is infinitely small, where it can reduce no more, where it reached its ultra limit, but at that point it cannot be zero, because the point is there notwithstanding that it is at a location beyond our Universe. But the spinning object **will have a middle point**, a very specific **centre point that does not spin** and only holds Π as a specific value. One value such a line **cannot have is zero** because **zero does not start any** line and therefore the **value of the line must be infinite**, just as described in **accordance** and by **the definition of singularity.**

From somewhere outside the Universe a line rises while remaining in a position allocated to space being outside the Universe. The line is activated by the rotation

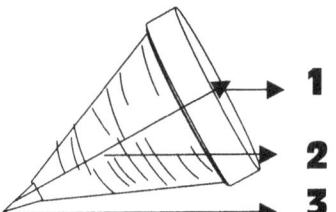

that sets the top in coherence with time and the motion grants the top individual status by establishing independence. The motion holding a dual action while being a unit lay down the ground rules for the Coanda effect. The four points turning around a fifth centre also charges a line to place the space-time the top holds to an erect status of independence. The line is at the centre and it is motion by rotation that activates the line. The erecting of the top underline the status the top receives as an independent Universe, which is maintaining an individual singularity. Even when the motion no longer finds the ability to charge the line in singularity and the top stumbles before it falls, there is still a desperate fight for keeping the motion and with that the independence active. The fight is not a fight for balance but a fight for survival.

Space parting eternity and infinity activates the line and Newton was of the opinion that rotation brings about no work. The line runs from the top of the top down to the bottom of the top without ever being present in the space of the top. The top forms the space being fully independent and it is most critical to consider that line that the motion activates also activate the third dimension of time in space. Space in time and time within space separates by finding separate identities in the cosmos and this gives singularity independent identity. It parts eternity in time from infinity in time. A universe is born through the rotation of the top in spin. By parting infinity from eternity the space in between fills with material that allows the top the position the material in the top has while the top is spinning. When the motion falters in sustaining the singularity it requires to remain independent the point holding singularity goes cold and looses the acquired independence it had.

That point albeit hypothetical, is also as much a reality none the less and is placed where that point must be standing still because every line running from that point in opposing directions are also in opposing directional spin the other or opposing side. Move the rotating line progressively to the middle by reducing the length the line have from the edge to the middle. At one point all further reducing ends. In considering the spinning motion in the fraction of time in the detailed instant every aspect of rotation will turn in every instant of change in time. Although the points had the same characteristics only one instant before, they oppose the characteristics it had just before and just after the very instant in which they are and to which they relate by similar points also in rotation. The fact of the graph proves my point in quarterly opposing dimensions and values.

This only applies in relation to time because time is the square or then if you wish time is the flat to space being the cube. Time in the square draws space in the cube flat and that is the why the Universe holds the sphere in place.

Understanding all the following is connected intimately and all conditionally to the fact of accepting that all individual particles in the Universe use motion and therefore spin.

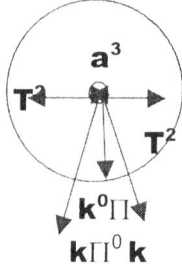

In dimensional terms, which I explain later on the value of **2k** relates to T^2. That relation extends to the next value where T^2 relates to **k**, which relates to T^2. The first space in the circle will then be T^2 **k**. From the centre being in infinity, one can realise by applying mental power the single dimension factor not seen but present all the same. Extending that into the 3D comes six **k** and any one of the six will further extend to form a seventh point as T^2

All this is a multiplying of $k^0 = a^3 / (T^2 k) = 7$

From this line of reasoning I dismissed the theory of the presence of a force being gravity but rather consider it as a dimensional changing contributed by the spin of the Earth and the spin comes from singularity located in the centre of the Earth. It is all about dimensional changing that influences space as a factor of ten to reduce to Π^2 on a continual basis from point forming new dimensions through billions of such points.

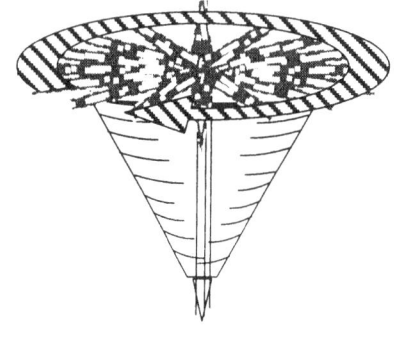

In conditions found in the Universe spin can only come about from heat that is concentrated around singularity. Dropping water on a red hot metal will lead to feverish motion as the singularity in the water absorb the heat and expand the space by accelerated duplication. The singularity find the time differentiation there is between eternity and infinity excelled and space excelled brings about motion in duplication amplified. That what we see in the top results from the interaction of life that in principle is some mistake the Creator allowed to happen in a very small region and as far as man is concerned taking into account all the proof man has to his disposal the phenomena of life is no where else in the Universe. By having T^2 overheat space will reduce in ratio that brings about linear motion k being shorter per time unit but much more frequent per time unit accelerating because $\mathbf{k} = \mathbf{a}^3 / \mathbf{T}^2$. However in accordance to Newton's law on motion there will be a reaction $\mathbf{k}^{-1} = \mathbf{T}^2/\mathbf{a}^3$ and with a reduced space the time in ratio will increase allowing for the heat rising and the amplifying of motion.

More spin increases both lines that force gravity by the increase of T^2 that extends the influence of $\mathbf{k}, \mathbf{k}^{-1}$ in the formula as factors because it reduces the moment of $\mathbf{a}^3$. The extending of the liquid heat will increase the motion and increases the contracting gravity $\mathbf{k}^{-1} = \mathbf{T}^2 / \mathbf{a}^3$ and the reaction to that is that the space reduces by a larger time contraction $\mathbf{k} = \mathbf{a}^3 / \mathbf{T}^2$. The space then has to duplicate more vigorously in order to cope with the rise in the time aspect. Therefore the space

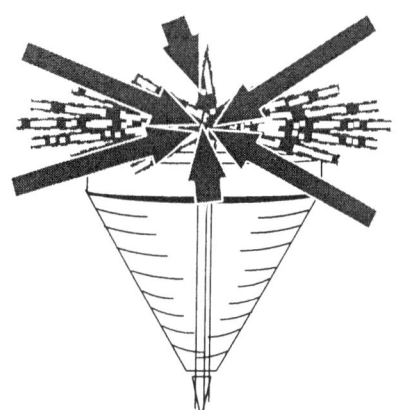

wants to exceed the boundary time slaps on space as a result of the Coanda principle by applying more contracting motion because of the rise in the liquid heat levels in time while space has to extend because the rise in heat produces a need for countering the overheating. The duplication of space by an increase of the number of $\mathbf{k}$ in the time unit $\mathbf{T}^2$ will allow as far as $\mathbf{k}^0 = \mathbf{a}^3 / \mathbf{T}^2\mathbf{k}$ will permit. In this there is a living up of standards in space and in motion. However not one of the mentioned normal aspects will apply to the top since it is the manipulating abilities of life that charged the top into action.

In the circle using $r^2\Pi$ the r has to have distinctive qualities placing it as a factor apart from Π. Where the growth shows no separate distinction but a continuous flow from the precise centre to the precise edge the flow would become in relation with Π depicting the circle and Π replacing r as reference to any point on the circle. By using r as a distinction in the circle division is possible but by using Π there is no distinction possible making it a solid flow. Any object being in outer space floats and such floating is seemingly random with no specific detectable interfering favouring a movement in a particular direction. Such a devise is depending on influences not in our scope of detection. But then the object comes closer to the Earth and reaches one specific point where the six dimensions that influences the object suddenly changes. At one point, one of the six dimensions falls away as it disappears and the object quite latterly falls to the Earth. The support of one side disappeared and the centre point of the sphere took over the control. At that point the object is under the influence of one centre point in the sphere and we all also know that in such a centre point one will always find the strangest or the controlling gravity.

Space-time is a four dimensional position of the Universe where the position of an object is specified by three coordinates in space and one position in time. This evidence we find as matter grows into the dimension we now share with billions of stars in the cosmos.

With the dimensional change from space in the cube to space in the sphere a relation of 5 to 7 comes about depicting gravity on one side of the divided Universe. The principle of 5 sides in space relating to 7 in the sphere holding matter forms the basis of the Titius Bode and the Lagrangian principles.

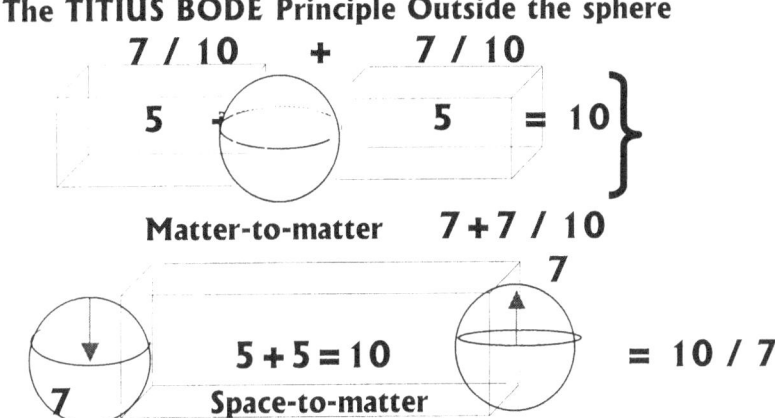

The Titius Bode law is an extending dynamic deriving from the law of the gravity dimensional factor where the space factor in a square of ten relates to a matter factor in the square by half (half since nothing can be in two places in the Universe simultaneously) of the matter factor of π^{7+7} or the square of space (10) relates to the matter factor of 7. From such a point every other point will be opposing any other point not pointing in the direction to which the first point is pointing, whereby it extends the direction it holds. No matter what the point is or where the point leads, such a point holding a specific direction will be unique in the direction it is rotating because at that or any other specific point wherever, it will be directing not in the direction it spins but in the direction flowing from the centre point outwards.

When the foundations were laid in place with singularity expanding even before it was growing The Roche limit became one condition. But while that was taking place another principle came about which is as secured in the foundation of the third dimension as the sides supporting the third dimension. Sides came about through the dimensions that are framing the dimensions, as we know them.

There was the dot. The dot had no borders therefore there was no separation and still we know there were more than one in a group of one. The evidence of this is very present in the cosmos at present and one can find such evidence all around us. The dimensions personify the Titius Bode principle and understanding the relevancy between the dimensions will also mean the understanding of the interlinking values of the Titius Bode law.

Everything is space-time by confirming space in establishing time

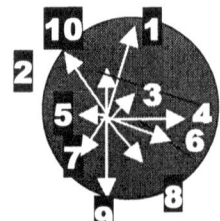 When the Universe was in the beginning with the entire cosmos still in a single dimension there were no limits as we know limits to form in the Universe we use and no borders indicating limits because after all it is the single dimension where there is only one dimension holding so much diversity. The dots referred to in this case have no space but were as close as singularity is when singularity has no sides but only shapes and the lot were the same, the very same one with a time delay parting them. The borders were part of development because we can witness the legacy of such borders in the present day holding the 3D in place. There will forever be smaller particles that combine to produce larger units. The forming of particles start at infinity and there no human can reach except with his understanding, and his mind power.

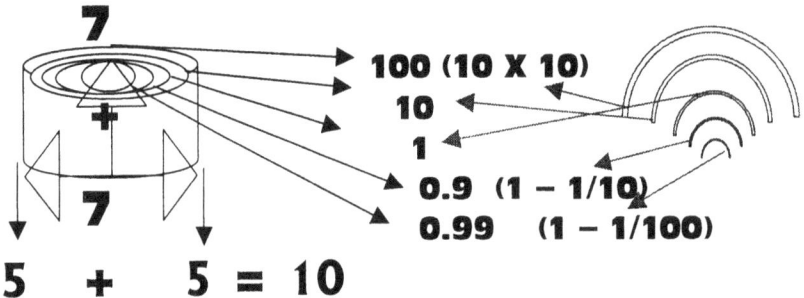

The normal flow will allow singularity extending to 10Π but when singularity blocks another sphere in singularity the two will form a joint value and by this joining the larger will dominate the space as well as the time of the lesser taking control of the surface and the atmosphere. Through this the Roche lobe comes about with all its other dynamics I describe farther on in the theses. The principle is the same, which we know as the conducting of lightning and Jupiter uses it extensively to implement this action. In the Roche limit the straight line forms part (1) and the half circle is part (2) and the triangle forms part (3) to singularity (4) Holding 5 points outside singularity. Every aspect connecting to the universe changes everything it holds totally and becomes the anti-matter to which it was matter 180° previously.

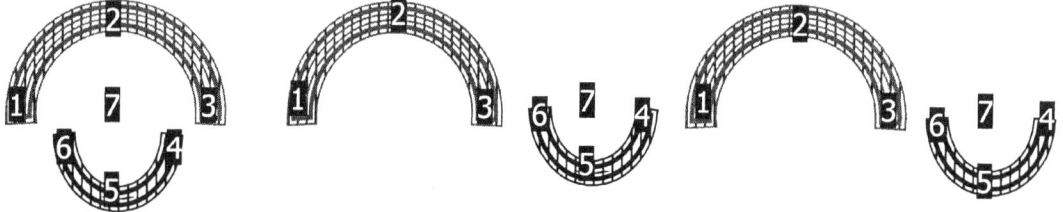

It starts where the first seven points serving singularity meets three points holding time. At present we named the proton combining with the neutron and served by the electron as the atom and from where we gauge the Universe to us the Universe is the entire atom.

Past   Present   Future

The atom holds seven points $(\Pi^2+\Pi^2)(\Pi^2\Pi)3=1836$ as the Universe but that Universe is seven points in Π being 3 points serving dimensional time to form the Titius Bode law, and a law it surely is! The gravity extending from the Titius Bode law forms the entirety of the building of the Universe by constructing the Universe in the using of the atoms to form the Universe in the entirety thereof.

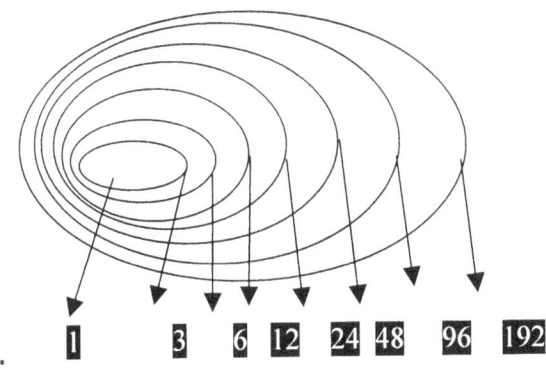

Bode's Law:

Planet	Mercury	Venus	Earth	Mars	Ceres	Jupiter	Saturn	Uranus
Bode's law distance	4	7	10	16	28	52	100	196
Actual distance	3.9	7.2	10	15.2	28	52	95	192

A numerical sequence announced by J.E. Bode in 1772, which matches the distances from the Sun of the six planets then known. It is also known as the Titus-Bode law, as it was first pointed out by the German mathematician Johann Daniel Titius (1729-96) in 1766.   It is formed from the sequence 0,3,6,12,24,48,96, and 192 by adding 4 to each number. The planets were seen to fit this sequence quite well – as did Uranus, discovered in 1781. However, Neptune and Pluto do not conform to the 'law'. Bode's Law stimulated the search for a planet orbiting between Mars and Jupiter that led to the discovery of the first asteroids. It is often said that the law has no theoretical basis, but it does show how orbital resonance can lead to commensurability. The importance that becomes known is the sequence the Titius Bode law saw in the number arrangement of 3; 6; 12; 24; 48; 96 etc. The incorrect application of the Titus Bode law lies in subtracting the figure of 3 from 10 leaving 7. The other way of reasoning is to add four each time to the first value of three starting with 3 and so on. The true significance of the Titus-Bode law is that it points directly to a circular growth of 7 stages. The 7 relating to 10 is a precise derogative of the Roche limit or the Roche limit is a precise derogative of the Titius Bode principle because the two systems interlink.

Gravity produces mass but mass is only the result of gravity. Mass do not produce gravity and the manner in which science uses mass can only apply when using the calculations in terms of the Earth. However applying it to stars as science indicates by their formulae used in their calculating of gravity on structure beyond the solar system is very inaccurate. Heat stored in motion produces gravity. Any one not in agreement convinces you by comparing the neutron star with the massive red giant. To calculate a Black hole they go and throw C^2 next to the dividing radius and throw the square onto the C that presents the speed of light. Then they sit back and feel smart in the way they manage to cheat once more to prove their incorrect views correct because after all who will ever fly down a Black hole and return to support or deny their calculations. The Gravity of the Black hole is a speed because the entirety of gravity is speed or better said it is motion. Then the speed that light has is gravity. The gravity of the light can be gravity as much as it at that very same time can be antigravity. What the hell has C^2 got to do with a Black hole because you can pop what ever nuclear device far away from a Black hole and it would be at the most and at the worst very much insignificant. The light will not even escape form the gravity of the Black hole. When this became apparent that the radius of stars reduces as the stars develop through progress. It is some time ago that someone was supposed to say: hey there is a dead rat I smell. For my saying so I am the clown in the courtyard, the one with the two dead brains cells and have no more to use as spare.

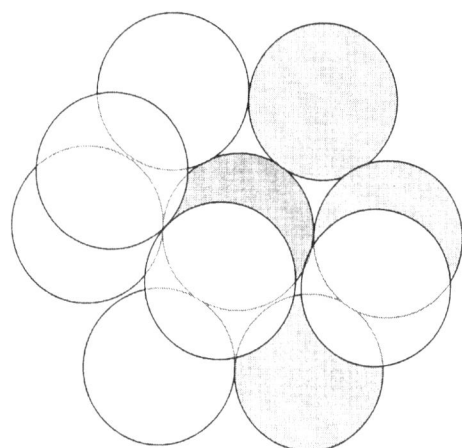

On the one side of the Universe in relevance to all the dots that came before, three dots landed forming one side while three dots formed the second side and three dots formed the third side, all relating to a centre dot which in turn related to the original centre dot from which all the dots came and developed.

Space generates the mass where the space has to reduce the size by becoming more intense and concentrates space-time to the time of 1836 time more when entering the point of the proton being on the verge of singularity. In single

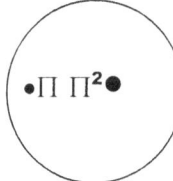

dimension seen from one aspect, with single dimension contacting the edges forming the sphere it will still keep the seven positions because the sphere remains a unified structure though apart because of singularity. In the core of the sphere the proton connects in alliances as $\Pi^2 + \Pi^2$ with the solidity of the neutron holding Π^2 as a second forming a π value. That brings the atom unit of π to a number of seven.

By reducing the space-time the lesser singularity is claiming singularity independence by offering reduced space, which will result in promoted time with heat being the net result. That heat is filling the space, which should then be entered by the independent space in motion if the motion of duplicating is brought closer in a relation to what the matching tempo requires.

It is clear that the density of material in motion is $\Pi^2 + \Pi^2$ but since that is k, which extended we know that that extending cannot sustain the initial speed. Since the speed is reduced the space in motion will value less. Taking these atomic relevancies into account, we can detect what relevancies brought about the atomic Universe of $(\pi^2 + \pi^2)(\pi^2\pi)3 = 1836$. The first substance that formed from singularity was solid and if that were the case the contra substance would then be a fluid with less motion filling more space taking shorter time duration in duplicating. The fluid substance that then formed was one less than the proton in motion which makes it slightly more in mass since it duplicates more with less space that then has to form $\Pi^2\Pi$, which has one Π less from $(\pi^2+\pi^2)$ which is resolved becoming a fluid like substance relevant to the first solid substance which is the proton. The loss of the one Π then became the factor claiming more space that is holding less substance. In this fluid state the neutron has more duplicating of the substance than is required of the proton. That what we find in space we also must find in the atom because the cosmos is not keen on inventing but is passionate on duplicating. This fact will also apply to space-time in many forms. That means investigation must prove the same results and what we find in the atom then also has to present in the cosmos at large.

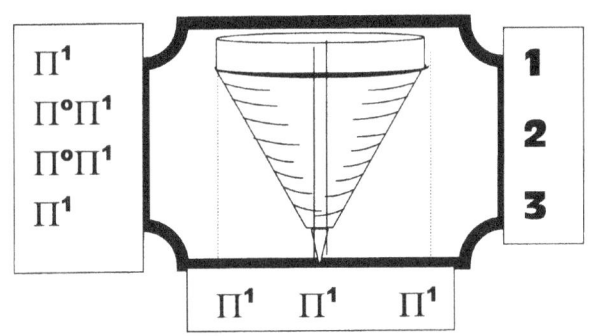

The overall picture resulted in a ring and all rings hold Π to secure the form. The only form that existed then was Π and therefore even today the borders use Π to indicate positions. But in the single dimension such definitions were far from clear and the only distinctions came from securing singularity in preserving the position of singularity to apply gravity and thereby absorb all anti -gravity. But anti gravity could not control expansion by counter acting contraction through gravity so the overheating continued forming non-existing borders in some thing infinitely solid just as Einstein predicted because this took place before light came about and therefore before the speed of light became part of the cosmos. The cosmos formed a partnership with one side overheating forming antigravity by expanding into space through the applying of the overheating and the other side formed gravity or contracting of space.

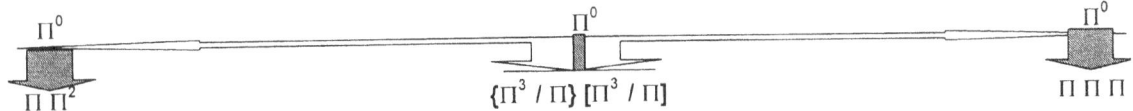

Singularity split the Universe into two parts that under no circumstances can ever meet. The one side of the Universe performs a balancing act to the other side of the Universe that duplicates but never double. The dot started overheating while the dot remained cool by activating gravity; the dot duplicated forming a sequence while the dot redefined the position in control.

Space

ΠΠΠ

    ΠΠΠΠ **Matter**

ΠΠΠ

Time

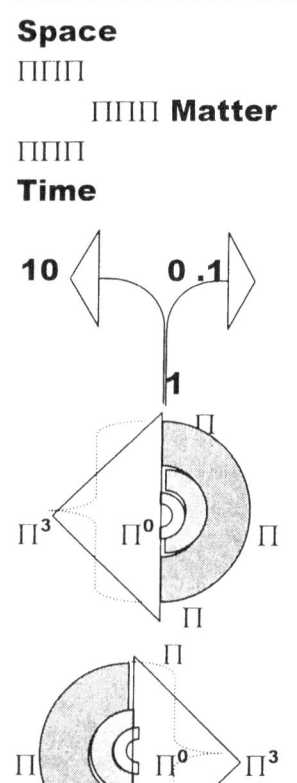

With the first dimension came matter, but also came space and came time splitting the universe in segments of matter relating to space filled with matter and time influencing the spinning matter.

Taking the queue from the numbers line that runs in opposing directions singularity by Π^0 is always going larger as well as smaller. However, the centre takes a value of one. It is a private choice preferring $k^0=1$ or $\Pi^0 = 1$ but that splits the Universe into two parts, being smaller and being larger in relevancies.

It is apparent that one cannot substitute the correct formula used to measure the area of a circle by using $a^3 = \Pi r^2$ because if **k** is the diameter then the formula must be $k^2 \Pi$. However, **k** cannot be Π because in Kepler's formula **k** takes the value of the radius. In that case, what will the value be of T^2? That places the formula outside the normal use of mathematics practised in the normal sense of $a^2 = \Pi r^2$.

By using the Kepler Formula $a^3 = T^2 k$ it is good to change the values to Π and see what pans out. If **k = 1**, **k** at the same time would be k^0. By replacing $a^3=\Pi^3$ then on the other side of the Universe $k = \Pi$ and $T^2=\Pi^2$.

Space

ΠΠΠ   **Matter**

  Π   ΠΠΠ

ΠΠΠ

Time

However, to secure this k in the centre must be 1 leaving $a^3= 1(1X\ 1X1)=1$ and $T^2=1(1X1=1)$. That complies with Einstein's definition of space-time being: Space-time is a four dimensional position of the Universe where the position of an object is specified by three coordinates in space and one position in time.

If k is the middle being $k^0 =1$ then $a^3 = k^0 =1= T^2$. When time is in a shift freezing then $a^3 / T^2 = k^0 =1$. In order not to overstep my limits by changing valid formulas I changed Kepler's formula to $R^3 = T^2 = 1$.

The book being written in Afrikaans the R stands for Ruimte meaning space and T is time. From that I deducted that the space used in a specific location will equal the time meaning the density of the heat in space. That brings the proof that space equals heat and space is the same as heat. Heat deforming or exploding is the equal to the space created. Also it confirms the substitute between Kepler and Π is correct

In the way space and the sphere connects the sphere will have 7Π points holding a relation to 3Π points not within the sphere forming the 10Π that creation started with. This will mean there is a division forever, and such a division may run smaller everlasting. With fluids connecting it is simple to recognise the sphere as Π for the form will indicate Π as the form of the sphere. By gas forming the connection there are the three points of space being apart and not forming Π, but still holds a relevancy to Π^2 through the value of Π.

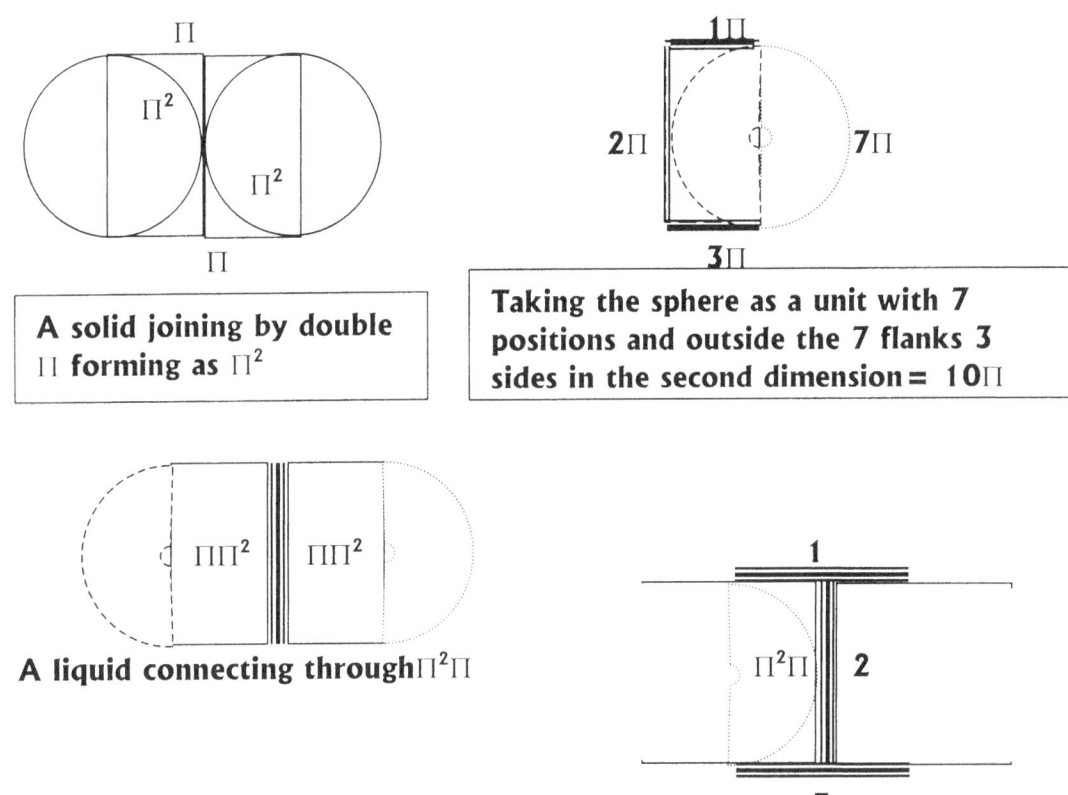

A solid joining by double Π forming as Π^2

Taking the sphere as a unit with 7 positions and outside the 7 flanks 3 sides in the second dimension = 10Π

A liquid connecting through $\Pi^2\Pi$

Total connecting relevancy of the sphere forming matter connecting to space = $\Pi^2\Pi$ 3

How many dots there was is a question no person can answer because

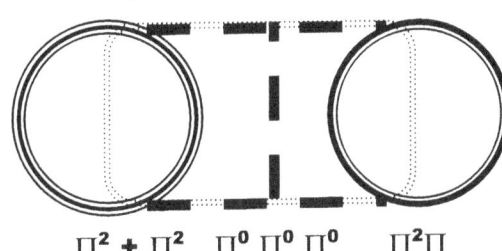

$\Pi^2 + \Pi^2$ $\Pi^0 \Pi^0 \Pi^0$ $\Pi^2\Pi$

everything was un-dividable solid and yet it did group together to form every atom located in the 3D. Individual singularity and governing singularity and group singularity enhancing the gravity every time singularity find an accumulation. The Universe came into position by deploying dots supporting other dots and some dots remained dots while other dots went on to become dots of hybrids as it was supporting dots through claiming dots of lesser density and pass that on to dots with larger density.

Space formed as motion came about through singularity overheating. Singularity k^0 produced motion at the point where k^0 became k and a^3 became T^2 by motion duplicating space. According to Kepler, a^3 is equal to k relating to T^2.

Matter formed where matter had to have ΠΠΠ = ΠΠΠ
space to occupy since it was to be in some space ΠΠΠ = ΠΠΠ

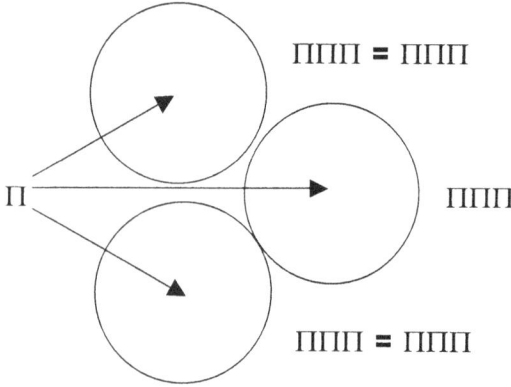

ΠΠΠ = ΠΠΠ

ΠΠΠ

ΠΠΠ = ΠΠΠ

therefore ΠΠΠΠ met with ΠΠΠΠ to form the proton in $\Pi^2 + \Pi^2$ because the matter is within the space it holds and another Π^2 employs Π as a representative of singularity. This then placed the seven positions of singularity as the ending of matter and the three squares ($\Pi^2 + \Pi^2$ and Π^2) of singularity as the limit of material. The last ΠΠΠ became Π^0 Π^0 Π^0 and that became the space producing heat without occupying matter in order to allow heat to be restrained inside the dome singularity provide.

It is all about relevancies applying the relations gained and lost through relations. If one place $\Pi^2 + \Pi^2$ on one side then $\Pi^2\Pi$ is the related form, where $\Pi^2 + \Pi^2$ is in the other side of the Universe being on the other side of the relevancy. Then Π^0 Π^0 Π^0 will again relate to the other two factors forming the "outside" of the other two being the "inside".

The Universe divides into two separate issues because of singularity. Nothing can be in two places at the same time the rest has to confine to the law applied by singularity. Objects can only be in one side of the Universe holding three parts or in the other side of the Universe holding three parts. From the totality three will be a double with six sides too shows, but that forms 3D. From singularity it is flat with three sides forming on either side of singularity as the formula used to measure the sphere indicates..

Newton said a sphere is $a^3 = 4/3$ Π r^3

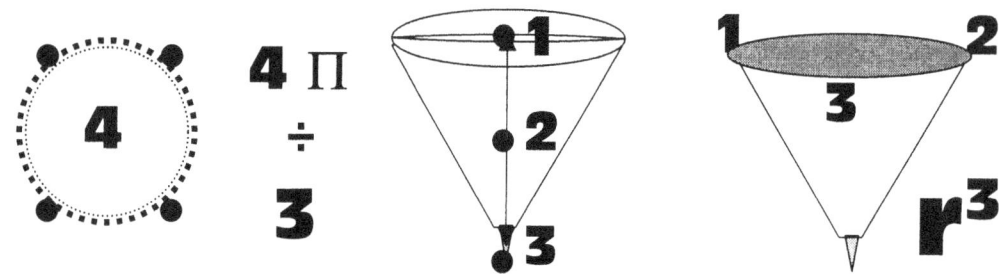

At first when material presented one side of the Universe matter had three sides to show. Matter had to have space to keep matter somewhere in some part of some universe and that made up three positions. Between the two universes **k** and **T**^{2} placed a value but since only singularity applied any values the value therefore was $\Pi^2\Pi$ where **T**$^2 = \Pi^2$ indicated time coming from 7/10 in relation to 10/7 and $\Pi^2/2$ (proof of that is somewhere in the

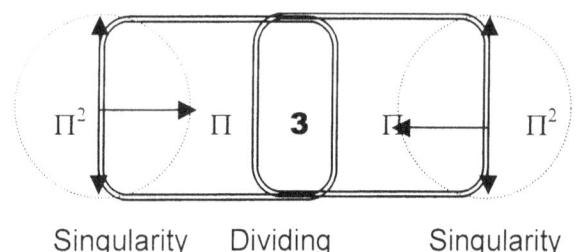

Singularity   Dividing   Singularity

book) and **k** = Π valued by singularity. When space-time developed 3D the dimensions falling outside the sphere becoming space-heat formed as $\Pi^0 = 1$. The electron holds a relevancy of 3 relating to the Neutron being $\Pi^2\Pi$ and the three keeps the electrons in different universes relating to separate or individual singularity.

The relevancy between the two particles secures individual positioning between the opposing particles, which positions the material that sufficient space secures cooling and preventing overheating.

As the relevancy between the particles promote overheating or applying antigravity (overheating) to the responding cooling or applying of gravity, the one repels material into space-time while the other is collecting material into space-time. The one loses material and sustain a model of preventing overheating while the other gains material and sustain a model of overheating. The one we named the Hubble constant where overheating produces space and the other we called gravity where gravity is demolishing space, but both phenomenon is at present dominating the flow of time in the Universe and will do so until equilibrium again comes about.

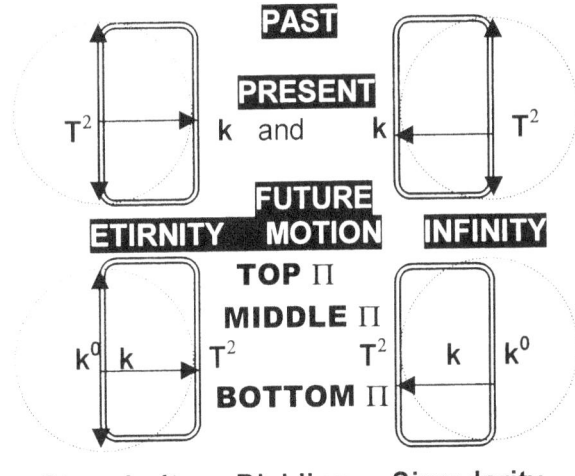

Singularity   Dividing   Singularity

The names I use in TOP, MIDDLE and BOTTOM must not be viewed as sides but merely as terminology using names to implicate divisions. Direction depends on positions and positions form a value only when the observer forms part of the cosmos and not part of the observing.

The universe divides into two separate issues because of singularity. Nothing can be in two places at the same time where as all the rest in the Universe has to confine to the law applied by singularity.

But when the Universe was in the single dimension, all values were Π, therefore every value related to $\Pi\Pi\Pi$ forming three of the same that was very different because it was where Universes met and formed relations. Every spot formed an individual dot or Universe and every dot was another new Universe.

In the relevancy where space divide eternity from infinity the three holding Π in relation to singularity holding Π^0 there are three points forming a square in relation to 90^0 which is implicating the law of Pythagoras while on the one side of the Universe the duplication is forming the same result and three points goes square. The result is that on the one side the square of space is ten and on the other side the square of space is also ten.

As the relevancy between the particles promote overheating or applying antigravity (overheating) to the responding cooling or applying of gravity, the one repels material into space-time while the other is collecting material into space-time. The one loses material and sustain a model of preventing overheating while the other gains material and sustain a model of overheating. The one we named the Hubble constant where overheating produces space and the other we called gravity where gravity is demolishing space, but both phenomenon is at present dominating the flow of time in the Universe and will do so until equilibrium again comes about.

Keeping these factors in mind it is clear that Π^2 are the choice of gravity and not r^2.

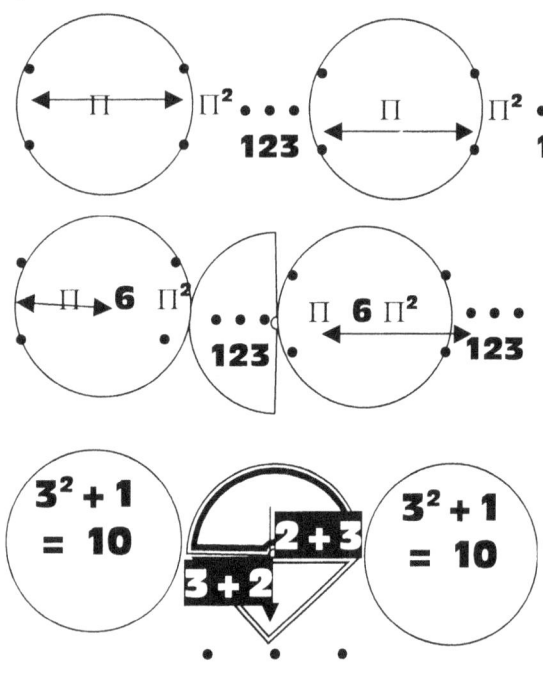

Material formed at a position of six points from singularity. That is three on the one side of the divide and three on the other side if the divide. It is one centre one on either side

In relevancy from one another material held five inclusive positions of two in time including the three positions as material. That made being in one quarter of time five in all. That makes the Lagrangian system dominant.

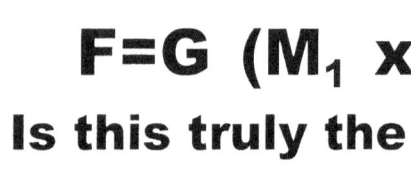  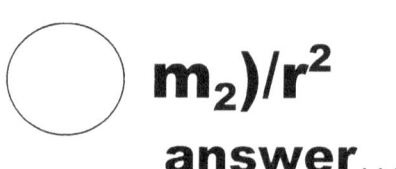

F=G (M₁ x ⬤ ◯ m₂)/r²
Is this truly the   answer...?

In the investigation of light and gravity and objects and gravity, the mathematical rule of the invert square law must apply without question. But according to the observation of Roche that is not the case. From what one gathers through the Roche limit implicating two orbiting structure the opposite is applying. One must accept that although k proves as an indicator it is also much more when complying the thin influences brought about by singularity in the values carried on by singularity.

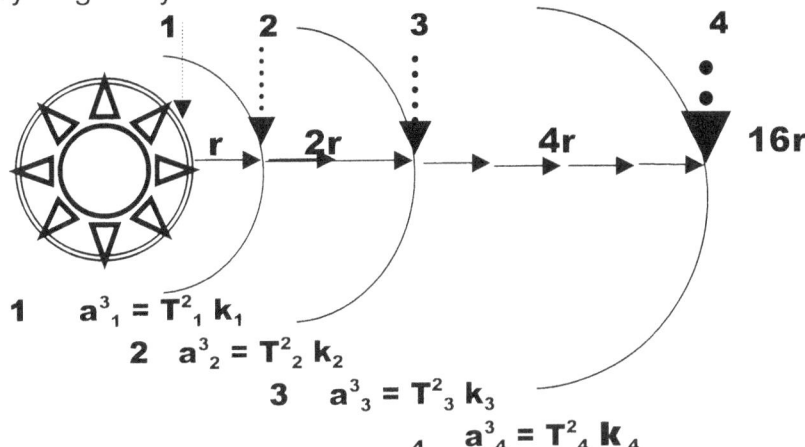

$$1 \quad a^3_1 = T^2_1 k_1$$
$$2 \quad a^3_2 = T^2_2 k_2$$
$$3 \quad a^3_3 = T^2_3 k_3$$
$$4 \quad a^3_4 = T^2_4 k_4$$

In drawing a most basic picture of light passing the gravity lines extending from any structure, I felt it was most insightful that the brains in cosmology was not able to see why light does not bend in the presence of increasing gravity. More surprising was that I found the mathematicians had to call on Einstein for advice regarding an ordinary problem. Light does not bend when passing large objects. It is Kepler's formula applying, and the evidence is clearly in front of the searcher for truth. But one has to go back to Kepler to re-apply what Kepler formulised and change the significant from Newton's significance.

As a^3 increases, so does T^2 as well as k increase and with that the influence of gravity per space unit increases with the concentration demise of a^3. But why would that be and what are we missing? Light shows there is an influence out there in outer space, that redirects light's route through space when passing large gravity fields. It is about the relevancy of k influencing the a^3 to allow the T^2 of light to divert in route because of influences established by k on a^3 and slowing down or increasing the line diverting. In this measure one may also find the Roche limit applying, but to truly understand how the Roche limit comes in place and how the Roche limit works one has to replace Kepler's factors with singularity and singularity extending being Π^3 $\Pi^2\Pi$ and **3.**

In the Roche limit the space factor provides space to a solid structure and therefore the value of r is replaced by the value of Π bringing about a square in half of Π. The cube holding 5 to either side removes allowing the extending of Π to indicate position to space.

Where Π extends to lock onto the next sphere's extending indicator, Π has

5/2

Five sides divided by two spheres.

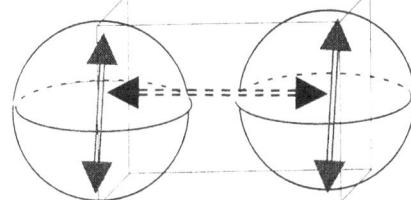

to connect to Π forming the square of space and translating that to the half of Π being $(\Pi/2)^2$.
According to normal mathematics the half of space should have been 5/2, but at

the time this divide took place, space was all in motion and motion was Π in motion Π^2 crossing the divide (/ 2) forming $= (\Pi^2 /2 X \Pi^2 /2) = 2.467)$

The space between the spheres divide in half, but because of the extending of Π and not applying r as ordinary mathematics will suggest where Π replaces r the singularity extending from Π^0 will be half of Π in the square of $\Pi = (\Pi/ 2)^2 =$ **2.4674.**In this lies the dynamics why planets have a positional (be it rather a dimensional) relation of 7/10

The Titius Bode law must not be seen as some obscure event that took place just before and / or after the Big Bang or when the solar system formed it fell into place. The Titius Bode law applies when the top is spinning, when an atom is spinning, when a motorcar wheel runs on the tar, when a jet engine fires up. It takes Place whenever the Coanda principle comes into effect and the Coanda principle is wherever there is motion in relation to singularity in a centre forming a centre.

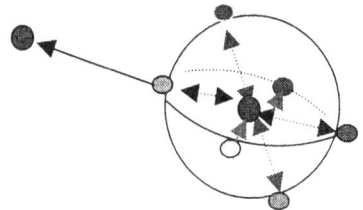
This is why we can use degrees measuring the circle by (6^2) (forming the square relating to matter through singularity) X 10 (square if space) $= 360^0$ however it is always in motion. That proves no point can be static or constant, though it may seem that way to outsiders. Although matter is matter, matter can also be anti-matter and moreover form its own anti-matter at the same time. This degeneration of structure is very likely to occur with overheating.

Revaluing Π to Π^2 will bring about a new contact point where Π meets r forming another relation in Π^2. Every time material swaps sides it also qualifies as anti matter to matter because if it goes out of orbiting rotation frequency. It has the ability to collide with the same matter it forms union with but is located on the other part of the spin. It then becomes in a situation where Π revalue to r.

Time is the changes in relation where Π contacts a different r not withstanding the many r points there may form because every r constitutes a different value to the Universe through other ratios and relevancies brought about by heat and light. Time is the duration it takes Π to rotate between any two given points of r and therefore must always amount to a square (T^2) moving from point to point through the cube of space (a^3) in that duration of time (k). With that it proves Kepler's a^3 (space) $= T^2 k$ (time in the instant of motion) but motion must continue through a specific value in space where the space-time is maintaining relevant equilibriums throughout singularity connecting.

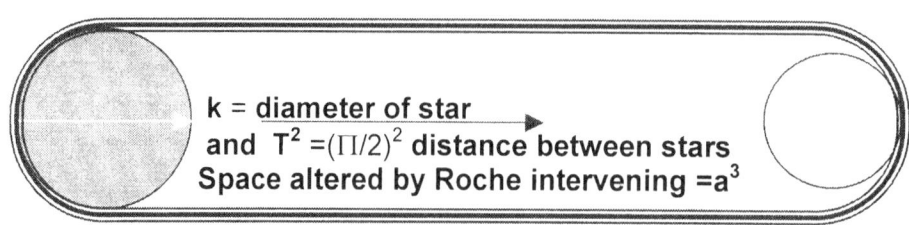
k = diameter of star
and $T^2 = (\Pi/2)^2$ distance between stars
Space altered by Roche intervening $= a^3$

The influence of singularity as the extending of Π into space links Π^2 to r and forms 2(5)+2(5) =10+10=20

From the position of singularity there are different values in Π where each indicate a position. The value it represents being $\Pi\Pi\Pi$, Π^3, Π^2, Π and Π^0

From there it influences singularity in the triangle flowing through to the half circle. It is an interaction between circular and linear motion as the value of Π continuous past Π^2 (at the end of the solid) and every cosmic structure holds an individual and specific singularity.

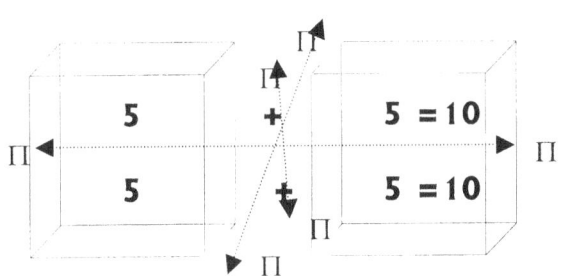

The field where Π extends we call the atmosphere having a value of 21.991 / 7, which is Π.

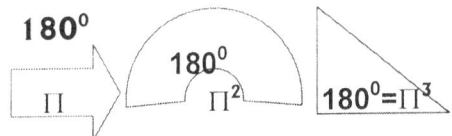

The triangle, the half circle and the straight – line has two things in common, they share 180^0 as a mutual value and they are part of singularity.

Using the concept that gravity applies Π as the circle factor Π as well as Π^2 replacing r^2 the replacing by Π brings two values as Π and Π^2. That I found is the case with gravity and will be apparent when explaining the sound barrier as well as the Four Cosmic Pillars. In order to create a distinction I remained using r as the indicator of the cube or non-circle that has vacant space and by vacant space I refer to non-solid structures. In the solid structure I use Π as a value for reasons that will become apparent in due time.

Gravity does not apply mathematical equations to the letter as we would like, but rather use Kepler's thinking by enlisting an average gravity applying through out because it never favours and is equal every where. In gravity one find the extending of Π implementing Π^2 on average as a unit and not the radius r as a specific.

Looking at the affect of gravity it shows the precise quality of no distinctive point as gravity never seems to end at a point but flows all over affecting all that holds a position in its sphere of influence. The gravity coming from China meets the gravity coming from America at no particular spot but intermingles without distinction. This takes mathematics back to another fact beyond normal explaining.

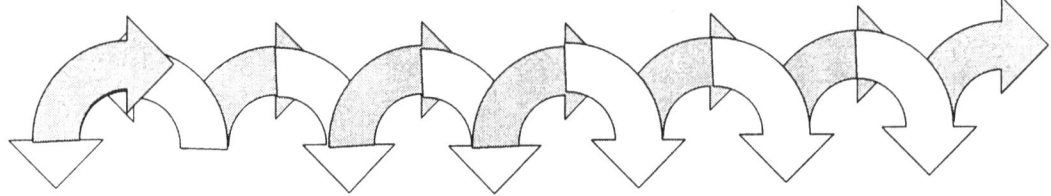 We take a line running between two points as being 180⁰ and the rest of the explaining is saved in the accepting part of mathematics. Any one of the two points the line starts or ends at is a point in infinity, The start and the end depends on the viewer putting the relevance to favour the side of choice. That puts the point of end or beginning in the spectrum of choice and not fact. Any direction is as equal as all other directions.

Following the flow of any line such a line is an extension of the previous dot in infinity to the next dot in infinity without any ability to skip or bypass any of the other dots in the connecting line. Any direction change including the remaining of travelling in the same direction is in relation to a line travelling all being the very same. Change does not affect the line.

A straight line, triangle and half a circle will always have equality in dimensional capacity providing equilibrium being 180⁰ because each one shares a common denominator in singularity to the value of Π. As the straight line averts a zero it holds another straight line in place to set about such an averting where the two lines will always carry a relevancy in relation to progress (the triangle) and a common denominator in the start from singularity.

This concept we apply as the graph or the vector. By going back to a line, any lines and all lines, the line is a connection of dots in infinity, running from one specific to another specific and avoiding zero or dots. At every point in infinity it dips into infinity coming out on the other side by choice of direction and the direction is unforced and change presents any angle including the straight line, which incidentally is just another angle.

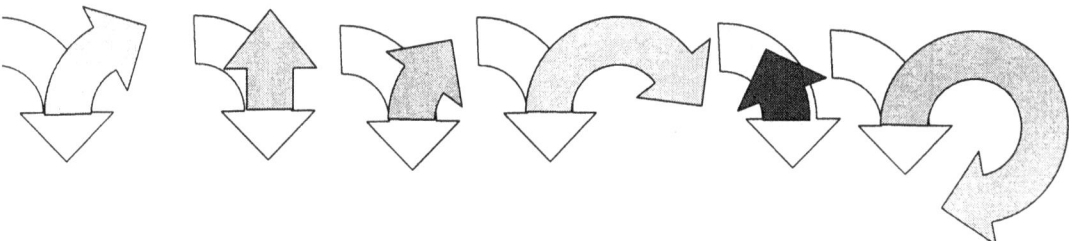

When connecting to the dot representing infinity the flow can be in any and all possible directions, including in the same direction. We all live in a graph, as the universe with all in it is nothing less than a three-dimensional graph flowing according to time. That means in the case of Pythagoras the mere fact that the line shows changes in direction does not implicate or affect the line as a tool of mathematics. Whether the line changes into a half circle meeting at the other end again or meeting in a triangle in forming a half square by joining the point where it began, the result still indicate a line flowing between points. Motion became an integrated part of space because motion is what establishes space. If motion redirects space and such redirection is not complying with a balance, there will be even further delay in time producing motion, which will bring about more time distorting and heat. Since the very first space from point to point was Π and motion produced a value of Π^2 while the four points indicated time, it is presumable that from that the Roche factor of $\Pi^2/4$ came into place.

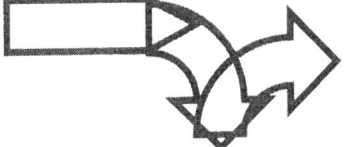  

The line dips into infinity every time it passes infinity when it cuts through infinity. The line going into infinity comes natural as the line progresses because all lines are infinite dots linking one point to another point. That means that coming out of infinity might slightly change the angle but that directs the route to the future and not the form because the form still remain equal whether the form is a triangle, half circle or straight line. The form remain a factor that confused every one in the past. When replacing the value we normally attach to circle being r with Π, the law of Pythagoras becomes quite meaningful and mathematical.

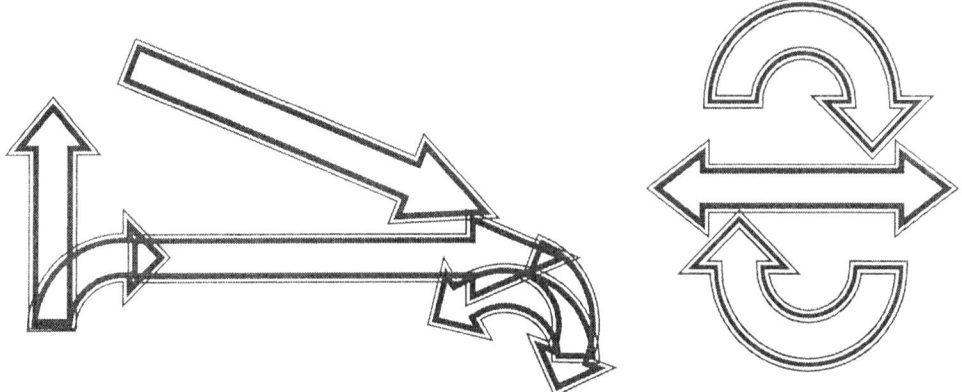

In that way a circle is a straight line following a loop as it comes out of singularity at a different angle and a triangle is a straight line that dipped into singularity but at three stages changed the angle with which the line then left to follow different directions at specific points. From the point singularity observes it still remains a straight line because there is no direction alternation in the first dimension and in that dimension it still remains a straight line in which we on the outside may experience as three forms but is in fact one single line. Only when the direction changes completely in reverse the line doubles in value but comes from multiplication for instance 2Π become Π^2. But the Lagrangian system proves much more than dimensional interlinking, it proves Pythagoras in principle.

LAGRANGE (-TOURNIER), JOSEPH LOUIS DE (1736-1813)
French mathematician, born in Italy. In celestial mechanics, he studied perturbations and stability in the Solar System. He examined the three-body problem for the Earth, Moon and Sun (1764) and the motion of Jupiter's satellites (1766). In 1772, he found the particular solutions to the problem that give rise to

the equilibrium positions called Lagrangian points.   Lagrange also studied the Moon's liberation.

LAGRANGIAN POINT
One of five points at which small bodies can remain the orbital plane of two massive bodies; also known as liberation points.   Three of the points lie on the line joining the two massive bodies: L_1 lies between them, while L_2 and L_3 have the two bodies between them.   These three points are unstable, slight displacements of a body from then resulting in its rapid departure. the fourth and fifth points (L_4 and L_5) each form an equilateral triangle with the two massive bodies, 60° ahead of and behind the smaller body in its orbit around the larger one.   A well-known example of bodies flying at the L_4 and L_5 Lagrangian points are the Trojan asteroids in Jupiter's orbit. Among Saturn's satellites, Telesto and Calypso lie at the L_4 and L_5 Lagrangian points in the orbit of the much larger Tethys. In similar fashion, tiny Helene precedes Saturn's satellite Dione, keeping 60° ahead of Dione.   The Lagrangian points are named after the French mathematician J.L. de Lagrange, who first calculated their existence.

LAGRANGIAN POINT:
*The Lagrangian points
are five equilibrium points
in the orbit of one body
around another, such
as a planet around the Sun*

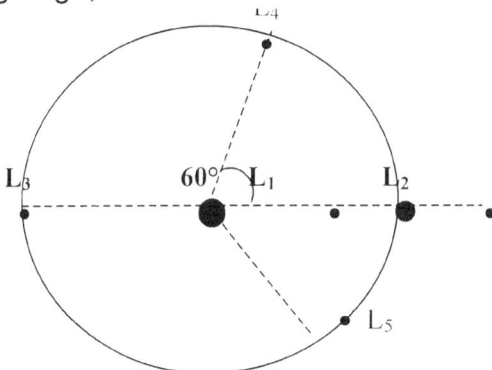

The entire concept of motion rests on the centre forming time and having one point outside time to be delayed or behind time. The delay parts motion in eternity from motionless infinity, which results in forming the Universe. Since the satellites are located as electrons the motion gathered from that falls in as a time delay. All motion is about time trying to cross that space to form a unity with infinity. That is the essence what keeps the top straight when spinning. The spin puts the outside of the top in another time zone than that the inside of the top is in and the four inside has to align with the fifth one on the outside where the fifth one is one in three positions allocated to the flow of time. By having time parted from time there is a flow coming from the fifth to the centre.

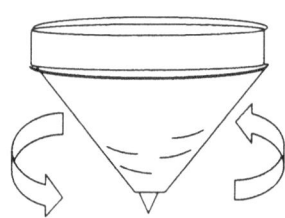

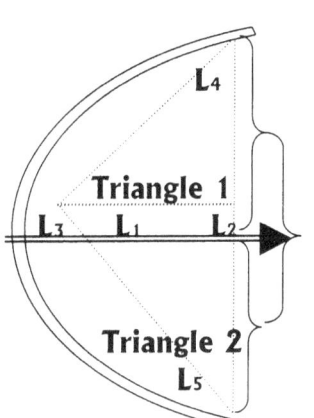

The Lagrangian System implicating the five positions extending from singularity
**Each triangle claiming a side of the universe
The half Circle = 180° combining as a Sphere when comprising
Singularity dividing the cosmos
1 Half circle   = 180°   L₃ L₄ L₅
2 Triangle 1   = 180°   L₃ L₄ L₅
3 Triangle 2   = 180°   L₃ L₄ L₅
4 Straight Line = 180°**

Singularity in the matching of the value of the straight line forming the half circle and combining as the triangle and all are equal 180⁰

The second one also fits in the singularity influence on the Universe.

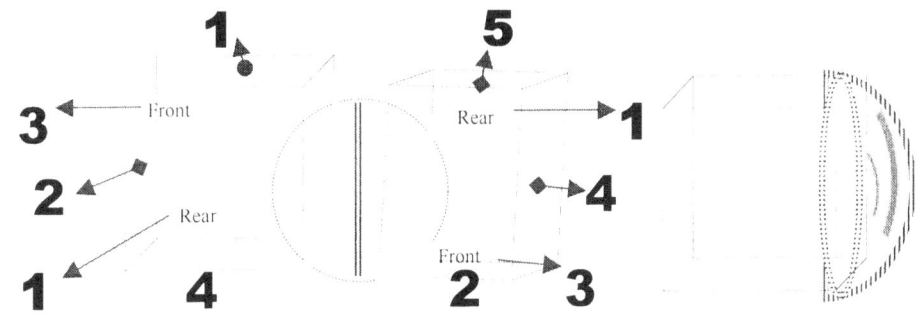

1 Relating to 5

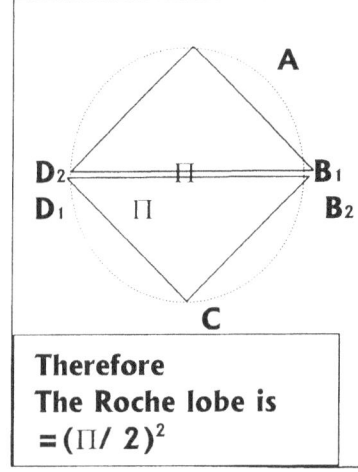

A

D_2 $\longleftrightarrow$ II $\longleftrightarrow$ B_1
D_1 II B_2

C

**Therefore
The Roche lobe is
$= (\Pi/2)^2$**

$(D_2 A)^2 + (B_1 A)^2 = (D_2 B_1)^2 (\text{PYTHAGORUS})$
$(D_2 A)^2 = (B_1 A)^2 (\text{EVEN SIDED TRIANGLE})$

$2(D_2 A)^2 = (B_1 A)^2$
$(D_2 B_1)^2$ (DIA. OF CIRCLE) AND ABCD EVEN SIDED SAUERE WHERE AB = BC = CD = AD

$(D_2 B_1)^2 / 4 = (AB)^2 + (BC)^2 + (CD)^2 + (AD)^2$
$2(D_2 A)^2 = (D_2 B_1)^2$ BUT $(D_2 B_1)^2 = \Pi^2$ (Replacing r^2)

$(D_2 A)^2 + (D_2 A)^2 = (D_2 B_1)^2 [(D_2 A)^2 = (B_1 A)^2]$
THEREFORE $4(D_2 A)^2 = (D_2 B_1)^2 / 4 = (\Pi/2)^2$

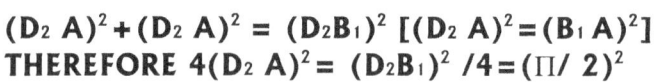

The value of singularity stems directly from the law of Pythagoras or Pythagoras is the result of the average of singularity. With the shortest line being a dot, all lines must start from a position implicating Π. A circle is a square without corners implementing Π and a half circle is therefore a triangle without corners. The corners are, an average of Π in the connecting line will come about. As both lines are the straight line forming singularity coming from one line being Π, the connecting line then must be the average of the two lines as Π^2. That is what the law of Pythagoras says.

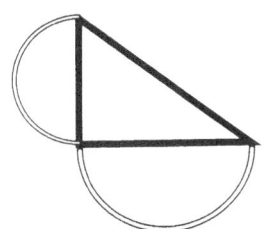

By placing a connecting circle on the sides of the triangle half a circle forms. By implicating Π as a relevancy and not the straight-line r, two values of Π applies to each circle, and the straight line is no longer r, but is Π^2. This will bring about that each circle holds half the square value implicated to the allocated conditions applying to Π in that specific instance.
By adding the two half squares forming the two half circles and then calculating the square root of the total that then forms the average diameter into infinity comes natural as the line progress because all lines are infinite dots linking from one point to another point. That brings about that coming

from infinity might change in angle bit that directs the route and not the form. The form is all the same

A STRAIGHT LINE, TRIANGLE AND HALF A CIRCLE WILL ALWAYS HAVE EQUALITY IN DIMENSIONAL CAPACITY PROVIDING EQUILBRIUM BEING 180^0 BECAUSE EACH ONE SHARES A COMMON DINOMINATOR IN SINGULARITY TO THE VALUE OF Π. As the straight line averts a zero going down infinity it holds another straight line in place to set about such an averting where the two lines will always carry a relevancy in relation to progress (the triangle) and a common denominator in the start from singularity. This concept we apply as the graph or the vector.

With the normal extending of singularity it will always form the triangle in a half circle whereby Π relates to the cube by 5 points to either side of the line singularity forms. Thus there are 10 standing related to seven and visa versa. By calculating the 4 squares in the circle with the dimensional changing of space (5) becomes the twenty

BC EITHER RELATE TO AB OR AC AT ANY GIVEN TIME OCCUPYING SPACE AS MOVEMENT DICTATES DRECTIONAL CHANGE THROUGH DIRECTIONAL FLOW

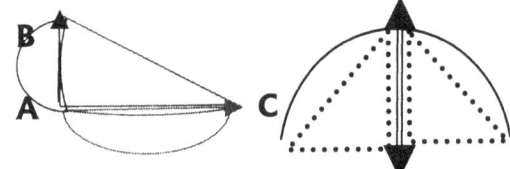

The normal flow will allow singularity extending to 10Π but when singularity blocks another sphere in singularity the two will form a joint value and by this joining the larger will dominate the space as well as the time of the lesser taking control of the surface and the atmosphere.

Through this the Roche lobe comes about with all its other dynamics I describe further on in the theses. The principle is the same, which we know as the conducting of lightning and Jupiter uses it extensively to implement this action.

In the sphere there are never only one direction implicated in movement. Movement are always in relation to the centre position because as a line goes up it also goes in or out. When a line goes north or south, it also comes towards the centre or going away from the centre.

There is always relevancy present in movement. As this moving indicates direction it also apply Π^2 for indicating value forming the time factor.

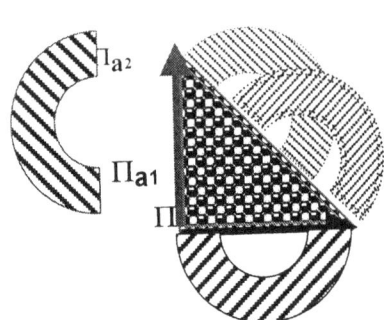

$$(\Pi_{a2} \textbf{X} \Pi_{a1}) + (\Pi_{b1} \textbf{X} \Pi_{b2})$$
$$= (\Pi^2_a + \Pi^2_b) / 2 =$$
$$\Pi^2$$

= **gravity and that is proven by Pythagoras.** Gravity is the average movement of matter through space in time determent from the position where matter in the sphere meets space in the cube from a point of Π to a point of Π^2 In this the figures of 2(5) = 10 (space) stands related to 7 from singularity as (matter)

 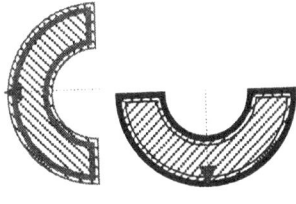 

From the star holding a dominant point or most valued point in singularity it affirm all three other structures, each holding singularity individually and in a compliment of 5.

The network of individual singularity not only provide spinning through governing singularity in the sphere but also provide spinning in the geodesic through out the cosmos linking all matter to matter in a network no one will ever come to understand in full. In the sphere the four squares forming the triangles linking the lines to the half circles holds space in time maintaining singularity of different assortments. In view of the matter-to-matter Roche factor where the factor consists forming relation between particles occupying densified space-time of where (Π / 2 X Π / 2) relating to the foursquare triangle the value of gravity Π^2 comes in position as Π^2 / 4 X 4 = Π^2.

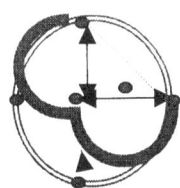 **Because every moving line represents one quarter of the sphere in relation to the rest of the sphere and the line also indicate the relevant position between the point indicated and the point in the centre it is a relevancy of singularity in progress. By connecting the line, as Pythagoras will suggest the singularity within the sphere become a specific value indicated representing one half circle.**

No object can be in two spherical quarters in the same time, but has to alternate in aliens to the space in accordance to time rotation.

To alternate in aliens to the space the relation of time in space has to alternate relevancy to the cosmos.

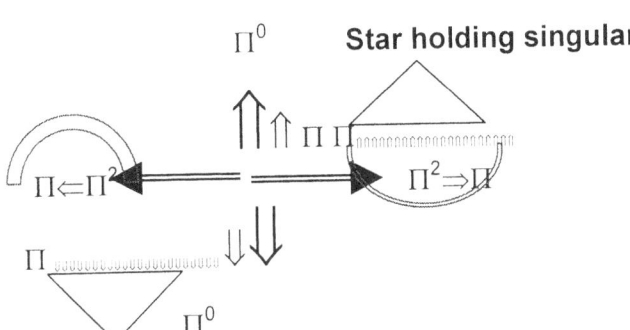

Star holding singularity

Singularity holds five dimensions inside and five outside singularity as matter and space forming space-time. The ten dimensions I named the atomic relevancy is also showing the double value of singularity as singularity extends into as well as beyond space. The atomic relevancy is **(Π^2+Π^2)(Π^2 X Π X 3) = 1836** that is the mass relation between the electron (3) and the proton. Proton = $(\Pi^2+\Pi^2)$ Neutron $=\Pi^2\ \Pi$. The atomic relevancy holds the dynamics

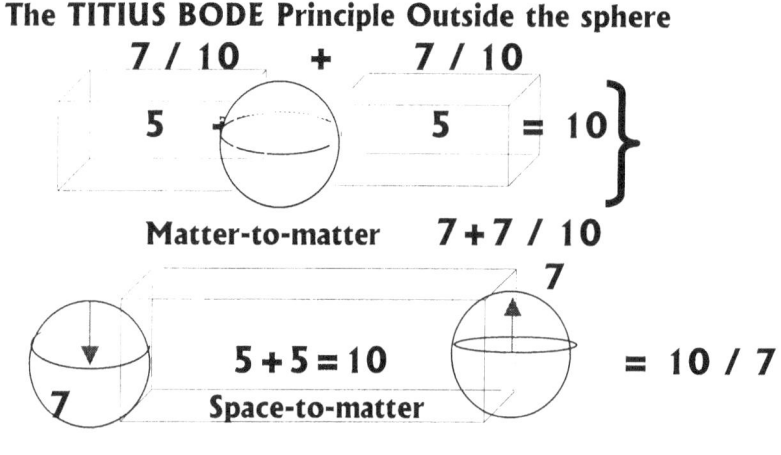

The TITIUS BODE Principle Outside the sphere

From the dimensional implication comes about, not only the Doppler's effect, but many more of phenomenon not yet understood. The dimensional relevancies formed between matter as six, matters end at seven and space at ten, comes the value of Π.

The process is all intermingled and stands in relevancy to one another. The relevancy compliment holds such attachment that none of the factors can even stand-alone. It is the way that science places every aspect in the cosmos as individual and not related to each other that launches the problems of miss understanding. The Value of singularity appreciates or demises by ten fold. For instance, the value of Π will increase by ten every time singularity applies another layer.

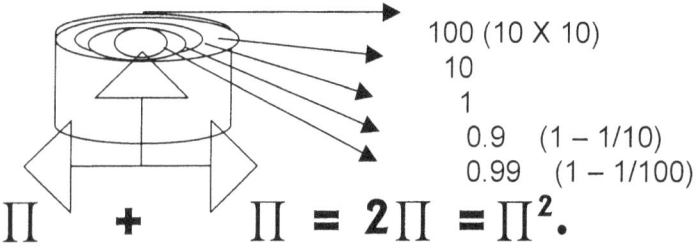

100 (10 X 10)
10
1
0.9   (1 – 1/10)
0.99   (1 – 1/100)

$$\Pi\ +\ \Pi = 2\Pi = \Pi^2.$$

The normal flow will allow singularity extending to 10Π but when singularity blocks another sphere in singularity the two will form a joint value and by this joining the larger will dominate the space as well as the time of the lesser taking control of the surface and the atmosphere. Through this the Roche lobe comes about with all its other dynamics I describe farther on in the theses. The principle is the same, which we know as the conducting of lightning and Jupiter uses it extensively to implement this action. In the Roche limit the straight line forms part (1) and the half circle is part (2) and the triangle forms part (3) to singularity (4) Holding 5 points outside singularity. Every aspect connecting to the Universe changes everything it holds totally and becomes the anti-matter to which it was matter 180° previously.

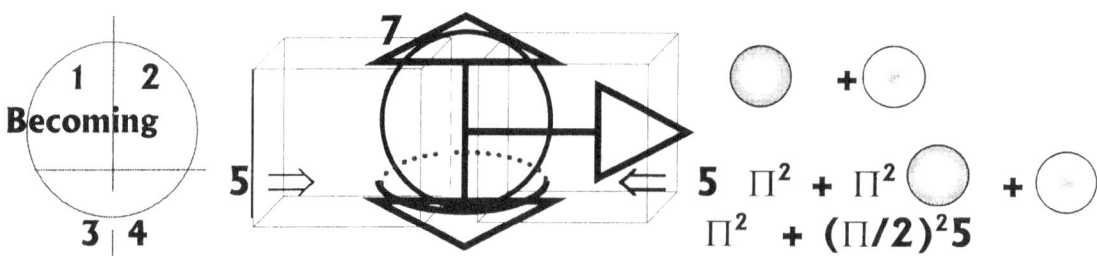

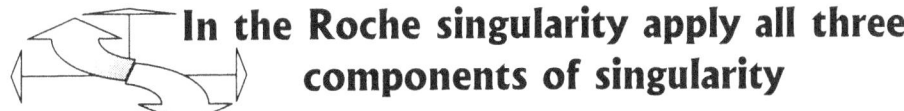

In the Roche singularity apply all three components of singularity

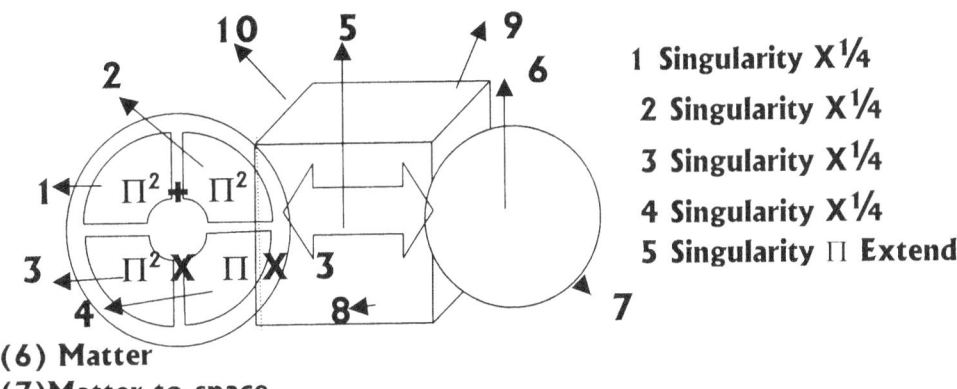

1 Singularity X¼
2 Singularity X¼
3 Singularity X¼
4 Singularity X¼
5 Singularity Π Extend

(6) Matter
(7)Matter to space
(8,9,10) Dimension1,2,3) in the cube's six sides

Gravity is about a relation established when time begun between particles we know as material and particles we know as free or unoccupied space. Gravity reduces space to apply to fit the form of the sphere and later accept the form of the sphere.

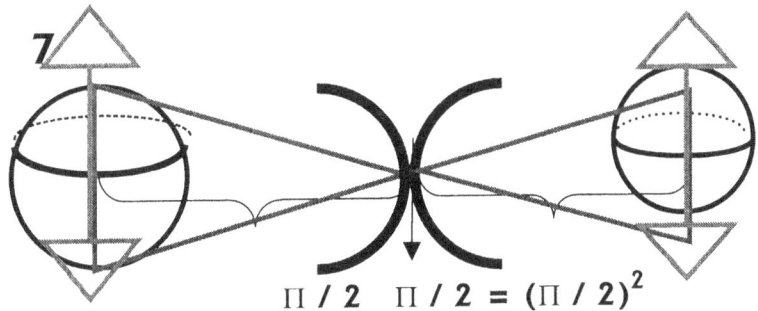

$$\Pi / 2 \quad \Pi / 2 = (\Pi / 2)^2$$

SINGULARITY MEETS AND COMPLIMENTS EACH OTHER.

The diameter of the cosmic structure holds the value of r and singularity holds the dimensional value of Π meaning that the radius or diameter (r) extends to become the diameter multiplying the value of singularity. But since r already consists of the square of space holding a definite positional relation with the value of singularity being Π the diameter comes into effect. Π extends each to an individual value to a point where the singularity on each side meets, bringing about a mutual Π^2 to the value dominance of the larger singularity control.

At this point the equality of the straight-line dimension to the triangle and the half circle holds prominence as a straight line, a half circle and a triangle is dimensionally equal. The common denominator will bolster all factors to an equivalent ratio.

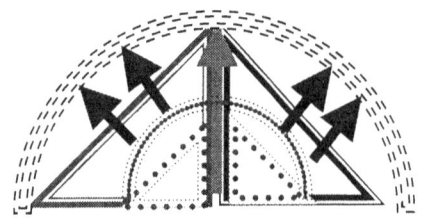

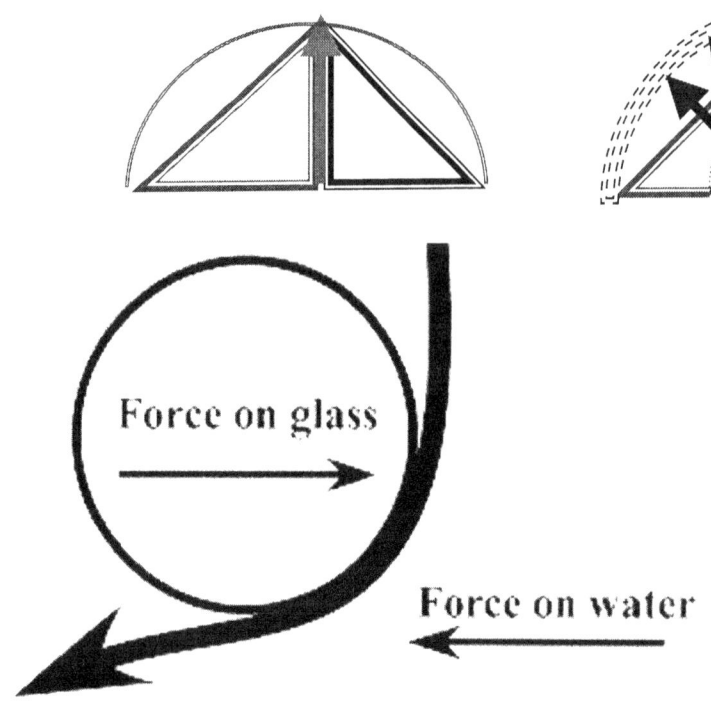

Force on glass

Force on water

The Coanda affect is proof of the functioning of gravity inside the atom. It proves that motion (T^2) of the neutron establishes a centre in line where the compliment of material forming the atom will secure a controlling singularity that is governing the entire atom. That forms the centre of the Universe. Singularity then finds a position at the distance of (**k**) and such motion claims the space (**a^3**), which is the atom by construction from a centre within that motion (T^2). The motion (T^2) creates a centre at the line of (**k**) and a centre of the space (**a^3**) the motion (T^2) establishes a gravity field all along the lines and at the distance of (**k**) in the space (**a^3**) that the motion (T^2) created.

When singularity by the straight line increases the singularity by the triangle it will also bolster giving equal potency in singularity by the half circle. As the singularity of the major component revives the lesser singularity to equality, the **triangle in singularity** will match the performance and so would the half circle respond in precise ratio setting equilibrium in order. The major partner's singularity in the straight line excites the minor partner's singularity in the straight line affecting all other aspects holding singularity in both objects to match equilibriums in all aspects of singularity. That is the Roche lobe.

From this the lesser partner will fill by the extent of the larger partner and as soon as equilibrium sets in the growth will duplex to matching in both accounts, normally to the fatality of the lesser partner, as the lesser partner will be capitulating under the strain of the dual. In that way the inner planets came in place as I explain in part 7 *of Matter's Space In Time The Theses*.

The Titius Bode configuration in accordance to orbiting formation holds a slightly different explanation to the explanation that applies to cosmic structure surrounded by space. It is moreover the individual singularity in maintaining the major singularity, which sustains the governing singularity providing equilibrium in space-time. Not only does atomic individual singularity maintain self-preservation, but in doing that it also sustains a governing singularity holding structural composition and form within a cluster of matter for example a star. As there is between stars so there is in the same manner a mutual or bonding singularity between atoms in stars, which we see as fusion. From this one may freely deduct that gravity is not forcing material closer but is destroying space whereby it converts the space to a density the senior partner has in the atmosphere of the senior partner. Where does all the information given thus far take us you might ask? For one it can help to explain something Newton science can never understand. To start with we have to realise that the Coanda principle is the manifesting of Kepler's gravity and we have to accept Newton's version of gravity is a load of rubbish.

Years ago it dawned on me why we all would be so egocentric. This was a problem that was eluding every thinker ever thinking. I admit as a thinker I am quite average but still we are all thinkers, what puts us apart is what we think about and in that I am then equal to the attempt of any other average person with the right also to think. There is something that makes every person in his or her eyes having the opinion that that person is the greatest there ever was. Let's call it a Jesus syndrome. Either the person frequents with Jesus or the person has a special prayer linking such a person directly with Jesus or the person may recognise Jesus or Jesus has come in person to meet with that person in particular and others just simply become Jesus. We all know what I am talking about. What is it that gives every person on Earth the idea that that person is superior to all other persons except those we regard as being more advanced than us? Why would every man walking on Earth think his sperm is just what every woman on Earth would give her front teeth for? Why would every man that walks this Earth do so with the idea that every woman is just waiting on him to impregnate her and that his her sole purpose in life...to wait on him to fertilize her? Why would we be so God damn ghastly superior in the way we see our status we have? Why would every person see him or her with the superior capabilities of reinventing life? Some would not eat meat. Others would bullshit through their teeth about health implications and the misery of death just to get the world to stop smoking. If we are that scared about death then we better ban the wheel first before any other thing because the wheel in whatever form is killing a hell of a lot more people than smoking can ever achieve. It is the thought that a person can impersonate God and that would allow and enable such an individual to change the course of man forever in all time to come... Some would go to war for any reason because only leaders that killed millions are worthy of the remembering by Historians. The more any leader killed off his fellow beings the greater role his memory has in the history of man. Others would not war for any reason even in the face of being threatened by death. Some would drop a Uranium bomb on others with the pretext that they did it to save lives. Others would drag a whole world into a war for the benefit of monetary gain, because lets face it, in the back ground behind the drawn curtains there are those bankers and industrialists that makes enormous profits from other fools fighting "for justice". Something is making every person feel horribly special. Something allows every person to know that that individual is in the centre of the Universe right where God should be. There is a very good reason we all feel that way because we are not wrong to feel that way, and we are in the centre, the very centre of the Universe.

Step outside into the night sky and the reason is in front of you. Every sparkle of light coming from where ever is coming to you honour. All the light that was released from any and all points in the Universe is coming to the place you stand. That makes you the most important person ever born because you are the **centre of the Universe**.

When any person is standing on any place anywhere, while viewing the Universe, that person is filling the **centre of the Universe**. Let's get more personal. When you, the person that is reading this, are standing at night and is looking at the Universe you are seeing the Universe from the position that one only can have if that person is filling the specific spot in the **centre of the Universe**. All the light, every single beam that ever left any destiny at any time acknowledges this fact. You are the most important person in the Universe because you are holding the most important position in the Universe. All the light that come across and travelled all of the vacant space from any and all possible positions in space runs directly towards your position using a straight line towards you where you are filling the **centre of the Universe**. Not excluding the effort of one photon, all light is heading to meet you where you are in that centre spot and not one photon will pass you by. Not one photon dare miss you because if they do they miss the effort that all light has to accomplish and that is to locate you as the person filling the **centre of the Universe**.

Should you decide to shift your position to any other place in the Universe, you will shift the **centre of the Universe** to that location as well. If you install a camera on Mars, the light is obliged to acknowledge your relocating the **centre of the Universe** at your will to reposition you're being that **centre of the Universe**. All the light that ever left its destination crossing the vast spaces of the Universe, excluding no particular light, travelled all the way just to find you filling the **centre of the Universe**, right where you are. By you're standing anywhere, you fill the **centre of the Universe**, and the entire Universe admits to that because all the light comes to meet you there. If you shift from the North Pole to the South Pole you will shift the **centre of the Universe** because all the light travelling throughout the Universe will find you where you then moved the **centre of the Universe**. The light left its destination billion years ago as it travelled through space at the speed of light anxious to acknowledge you're being in the very **centre of the Universe**. No photon will be able to pass you by where you are in the **centre of the Universe** because all light is heading your way from their starting positions. No wonder every person born has the idea they were born to fill **centre of the Universe**, which we do fill. The Universe is spinning around you or I, which is filling a centre where all motion is connected. That is the Coanda effect on the utter-most grandest scale imaginable; nevertheless it is only a manifestation of the Coanda effect. It implicates gravity as wide as can be… Some things mathematics is able to explain but other explaining goes beyond mathematics. Try to explain mathematically the colour of the sky being blue in a clear Sun ny day and changing to black when nighttime falls. Do the explaining in mathematics to a blind person that had no vision since birth in such perfect mathematical detail that would allow the person afterwards be able to explain the difference between blue and black to other blind persons by using only mathematics. Some aspects of the Universe go beyond mathematics and some even go beyond words. It is our task to find space, to find time and moreover it is our optimal task to find the Universe. We have to see what is solid, what is liquid and what causes gravity. Please keep this aprt in mind because in a short while I am returning to thsi to show how this becomes a cosmic reality.

Gravity **is to move or apply the intension to move** space a^3 **at the** distance or relevancy of **k** while T^2 is the time it is going to take to **apply gravity** or move the space filled with material space a^3 at the distance of **k** in the time period of T^2. That confirms Kepler's attribution to gravity where according to Kepler space a^3 is equal to the movement T^2 (time it takes to move) at the distance **k** from the centre specific.

Then the I took Human nature and science and combined the two which gave me the vision on the findings Kepler received from the Cosmos. It puts all aspects of gravity in the Universe in new dimensions. But the visions formed the beginning because the visions unleashed many new questions. If gravity is motion, what causes motion? What stops motion? That answer is in the Black Hole. In truth the explaining of the Black Hole is as complicated as the Universe may represent and as simple as the cosmos truly is. If a star is about fusing atoms and with such fusing of atoms is thereby growing, what happen when all the atoms fused into one all collective atom in one already all—atom-accumulated star? What is the gravity if the star has melted all atoms it had into one all-inclusive atom and this all-inclusive atom is providing all the gravity that the star had when the star still had massive volumetric space? If all that space that once filled an entire giant star fused into one specific space less centre holding singularity 1^0 then the enormous gravity is applying to the centre of such a non existing space-less atom and that entire enormous force has been secured in the space less than that which one atom holds. In that case the atom would then show a force that would pull the surrounding Universe flat. The purpose of fusion is to reduce space and magnify space less ness inside the sphere. Where does the gravity of the star end when all the atoms in the star became one giant atom by fusing all atoms into one nucleus? Gravity is smallest where space is least. Where space of an entire massive star is left in the size of one atom the gravity coming from that will pull the Universe flat at that point.

Newtonians have the opinion that it is energy that keeps the planets in rotation and the system is equal to the rotation one will find in Earth. There is one slight problem and that is that all the mass used in the calculation is not worth a penny in practise. In nature all the planets orbit in an equal ratio while in their opinion the mass is the key factor, which implicate all aspects of the energy requirements in the planet orbit.

They say that $E = -(GMm) \div 2r$ and the gravitational constant (G) is one factor of three where the product of the three factors holding the Mass of the Sun multiplied by the mass of the Earth (or what ever planet apply at the time) giving the Mass X the mass X the Gravitational constant and this is in division of the radius (r) from the Earth (or what ever planet apply at the time) added (2) from both ends. There is a problem looming on the horizon...

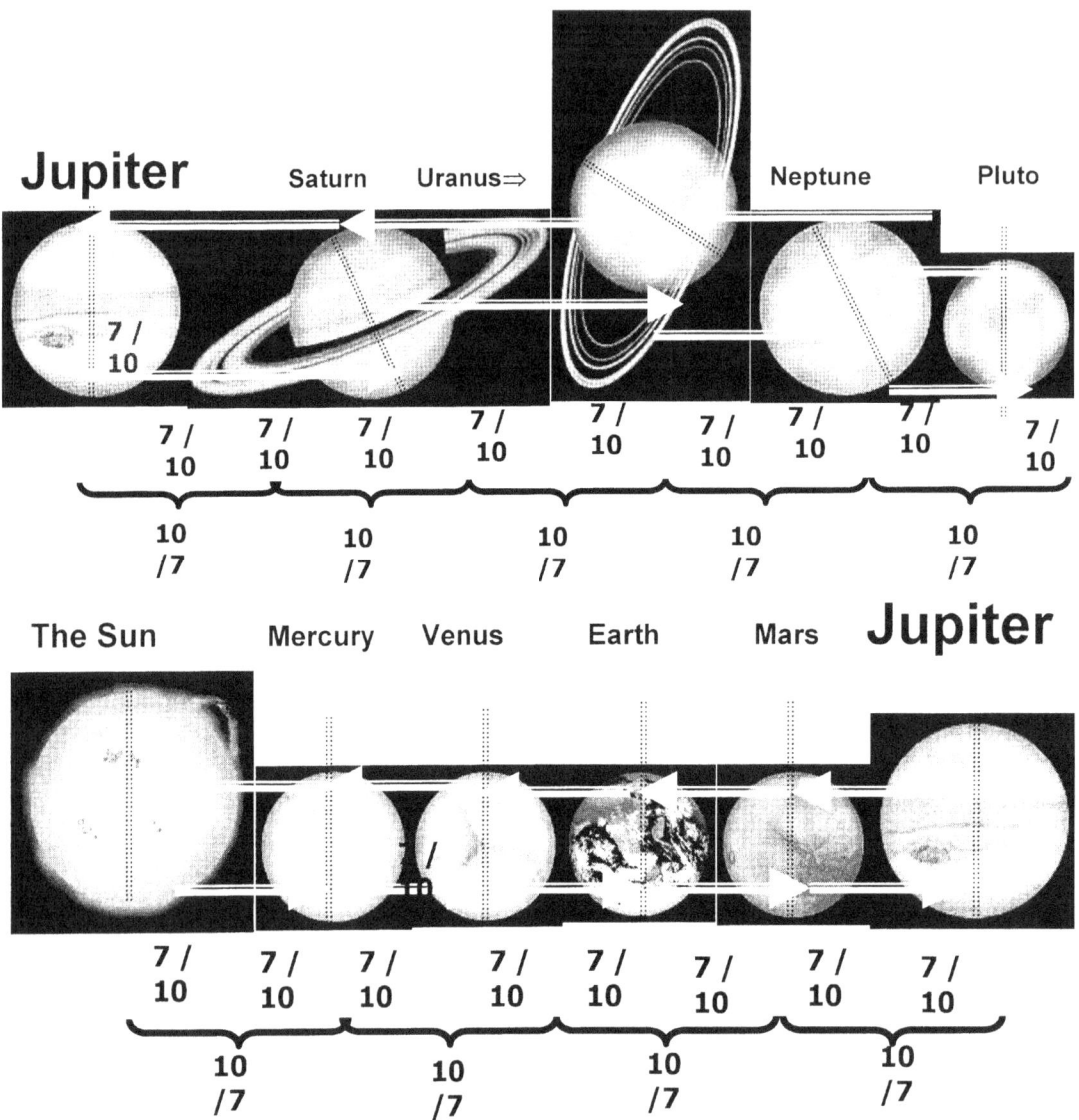

Notwithstanding mass differentiation and mass discrepancies of the large planets in relation to the small solid planets all the planets are in a similar ratio in space and time around the Sun . That means big or small, they travel alike. You can say what ever you like about Newtonians but stupid they are not. They know how to think and think they can…fore instance try and beat this:

Mercury	Venus	Earth	Mars	Jupiter	Saturn	Uranus	Neptune	Pluto
0.055	0.86	1.0	0.11	318	95	14.5	17.2	0.002

Notwithstanding the enormous mass discrepancies we see illustrated in the table, all the planets orbit equal in ratio. That means we can ignore the fact that Jupiter is 318 times more massive that is the Earth because they use the same time to space ratio. One might think that if the one mass (the smaller mass) in the case of the Earth stands to be used in the formula $E = - (GMm) \div 2r$, in comparison to the case where Jupiter is 318 times more, or in the case where Pluto is 0.002 times that of the Earth, the mass will bring changes. As I said, one thing you may not call the mathematicians is that they are stupid. They did notice that all the planets orbit equally and at the same ratio. That did not stop them from implicating mass, no they just went on to blame the gravitational constant being guilty of eliminating the mass discrepancies.

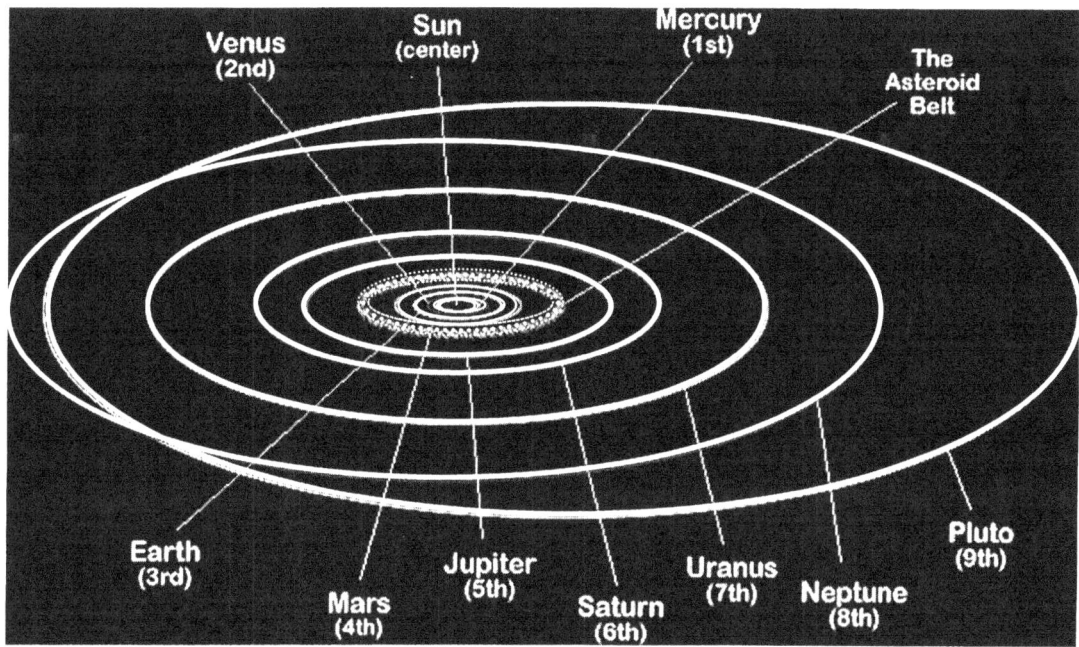

If it were true that it is the gravitational constant that is eliminating the supposed effect of mass on the potential gravity of a star then it would be that the formula would read as follows:

$F = (M \times m) \div (G \times r^2)$ where $(G \times r^2) = (M \times m)$ because that will mathematically show that the Gravitational constant (if there were anything of that nature applying) cancels the effect the mass factors has on the orbiting structures. That would mean that the gravitational constant eliminate the mass factor on bother ends of both the radii and not as it is at present where the gravitational constant incorporates the mass as the mass on both ends incorporates one another in order tot compliments gravitational constant to calculate the required planet orbit. As I said, they are not stupid, they will use any bullshit to wiggle them out of a loop. They do with that problem just what they do with me as a problem they pretend it never was a problem and ignore the problem.

In another pert of the book I went into the criminality of falsifying evidence in order to colour a picture to the likings of the person acting criminal or to falsify in order to bring about purposely an incorrect situation. In this part I wish to elaborate on the incorrectness of this approach and the magnifying of I\the intended incorrectness. It is acceptable that there was no one in the past that saw the Titius Bode law for what it is but in the same manner if there is deliberate protectionism of the corrupt and a deliberate effort to cover falsifying evidence and statements, then it will be a natural tendency to over acclimate the process where further investigation is required.

In the Titius Bode law on find the distribution of planets in response to the allocation of singularity respectively. By having a distribution of twice time seven divided by ten in relation to ten divided by seven is the location or position that serves the outside planet. Where the one is twice the other we find that the distribution is coherent with one marker and one planet. The location of the other planets has no role in the position the outside planet has since to the outside planet only its immediate inside planet is a seven. Any planets closer than the immediate inside planet is to the Sun or farther away from the Sun than what that outside planet is, has no function in the allocation of the planet distribution. To every planet that planet is forming the outside and all other planets except the one to its immediate inside is of no significance to the planet forming it's most outside border. This results in the way the distribution uses (7+7/10) in relation to 10/7.

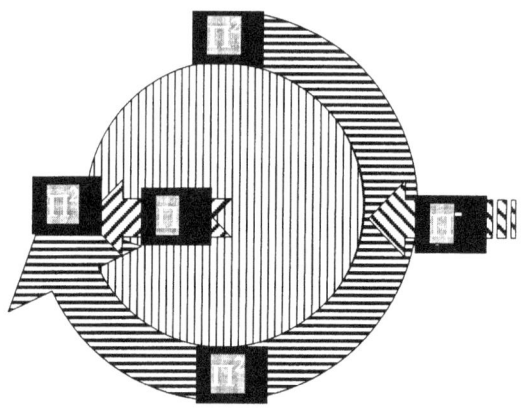

It is so obvious that mass plays no part in the orbit of planets. I just cannot believe any reason or excuse put forward why the worlds most intelligent will hide the truth about mass not playing any part! Yet where the Titius Bode is so overwhelming in evidence of being the process used to form the allocated orbits of the planets, there is such a strong and deliberate attempt to by pass the issue. The blatant misleading reasoning about why the mass will be illuminated by the gravitational constant without having that reflected in the formula used is shocking but even much more shocking is never having one person investigate (in earnest) the Titius Bode law.

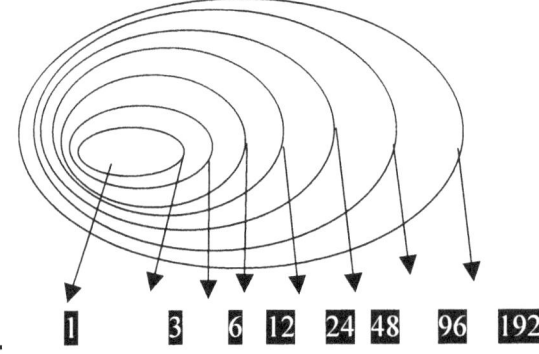

Bode's Law:

Planet	Mercury	Venus	Earth	Mars	Ceres	Jupiter	Saturn	Uranus
Bode's law distance	4	7	10	16	28	52	100	196
Actual distance	3.9	7.2	10	15.2	28	52	95	192

That brings us to another Newtonian problem that they deal with in precisely the same manner; they ignore it and declare it never existed in the first place and any one mentioning it must first prove that it ever existed by proving that it never was a coincidence to start with.

One can clearly see how the singularity of the atoms form the building form used to increase the space –time growth. The seven that material holds are in double relation to the ten that time holds. By valuing the atom as $(\Pi^2+\Pi^2)(\Pi^2\Pi)3=1836$ we find that the seven reflect as the material component and the seven on both sides of the Universe is in regard to the five it is in contact with. But on the other hand the five doubles to ten on every side of the Universe since no one can determine precisely where the five begin to form seven and the five will always be a square to the seven it is in contact with.

The square however dates back from a time when the square still was just a doubling to bring a duplication of one to the other side of the Universe. For every seven in singularity holds relating to material (7/10+7/10 = 1.4) the time doubled by remaining the same ratio (10 / 7 = 1.42) That allocates one line in singularity in space holding time to twice the ratio of time holding space while the ratio remains the same. That means the radii (if one could call it that) in distance doubled (.7 + .7) by allocating one time unit in relevance (10/7)

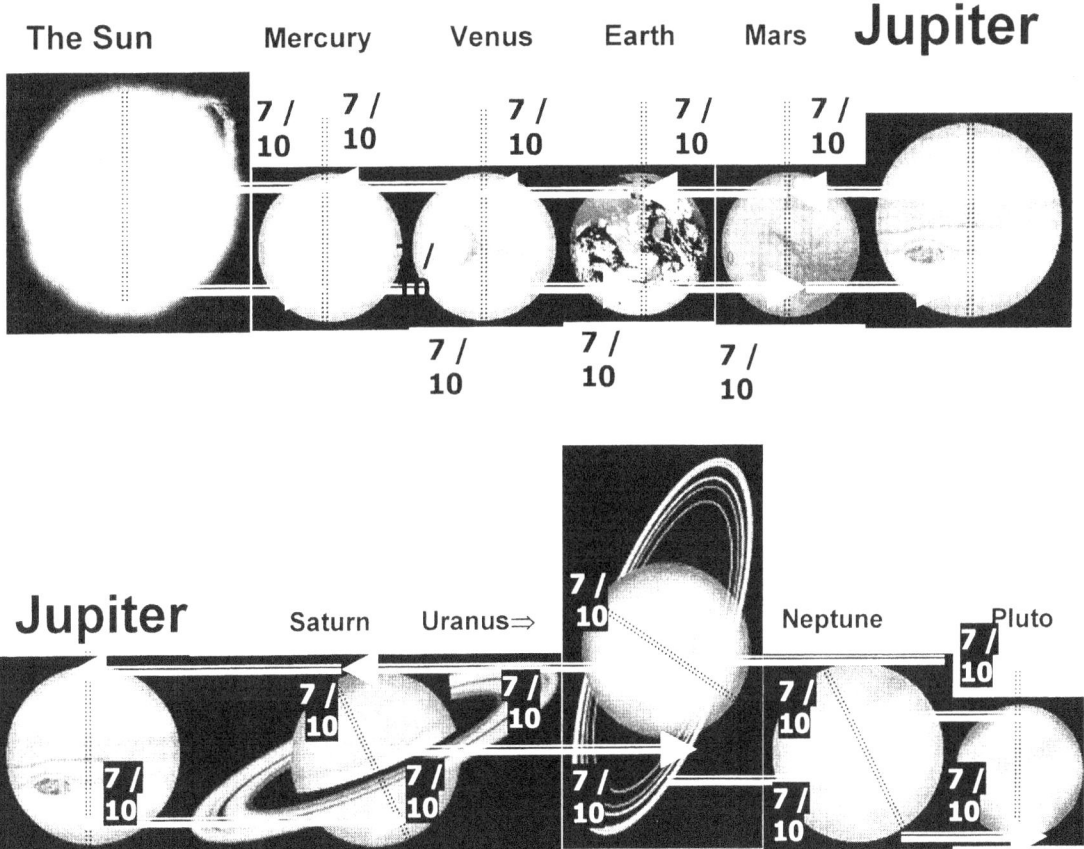

There is no one of the seven directions we move in because time takes the seven directions and move infinitive number of sevens in three time positions from the future to the past. The direction we see is the Universe coming towards us and disappearing into the infinity we have. We are moving from eternity in the direction of infinity.

That easy part explains the frequency Titius and Bode mathematically could interpret. The outer space region is the neutron. The neutron provides gravity by producing motion. Motion is (1.4 / 1.42) X 10 = Π^2 and that makes outer space the compliment of motion Π^2 going to Π. That is way the location (Π) is in double the time (Π^2).

Another bone of contention I fail to see is how does Newtonians compromise logic in order to justify Newton in terms of Galileo. Yet I have been, to put it very frankly insulted on more than one occasion because I fail to see how Galileo says mass plays no part in the falling and Newton formulate that the whole affair is mass orientated. F = G (Mm) / r^2. On one campus in particular there was one professor that truly got nasty about this and he insulted me in a way I cannot forget. However that same professor failed to show me how Newton's mass brought any object faster to the ground since (GMm /r^2) = mv^2/r which suggests that the square of the velocity multiplied by the mass is the same as the gravitational constant multiplied by the product of both the masses and then divided by the square of the radius.

That means the mass m has to multiply X with the velocity in the square (v^2), which then will reduce (demolish) the distance (r) there is between the Earth and the object on a continuous basis until the distance is reduces. That's rubbish. How do they console this statement with that of Galileo where Galileo said all objects fall equally to the ground! Galileo said that notwithstanding mass discrepancies will all objects hit the ground at the same moment when dropped the same distance and at the same moment. Newton insists on mass while Galileo insist on equality of mass during the fall. The biggest bogus part of the lot is that I have not come across one Newtonian that was able to see this distinction. It is as if they all have an inborn blind spot.

Galileo said that the atmosphere is a neutron that is providing unrestricted mass in the time period that the earth set. Galileo unwittingly suggested 7 / 10 and that is what gravity is. I found the sound barrier as 7(3Π^2) = 207.2616km per hour. That is applying to what ever is falling whether whatever is falling or intending to fall at that moment. That is the neutron state of a body in the atmosphere.

A while back I indicated how man's senses evolved around his view that man (every one alive) is in the centre of the universe. Everyone and I can see how all light coming from wherever is heading directly towards me. By standing outside and gazing into the dark eternity that never ends I see from eternity light flows towards me and that places me in the centre of the Universe. That is a cosmic reality.

The atom holds seven points ($\Pi^2+\Pi^2$)($\Pi^2\Pi$)3=1836 as the Universe but that Universe is seven points in Π being 3 points serving dimensional time to form the Titius Bode law, and a law it surely is! The gravity extending from the Titius Bode law forms the entirety of the building of the Universe by constructing the Universe in the using of the atoms to form the Universe in the entirety thereof. That puts the atom in charge of the Titius Bode law since the atom forms the Universe.

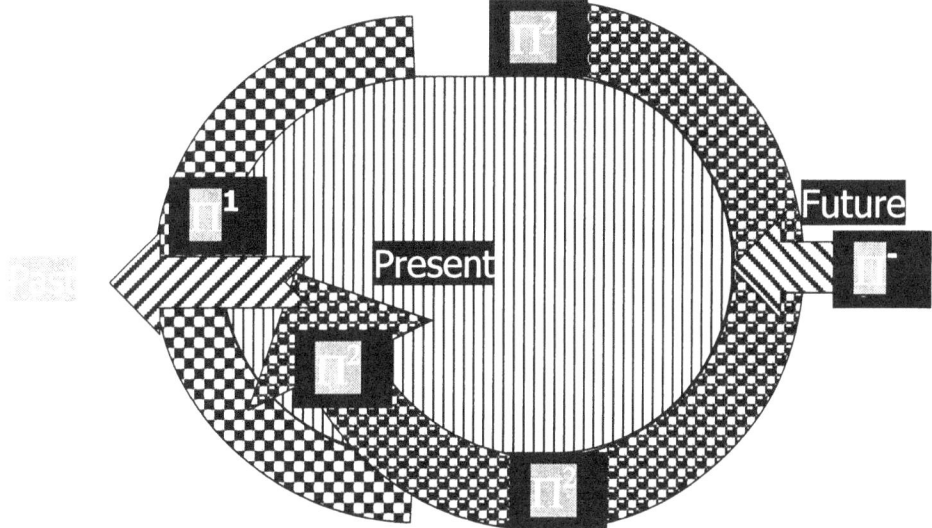

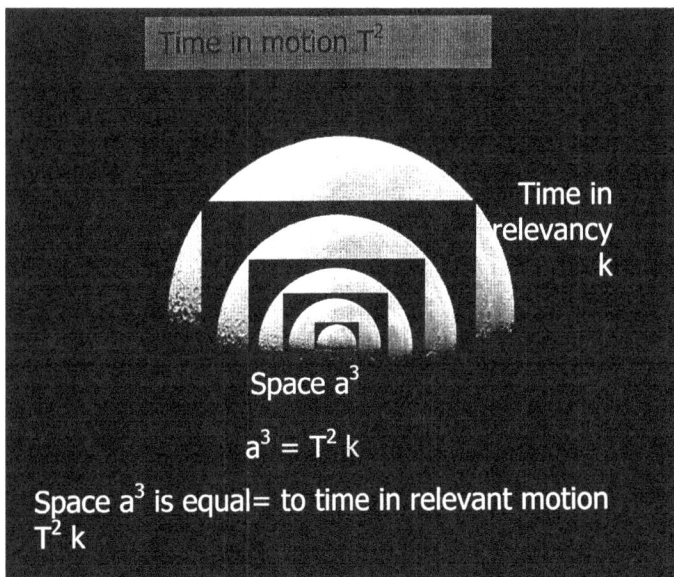

The three points we find time to be moving in is a direction unlike what we in the past thought about as a direction. There are seven basic directions being front and back, north and south, top and bottom and in or out. To our view that is the only way anything can move. It is either one of the lot or a compliment of two forming one of the lot.

By seeing light travelling towards me I am seeing time travelling. I am the direction that time flows. I can see where light was. I can see where light will be. I cannot see where light is going because that is within me and my singularity presents the future. Any one in disagreement should just go outside and see the light coming towards you. See how the light meets from all over the Universe precisely where you are.

The flow of time must never be confused with any part of the seven dimensions in space-time. The flow of time is away from the structure towards the structure then into the structure, through the structure where time disappears into singularity by measure of infinity. The flow of time is the motion that concerns the part science at present think of as outer space but which in essence forms time in eternity. Because Mainstreams science has the name incorrect they also have all attachments they connect with time in eternity incorrect for instance that the Universe has an edge and they give the cosmos a place where the Universe ends. That part on the outside of the atom never ends because that part is continuous to the point where time ends and time cannot end in the part holding the seven dimensions of time in space. Therefore time ends within the atom or as I have life my time ends in me. Therefore I have my position where I am in the centre of the Universe. Time comes from the outside as far and as wide as things can go but time ultimately ends within me.

Time is the outside of the atoms. Time is the inside of the atom where time is excluded from eternity by giving time specifics in motion and a defined value confirming space while the space is conforming time. Then time is taken to infinity where infinity absolves time into a unity once more. Time is what is between infinity and eternity and while eternity is parted from infinity time in eternity as a unit is also standing between eternity and infinity. Eternity is part of the part that is standing between eternity and infinity and therefore I am able to see eternity as a reality.

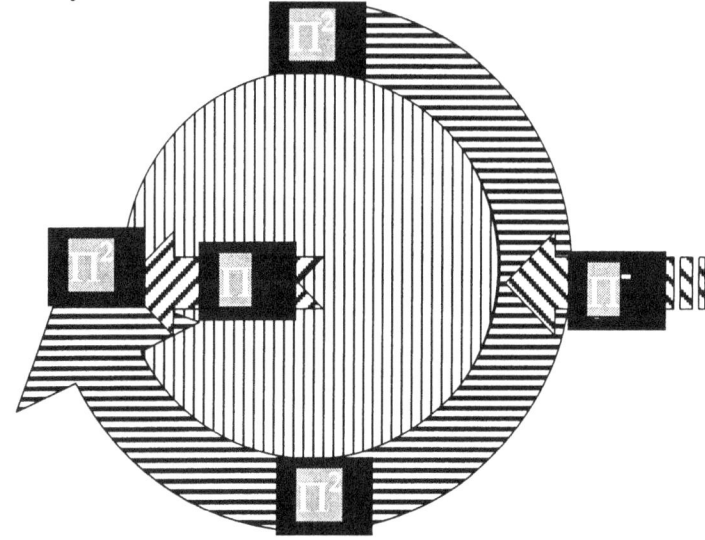

The light coming toward me is going whereto after it is upon me. It is going to the past. But from where I stand the light is representing my past so I am taking my past down my infinity into the future. That is why the Universe is shrinking onto the oblivious.

That is why everything into my future is shrinking into the oblivious as time engulfs material into the future.

You were in eternity because the light is coming from eternity towards you. You are where you are because I can see where you re plus the time it takes the light to come from you to me added to where you are in time. The light is going to disappear into where you are but that is not true. You are dragging the light that reached you into infinity berceuse light tries to escape time by going infinitive.

Infinity that which has no start is in you and you with your eternal life is generating time that parts infinity and eternity. That is not religion because that is raw physics. I have my doubt that any Newtonian will understand this concept since they can't even see that mass has no application on objects in orbit in outer space. If they are incapable of seeing the obvious how the hell will they be able to see what only those with intellect can see. That is why they can see no God. It is because they see mass applying in locations where there can be no mass applying.

Where you are and where you hold your body is the closest Black Hole to you because at that point time converts to space and space disappear into the gateway of singularity. Time ends where you stand but that only applies to your

Universe and while your Universe is in contact with the rest of the Universe your Universe is solo and alone in time.

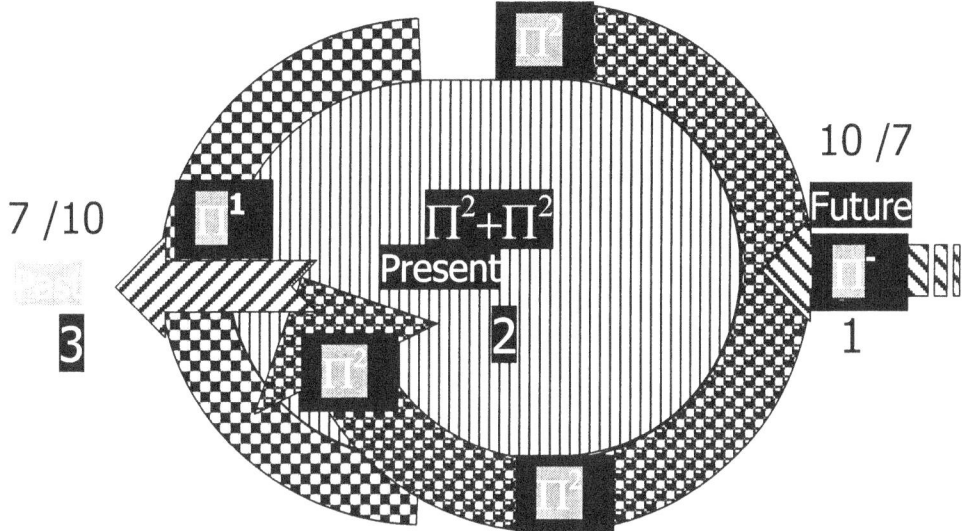

Time is taking the seven that was into the seven that is through singularity (.0999991) onto the seven that is going to be and that (3X7 = 21 + .99991) / 7 of material to which I relate I can be sure the Universe having time forms a sphere. By forming a sphere it gives meaning to the growth we see as the Hubble constant without Newtonians trying to rape any common decency out of it by their 13.5 X 10^9 years. God how could or can any one be that crude? The Earth alone is one million times older that that because what they use to measure time is the readjusting of the atom in relation to the factor the space represents. That is how the star inside accumulates the liquid by freezing the star, however I put more on this in another book where that belongs.

One thing we must not forget is that outer space is what material that is orbiting through outer space is allowing outer space to be. The Universe is the proton. The Universe is 7 / 10 in relation to 10 / 7.

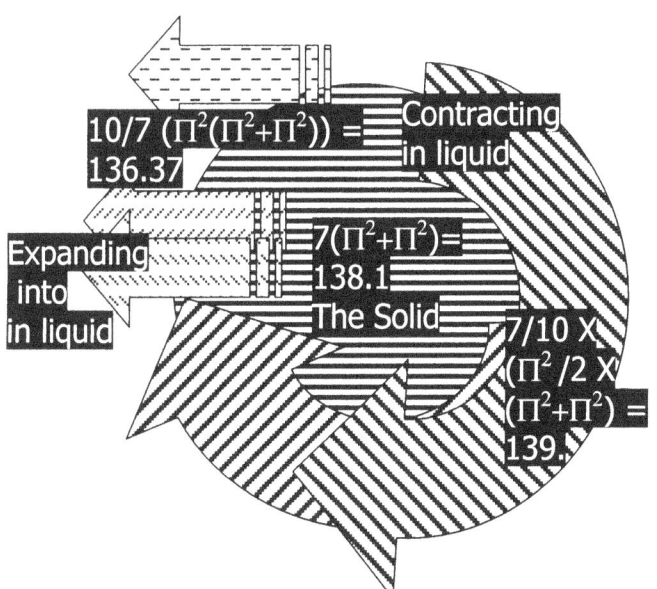

The Universe was what we now have from the first instance but in our perception that which was then does not apply to what we see in the Universe. We have an individual Universe from the one that will apply one day when one hydrogen atom will be a full star at an era of 7/10 Π/ 2 =1.09955. According to my opinion and that is my opinion, what we see as the Universe first applied when liquid and material stood apart from singularity. It was when liquid transformed space to combine again. That was when the neutron as we know the neutron first found a measured value in the Universe. Before that it was a factor but motion in time was at that point only a definition in our standards we now apply. It was when $10 / 7\ \Pi^2(\Pi2+\Pi^2) = 136$ formed the one wall of the then applying Universe while $7(\Pi2+\Pi^2) = 138$ formed the solid and the material was $7 / 10\ (\Pi^2/2)(\Pi2+\Pi^2) = 139$.

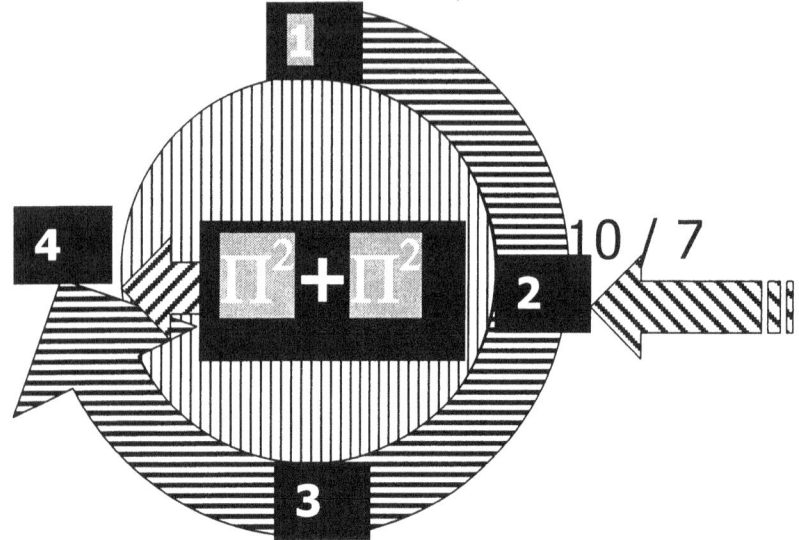

 Today in our Universe we have the wall of time at $10/ 7(4(\Pi2+\Pi^2)) = 112.\ 8$.
That is from where liquid flows to singularity. That from where gravity is generated by the iron core of the star. The core must have a relevant displacement of $7/ 10(4(\Pi2+\Pi^2)) = 55.267$ in proton displacement to have gravity establish the concentration of heat. That puts the Universe within the borders of the Titius Bode law at $10 / 7$ and $7 / 10$ in relation to the proton $(\Pi2+\Pi^2)$ forming time (4).

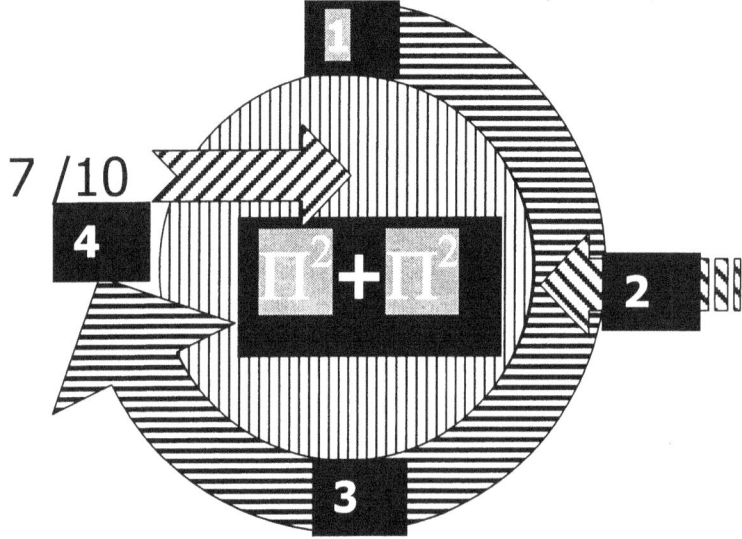

That is where liquid ends at material begin. That is where contraction of gravity begins within every structure that in our era has the ability to generate gravity. It therefore has to have an Iron core.

At the point where the neutron disengage from the atom we find our Universe catch up with time as time then takes control of space once more. The neutron is the lagging of time between 7 / 10 and 10/7. When the neutron removes as a factor that influence the displacement from the atom at $3(\Pi^2+\Pi^2) = 59.217$. As one can see the neutron removes all influence from the atom and when that happens we have a neutron star' which is no longer valid in out Universe. Outer space is not mass implying the gravitational constant. It is not mass that is producing the product by multiplying mass. Outer space is the Titius Bode law. It is gravity or motion or the neutron or movement. It is what the Titius bode law says it is. It is seven where four relates to three. It is where the building blocks of the atom leave their layers in the forming of time.

Our Universe is the flow of space-time in the form of retarded heat coming from the region 10 / 7 $(4((\Pi^2+\Pi^2)$ where gravity is generated by a revolving planet similar to electricity being generated by a spinning armature. There is no difference between electricity and gravity except that electricity is more concentrated in are of distribution. The gravity flow is directed by the iron core within the structure, which has to have, a displacement of is 7/10 $(4((\Pi^2+\Pi^2))$ in order to establish a flow direction. Just as electricity is charged by a directional flow between Iron 7/10 $(4((\Pi^2+\Pi^2))$ and copper $(\Pi(\Pi^2+\Pi^2))$ the flow is between 10 / 7 and 7 / 10. By collapsing the space-time within the core of the planet, there comes room available and with that reducing of space-time it starts a need to fill the collapsed space-time

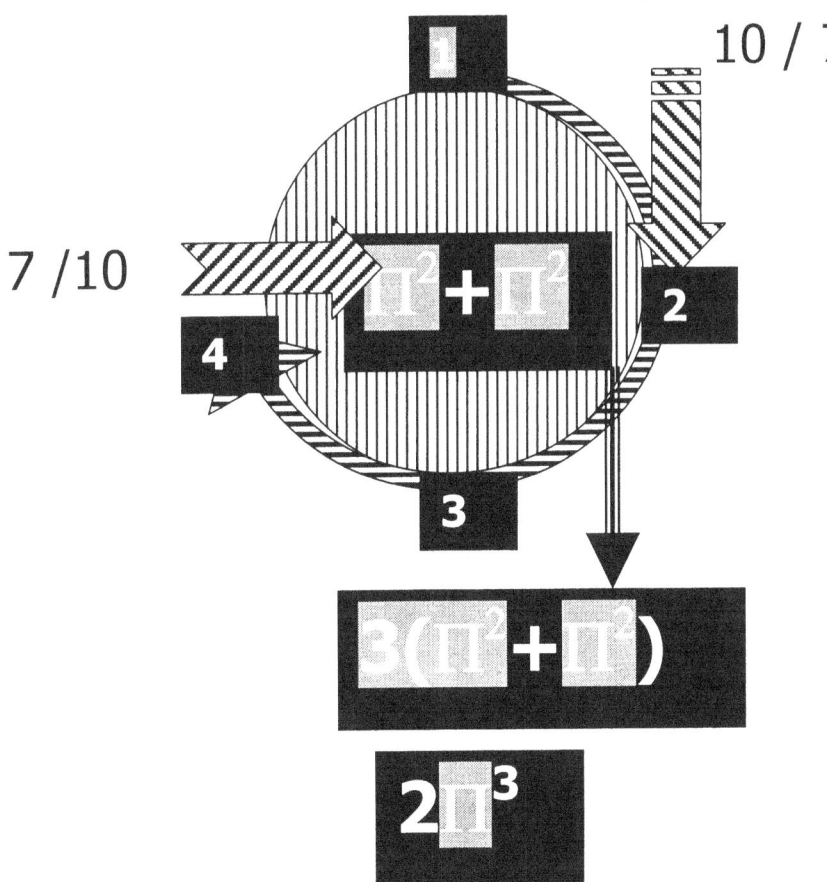

where that flow in space-time or heat contracting is gravity. This can only be when the atom freezes into a position where there is no required motion available the host the neutron. As soon as the proton$(\Pi^2+\Pi^2)$ links with the electron 3 by forming $3(\Pi^2+\Pi^2)$ the star is going outside our Universe and then becomes a proton star. By further cooling the atom will then directly link singularity outside the atom the proton $\Pi(\Pi^2+\Pi^2)$ and in that the space catches up with the time. The space goes double $2\Pi^3$ and eliminates the requirement for motion. Singularity feeds itself.

Time forming space = Π^3 =31.0061

Outer space is 10 / 7 $(4((\Pi^2+\Pi^2))$

Time collapsing space= $2\Pi^3$ =**62.01255**

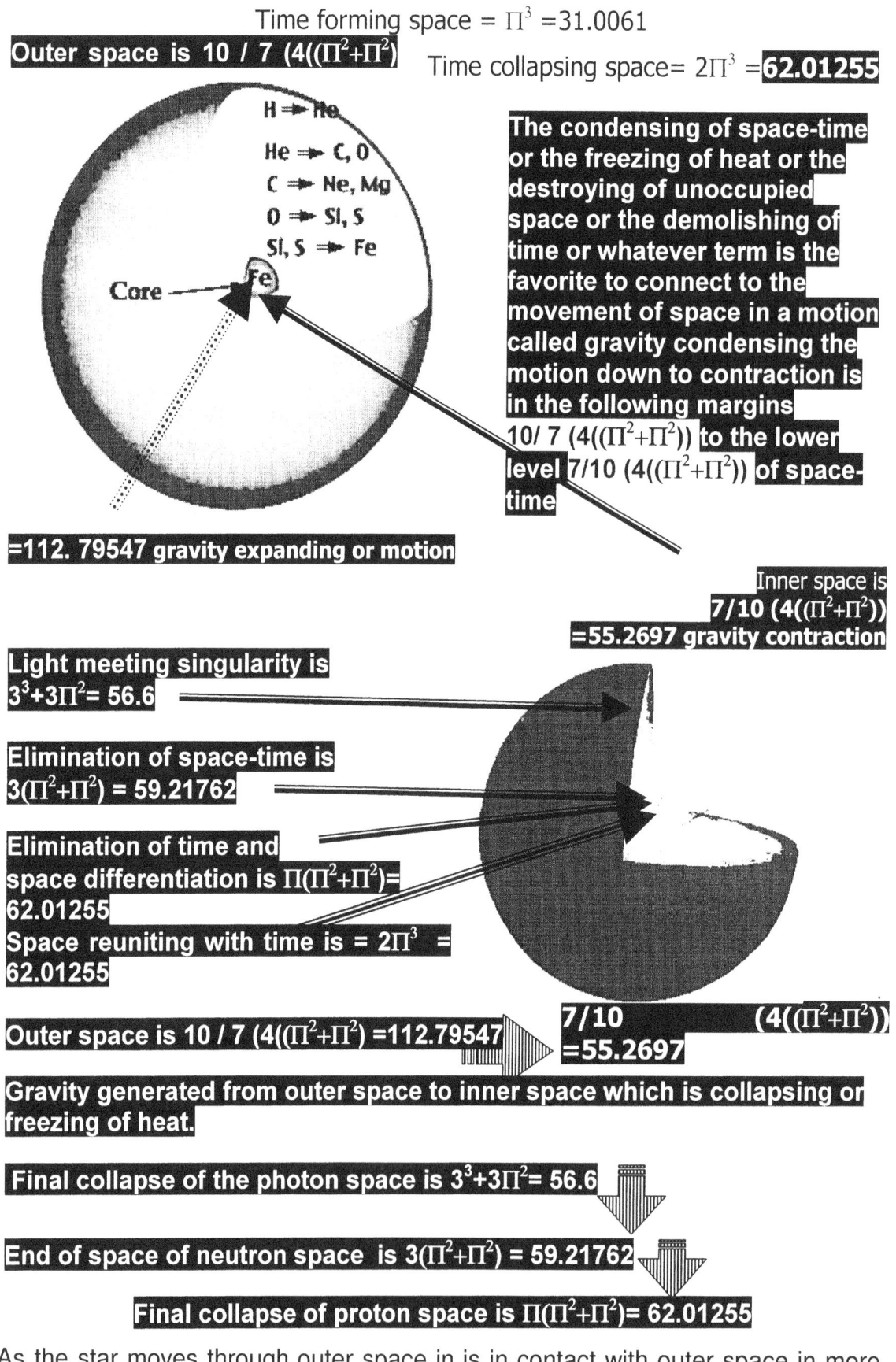

H ➟ He

He ➟ C, O

C ➟ Ne, Mg

O ➟ Sl, S

Sl, S ➟ Fe

Fe

Core

The condensing of space-time or the freezing of heat or the destroying of unoccupied space or the demolishing of time or whatever term is the favorite to connect to the movement of space in a motion called gravity condensing the motion down to contraction is in the following margins 10/ 7 $(4((\Pi^2+\Pi^2))$ to the lower level 7/10 $(4((\Pi^2+\Pi^2))$ of space-time

=112. 79547 gravity expanding or motion

Inner space is
7/10 $(4((\Pi^2+\Pi^2))$
=55.2697 gravity contraction

Light meeting singularity is $3^3+3\Pi^2$= 56.6

Elimination of space-time is $3(\Pi^2+\Pi^2)$ = 59.21762

Elimination of time and space differentiation is $\Pi(\Pi^2+\Pi^2)$= 62.01255

Space reuniting with time is = $2\Pi^3$ = 62.01255

Outer space is 10 / 7 $(4((\Pi^2+\Pi^2)$ =112.79547

7/10 $(4((\Pi^2+\Pi^2))$
=55.2697

Gravity generated from outer space to inner space which is collapsing or freezing of heat.

Final collapse of the photon space is $3^3+3\Pi^2$= 56.6

End of space of neutron space is $3(\Pi^2+\Pi^2)$ = 59.21762

Final collapse of proton space is $\Pi(\Pi^2+\Pi^2)$= 62.01255

As the star moves through outer space in is in contact with outer space in more than one way. By moving through outer space the star is disturbing outer space. Outer space is pushing against the star by measure of the star moving through outer space. The maximum displacement in duplication and contraction that outer

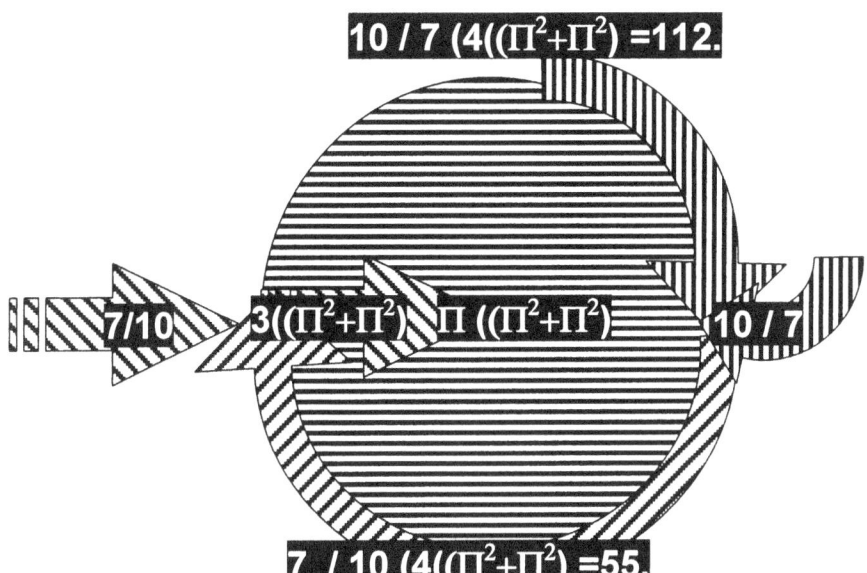

space can accommodate is the total sum of the atom in relation to singularity which is

and with that in relation to singularity it forms the atomic displacement limit of 112 protons to one cluster. With the motion that is the maximum expanding there can be when duplicating. However motion stands in relation also to contraction. The contraction is freezing of heat into a state of liquid coming from a gas. At 112 the state of singularity is expanded at a maximum and cannot cope with more heat than 112 protons will manage to control in one atom cluster. But relative to that must be a cold where such a cold will not hold space under a specific level of freezing. Beyond a specific limit the cold of space freezes time into singularity. The flow of electricity is not the transporting of some unattached electrons lazing around and then put to labour. That is Newtonian thinking. The shifting of electricity is involving motion, which stretches the neutron that then is running space-time all the way from (10/7) to (7 / 10). That also is gravity and electricity and gravity is the very same thing. It is the condensing of heat through motion.

That what is between unbridling expansion and total collapse of space is the Neutron. The neutron stretches fro 10/7 to as small as 7/10 from where it can reduce little more before abandoning the atom altogether. The Universe is gravity and gravity is the neutron where the neutron can have no mass because the neutron personifies motion of space-time.

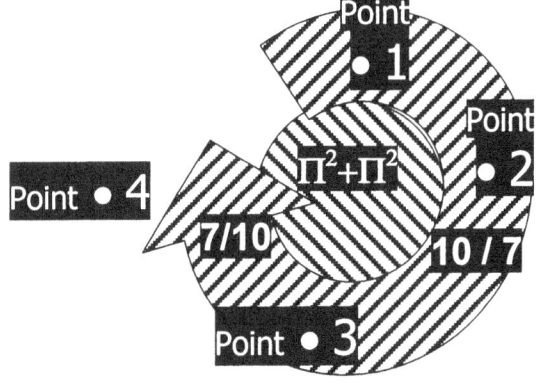

The motion is contraction but the contraction is not directly reducing. It is more a filling of vacant heat and displacing the heat in order to align the occupied heat with unoccupied heat. The expanding on the other hand is not expanding by going bigger but repositioning in order to duplicate. Such duplicating is exaggerating of space by instant reducing of space. In the reducing of space the virtual contact grow substantially more. In relevancy the star moves while outer space is motionless but because outer space is motionless the star is putting the friction of motion onto the account of outer space. The point of contact between outer space and the stat atmosphere produce heat as outer space e is reduces as well as accelerated. That spinning produces the light, which we see as photons. It is cooling the gas of outer space by reducing outer space to the point where outer space holds friction and the particles spinning in friction that comes across as photons. That is at a displacement level of $3^3 + 3\Pi^2 = 56.6$ the photon is the product of intense cold coming into contact with intense heat. By motion the atom is removing heat from inside the electron orbit to outside the electron orbit. The motion the atom is subjected too becomes more intense with every time there is a duplication of the space-time.

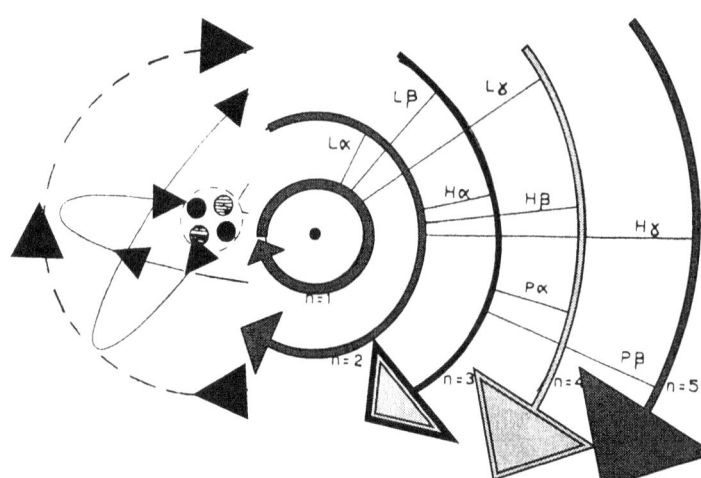

In this duplication presented as cooling the star puts the atoms of difference under the same state of affairs. The one duplicate more then the other does because the one consists of more protons in the cluster as the other does. The one has a more frequent cycle of time repeat that the other has. In this array of possibilities confusion sets in, in terms of duplicating while the other is contracting and fusion takes place where in motion the one incorporates another atom because the cycle period does not match. As I have indicated how one serves as a liquid to another particle serving as a solid the same process apply within the star. The contraction is presented by atoms with a larger indication of consuming space that the other atom can. This is what Newton got confused as mass. Because the one has a different setting in relation to heat one would stand prone to duplicate more and the other would rather contract. We see this as one being a gas while the other element is a solid. It is the measure in which the element favours duplication above contraction or the other way around. The fact that neither element of different standings show identical patterns in behaviour the one will try and incorporate the lesser developed particle by duplicating at the same pace while also duplicating considerably more at the time of duplicating. It takes considerable more duplicating time to contract an element holding say fifty proton displacement duplication than it takes an element with a duplication displacement of say 1. By duplicating the one element holding one proton in the nearness of the element duplicating 55 protons it can quite easily become a question of finding the element of one proton to be liquid and by the nature of the Coanda effect the space also extend to incorporate the additional liquid supplying the motion. The Coanda effect works on the basis that the extravagance of the liquid providing the motion suppresses the space when the motion is absorbent in nature at that point. There is a hot spat in all the cold and the hotspot causes sudden motion acceleration, as heat will do. The surge in heat has nowhere to go although the surge in heat at that spot insists on having more space. We know a liquid that heat takes up more space because it expands. If the heat at that point has no where to expand, but the heat is there altogether and the same, then the only expanding must be to incorporate the one proton element into the element holding fifty fife protons and that takes the total up to 56. But in this one must see the liquid surging in space as heat level rises but at that very point the space will reduce because the atomic relevancy will freeze the atom into less space. If the liquid becomes more heated and surge for more space but there is no more space to supply in a star that is predominant and overall engulfed with liquid the reaction of the solid will be to become colder. Becoming colder is also freezing in the face of the liquid heating and with the liquid heating the liquid will have much more motion. The more motion comes with supplying mot\re gravity where more gravity is pushing time longer in the face of space reducing. The situation is running at that point back in the direction of the Big bang and when the motion bridges the Big bang era fusion comes about between the two particles. In the end the particles joining space was the result of motion differences and mass is the result of motion difference. At one point the motion differences extended to a point running into eternity leaving the space at that point in infinity and infinity joins eternity at that spot where the element grows by one more proton. The mass of motion discrepancy did not create the enormous space - time deficiency but the moment the mass became eternal and infinitive the element joined space while enduring eternal time. That is the use nature has for the principle Newton named mass. It is a motion discrepancy whereby atoms would then comply to combine space should the motion applying validate such a drastic step.

Once more I have to return to the sound barrier to explain my point I wish to bring across. Where the sound barrier becomes evident there is a particle at one point displacing space-time at a rate of $\Pi^2/2$. In reality without the assistance of life to intervene it would then be the entire atmosphere that was moving at that pace having the particle maintaining such a motion in relation to all the liquid. In the case of the aircraft there is expanding without the earth compromising. But is the entire liquid atmosphere reaches such a point it could only come as a result of the earth in relevance duplicating at such relevance. The motion refers to the solid, as the liquid is motionless. Therefore the duplicating of the earth crossed the $\Pi^2/2$. By reaching $\Pi^2/2(\Pi^2+\Pi^2) = 97.4$ and that takes all material into the cosmic state of liquid. That would make the earth and the aircraft both having a state of being liquid where both can join. But also the earth would duplicate by reducing to the tune of $\Pi^2/2(\Pi^2+\Pi^2) = 97.4$ in order to reach such a state. In cosmic terms the earth is an atom and the aircraft is a lesser-developed atom but both adhere to the same atomic law, as would hydrogen and an Iron atom do. It is about solids and liquids and motion.

Then a point is reached where there is no more room to allow motion within the Universe. The atom in the star reached the cosmos limit on motion of $7/10(4(\Pi^2+\Pi^2))$ and has surpassed it. The atom has cooled to the point it could no longer sustain the neutron at a point of $3(\Pi^2+\Pi^2) = 59.217$ and the neutron at 7/10 is no longer part of the atomic space. It is where cobalt is and for that reason we find cobalt radioactive. In that region the neutron is no longer any part of the atom. Then more growth increases the atom to a displacement value of $\Pi(\Pi^2+\Pi^2) = 62.0$ which is a point where all space totally collapses. That is the end of space-time in an atomic environment because this is where the atomic space-time relevancy reaches $2\Pi^3 = 62$. At that point in the Universe the Universe at that point became a Black hole and as development evolves the dynamics of the Black Hole will eventually consume the entire star.

That is not the road that serves the Earth situation.

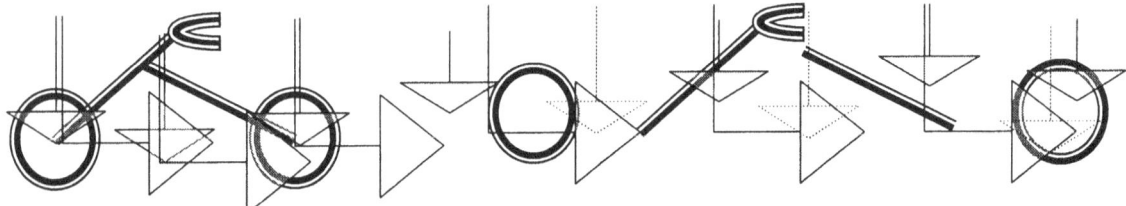

Motion distributes space and therefore decreases heat. By spreading the space over a larger area the heat in the space is reduced because the density of the heat allocated to the space reduces. Motion decreases the heat and therefore the Sun is the coldest place in the solar system while outer space is the hottest part of the Solar system

Let us first forget the accuracy of the bicycle moving in cosmic terms and concentrate on deliberate motion. It is said that if a butterfly flaps its wings in China, a hurricane will hit New Orleans. That is not true because we have to see what does exist and what does not exist.

When a bicycle is motionless the bicycle is not part of the cosmos. It has atoms but the atoms forming the unit do not charge a governing singularity where that governing singularity responds to the motion of the Earth singularity and by doing so establishes the unit into independence. The unit is a unit within the Earth unit. The unit is in motion with the Earth as part of the Earth. There is no additional motion confirming the bicycle as a force that is promoting itself by promoting the

Earth motion in addition to what the Earth does to promote motion on behalf of itself as well as the motionless bicycle. The atoms spinning would form a unit as far as confirming the bicycle independence in the unit. That does not make the unit independent. That confirms the unit as a group of atoms forming a

unity in form. That confirms $F = \dfrac{r^2}{M_1 M_2}$ which is what Newton saw at first. That is absolutely correct when some of the sharp edges of incorrectness are removed. There is only gravity applying between the Earth and the bicycle and the gravity confirms the restriction of the radius parting the objects to the very limit. That is because the only independence the bicycle has is in form and without cosmic motion.

Then motion enters the scenario on the pat of the bicycle. When the bicycle was motionless the gravity the Earth developed restricted the bicycle to one line placing the bicycle in a direct line with the centre of the Earth. That is a value of $7(3\Pi^2)\Pi^0$. That is the motion the Earth bestows on the motionless bicycle. The line running through the bicycle to the Earth centre has a value of Π^0 while the Earth reserves the motion on behalf of the earth and also of the bicycle at a premium of $7(3\Pi^2)$. Should some cosmic miracle wonder come about such as life is. The bicycle might just find the opportunity to achieve cosmic independence from the Earth such as the moon has. The bicycle then holds its cosmic independence at a value of anything between $7(3\Pi^2)\Pi^0$ and $7(3\Pi^2)5\Pi^0$. The moon however holds its cosmic value at $4(7\Pi^0)$ days. And in return the earth holds the moon at a reference of $\Pi^0 / 2$ days. The one day of the moon is 28 days made up of one Moon day = $4(7\Pi^0)$ days in the life of the earth while the Moon is $\Pi^0 / 2$ = ½ days in the life of the Earth. There is a definite division of cosmic liquids that is time between the earth and the Moon and the moon holds a stronger identity of independence in relation to the Earth than the Π^0 that the earth offer the bicycle.

The liquid space surrounding the Earth confirms the bicycle as part of space which time in motion draws onto the space. If the bicycle has independent motion the bicycle side with the liquid time that holds the motion in the relation. If the bicycle has no independent motion the bicycle sides with the Earth putting all relative motion in the basket of time. One must keep in mind that although it is the earth having the motion the Earth projects the motion onto the liquid time since the Earth holds a steady point on the surface of the Earth, which is steady in relation to the centre singularity. The bicycle can be part of space $\mathbf{k} = \mathbf{a}^3 / \mathbf{T}^2$ that is confirmed by time or the bicycle can be part of time $\mathbf{k}^{-1} = \mathbf{T}^2 / \mathbf{a}^3$, which forms an extension of space. When I fall down a cliff I fall at a steady pace. It is the same pace that I would have when I fall down a waterfall holding a cup in my hand the water in the cup will not stay behind. The water in the cup will not spill. The cup will not fill with water. If I had to fill the cup with water I will have to supply motion in access to the motion with which we fall. The water and I will have the same pace therefore we will fall at the same gravity. My density will not leave me superior. The water mass will not have the water fall more forceful or less forceful. The motion considering all objects is not discriminating on any basic grounds. In such an event I will be submitted to $7(3\Pi^2)\Pi^0$. That is gravity and gravity is motion. Forget about Newton's mass controversy because blaming it on mass is instituted fraud.

Having independent motion requires more than gravity. One may even be able to apply $F = G\dfrac{M \times m}{r^2}$ under conditions about where what fits. This is not a cosmic

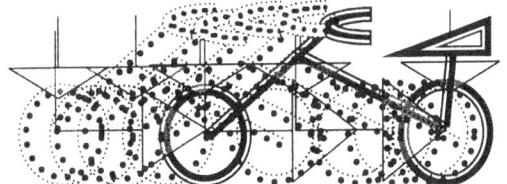

principle. It applies only to the Earth under conditions serving the earth and applying to all objects that submit to the Earth. The motion can be seen where the one M holds the motion the earth provides while the Other m indicate the independent object's independence while the G then will be the additional motion and the radius by square is the balance there is between being liquid and being solid.

In relation to the Sun the Earth is in motion and the earth is a part of the liquid that

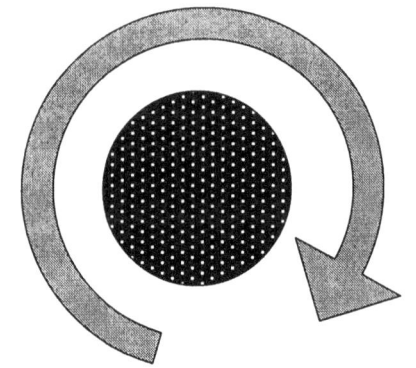

outer space provide. To the Sun the earth is a factor that flows with outer space and although the earth show the ability to counter the flow that outer space has, the Sun still regard the Earth as equal liquid to outer space. Being a solid or a liquid has no bearing on the state of the matter but it all depends in the flow in relation to a securing solid.

Understanding what the Universe is becomes the important key about realizing the dynamics of cosmology. The bicycle being without motion is a part of the Earth because of the Coanda and Kepler principle where the bicycle without motion sides with space a^3 and when achieving motion the bicycle sides with liquid by moving $T^2\,k$. This dynamic is the key in understanding what measures apply where in the cosmos. When not moving the atoms moving in the Unit the bicycle forms holds relevance relating too the general or governing singularity in the centre of the earth. It taps in to sustain the governing singularity by providing motion that forms part of the earth singularity. The unit uses the atoms to the advantage of the Earth motion and supplies the Earth with relevance in order to sustain as well as promote the earth moving. The bicycle is the Earth

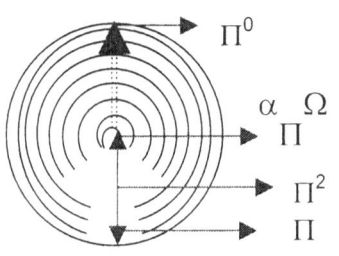

because it is space of the Earth within the space boundaries of the earth. It is a^3 and stands relative to T^2k.

Then when for some reason the bicycle finds the ability to move the bicycle split infinity an eternity. A universe is born. That which has no end parts from that which has no start leaving space-time generated. It is motion that puts the bicycle in three positions relating to time. That splits eternity and infinity just like it split eternity and infinity what moment-Alfa came about. There is just more heat in the backlog and less in direct relevance.

The rolling with time sets a differentiation between eternity and infinity and the

measure of the time delay forms matter in time. By providing motion the

bicycle no longer only keeps the earth singularity generated but also it supply a potential singularity by establishing the individual generating of singularity which sets out maintaining the individual singularity that is apart from the earth singularity while still being within the Earth singularity. The singularity it now generates is no longer Π^0 but forms an independent singularity by as much as $5\Pi^0$. It shifts the line of currently to at the most five positions in delay of currently. Any shifting further brings about serious conflict.

The Universe we have (not the earth filled with life that we have) but in the era we landed we find the Universe going from $10 / 7(\Pi^2+\Pi^2)$ towards $7/10(\Pi^2+\Pi^2)$ ending at $3(\Pi^2+\Pi^2)$ while eventually all space-time will form $(2\Pi^3)$ in the star limit. The proton disappears when the proton goes to singularity at $\Pi(\Pi^2+\Pi^2)$ which then becomes double space $(2\Pi^3)$ where space being double catches with time being single and loses it's lagging behind time quality. That is what they call a Black Hole or what I named a proton star. When the proton goes singularity then $\Pi(\Pi^2+\Pi^2) = (2\Pi^3) = 62.01255$

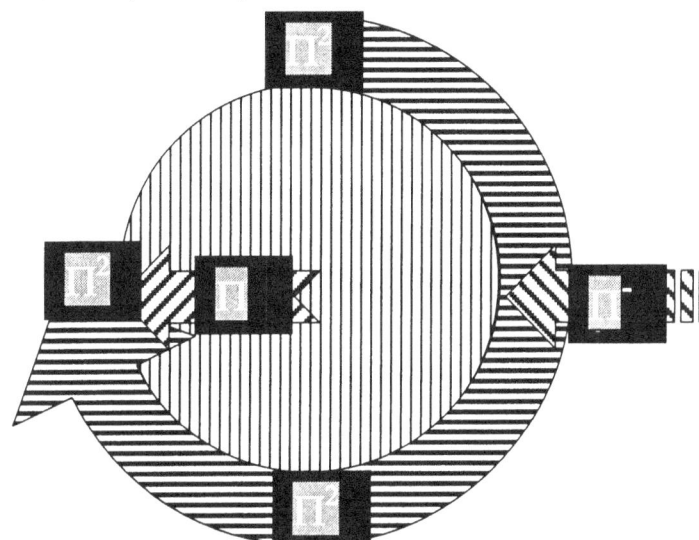

All objects are classed by heat either being in motion through duplicating (overheating and expanding) or being in motion through heat contracting (heat being reduced through motion removing space), but most of all is that all material is about motion forming the space-time and classifying the space-time. This is most pivotal in understanding cosmology notwithstanding Newtonian views.

The motion of the liquid which the neutron is proves to be the time (T^2) aspect because as it increases it claims the space (a^3) in the at a distance k of time (T^2) that the running has increased. The faster the motion is the stronger is the gravity that the motion generates in the space it claims by the gravity it generates

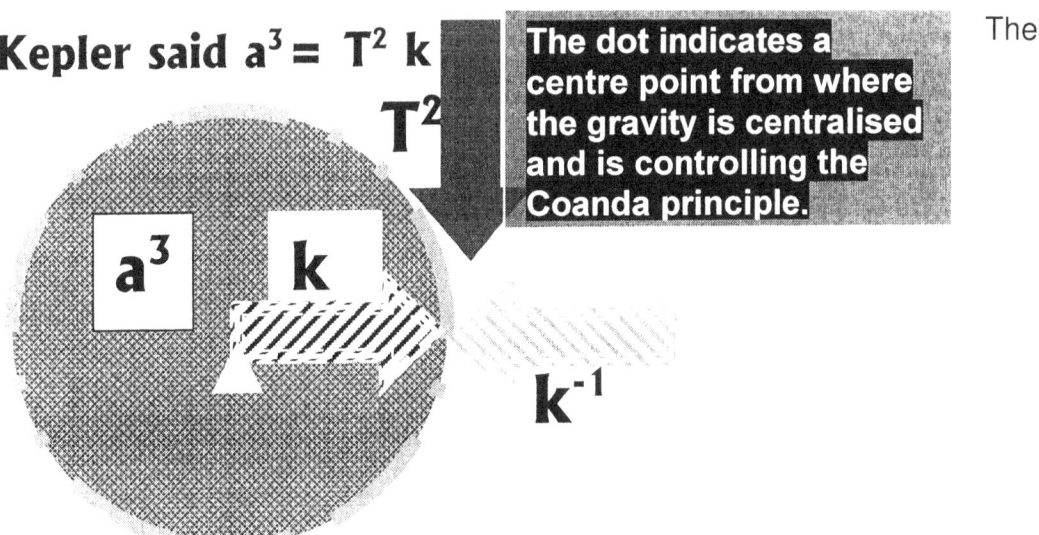

The dot indicates a centre point from where the gravity is centralised and is controlling the Coanda principle.

The

Coanda effect is proof of gravity coming about through space forming motion. In the case where water diverts the normal directional flow the space that translates to the motion is deflecting singularity with the flowing water charging the motion. In the centre of the object having the round form, singularity is duplicated and by transferring Π to form Π^2 and the motion of the water creates a line of gravity that pushes the flowing water to follow the direction that the newly gravity applies to the water. This again proves Kepler's statement of $k = a^3/ T^2$ that specifically states that space (in this case the object transferring singularity to a new position within the round object) and with the motion of the water redirects the gravity flow of the water to new space in new time. Only Kepler can explain the phenomenon but only when Kepler stands alone, correctly interpreted and divorced from Newton's opinion about Kepler's statements. There is a flow of time created by motion and defined by direction that produce expanding as well as contracting where expanding is contracting while it is on the other side of the Universe.

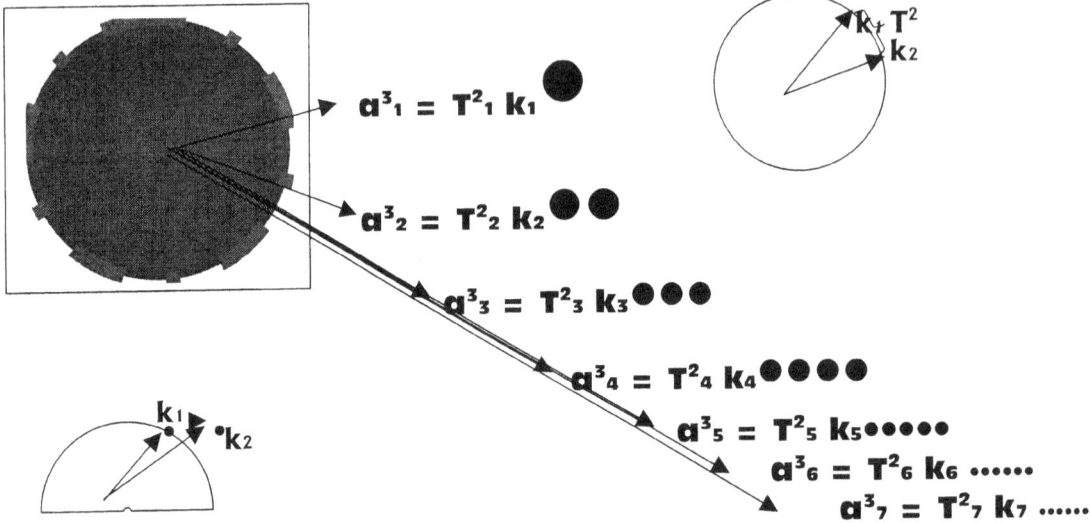

All objects are either cold and reduce space-time in relation to others being hot and expanding space-time. Being cold puts the object in the role of conserving space-time in contraction and that puts the object in a position of being a solid.

Then in relation to the first conserving factor there is the overheating factor, which brings into the relation the expanding, or moving away from the singularity.

The duplicating requires a repositioning of the aligning of **k** from a certain position to a more fore ward position in relation to and in that **k** will also have to extend a value when moving from k_1 to k_2. The essence of motion is to duplicate material that is in a process of overheating. By producing more than one of the same material unit the heat is distributed over a wider area and thus the heat has more space per time unit but less space in a time frame. That is motion. By applying heat to provide motion the Universe see that as overheating and the longer k_1 to k_2 is per time unit the lesser will T^2 be because a^3 is spread over a larger area.

This letter you are reading is my effort by which I hope to interest you in reading my Introducing letter an open letter To Selected Academics ISBN 0-9584410-9-X The book on offer has the title of an open letter To Selected Academics ISBN 0-9584410-9-X and is the actual letter I sent to various establishments.

What brings about the expanding?

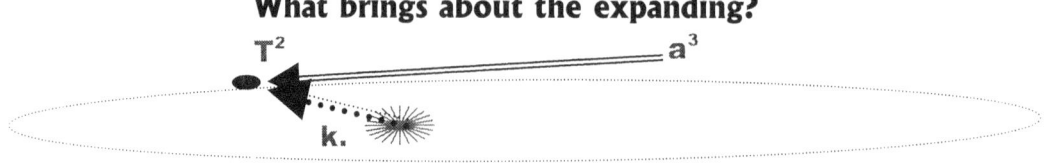

Kepler was the very first person to mathematically introduce space a^3 centre k and time T^2. Not only did he introduce space-time a^3 / T^2 but he also placed space a^3 and time T^2 in a relevancy long before Einstein did and placed gravity in space-time a^3 / T^2 even before Newton named gravity. Kepler was the person who placed gravity as the ingredient in the Universe that determines space a^3 and time T^2 and much more. Kepler was the first one that saw that gravity comprises of two factors being k or linear gravity and circular gravity or T^2 as gravity keeps space in form while all is staying together.

Since gravity also influences the space outside the sphere, the space we call outer space has seven plus three points bringing about ten positions of gravity influencing space.

The influence inside the sphere also captures the space outside the sphere.

This means that in the cube at the point of contact between the cube and the sphere the cube experiences such a contact point as if the "bottom falls out" of the cube and without a "bottom" to support objects they fall to the sphere as objects does fall to the Earth. Remember that a body "floats" in space, but at one specific point it starts to "fall" to the Earth. That is gravity and it is a dimension change much more than any force. I shall explain this last remark later on. That too is the Lagrangian system with five cosmic structures holding relevancy to the centre structure where the centre structure stands in for seven positions diverting from centre and the orbiting structures standing in for five positions in space.

Gravity has all to do with dimensional changing and reforming of forms to re-affirm alliances supporting the centre. It is the reforming of space converting space to more concentrated heat.

The Universe is in the three dimensions using twelve dimensions that is visible to us and indefinite number of stages in size differences ranging from the immeasurable small to the immeasurable large where mathematics becomes a short fall to the next and the previous dimension.

Up to now every one in science is normally acting as if gravity is a commonly explained factor, which every one knows every aspect about all principles that are involved in gravity down to the smallest detail. In truth, no one in science anywhere remotely knows what brings gravity about and I used Kepler to unravel this mystery called gravity. But no one in science will admit this fact about Kepler being the one who formulised gravity decades before Newton came and gave gravity the name.

Newton did not underwrite or define gravity and even today the most informed in Science at best can only assert their suspicion on a rumour presumed about what causes gravity to perform as the part interlinking the cosmos but no one can go any further by explaining the concept. Newton started this realising of gravity but it had and still has no more substantial proof than a rumour has and Newton admitted to it being a concept he could not explain. In Newton's ignoring to test Kepler's findings, Newton missed the opportunity to find what gravity is. Since Newton every person in science also ignored Kepler and every one is guilty of missing the opportunity Kepler maid available.

By my efforts of studying the implications that results from Kepler's finding I can now un- emphatically declare I know what gravity is. Gravity is the entire following locked into one compiling unit:

Gravity is not being some magic force found between particles grabbing onto everything. I mathematically explained the following phenomena:

Gravity is singularity as a factor forming space-time

Gravity is finding space-time

Gravity is proving space-time and aligning space-time with gravity

Gravity is the working principals behind all cosmic occurrences that pre dates the Big Bang period.

Gravity is the Roche limit.

Gravity is the Lagrangian system

Gravity is the Titius Bode law

Gravity is the Coanda effect

Gravity is the sound barrier

By being able to pin point prove what Gravity is that enabled me to unravel the other entire phenomenon that forms gravity. Each of the phenominon I mention above has one part or role in what is forming the totality that which we know as gravity.

Should you think this is rather a wild presumption I challenge you to spend a little more time and please think about what I say when you read about what I say in the next few pages. The first thing you should admit in private is what study did you personally so far made about the work of Kepler?

Still, to this day nobody in science at present will denounce the principle of gravity as vaguely researched. Gravity has never been explained as a principle. Even when one is considering what the importance of gravity is, gravity never yet has been understood. It is by now very clear that little if nothing of all objects is pulling closer in outer space. Comets are missing the Sun on a regular basis and no planet has come much closer toward the centre of the Sun. Still everyone in science acts in a manner as if Newton's gravity ideas are the best detailed proven fact and only occasionally does someone quietly admit that even Newton admitted not knowing what gravity is. No one ever comes to the front and boldly state that gravity is just a rumour spread by scientists pretending to know all there is to know and knows little to nothing about what there is to know. Newton admitted that much when he introduced the name (not the concept). Mistakenly Newton corrupted the concept he named as gravity.

Going according to what Newton introduced Newton's concept will by now have the moon much closer to the Earth than it was in the time of Kepler's studies, yet we know the moon is moving away instead of coming closer. By the same measure Kepler suggested that the space a^3 is content with the motion kT^2 as long as the motion T^2k is equal to what the space a^3 will allow. Kepler suggested motion of space remains in equilibrium as long as motion of space a^3 duplicated space a^3 by motion thereof T^2k. That is much more true than objects rushing towards one another by the pulling power of mass. Newton agreed that he could only declare gravity as a vague concept. This fact was at that time drowned by the man's stature and was relieved from the manner of requiring the proof that later Academic science became an absolute necessity. The proof one would demand now a day was never given to put Newton's rumour beyond doubt. When Newton announced a force he also admitted the force could be anything. No one ever came after Newton and proved the fact better. That still underlines the fact that the force to this day can be anything. Not once could one person in the past or present provide substantiating proof on gravity as reality by defining the very principles thereof.

That includes every one since Newton as well as including Einstein and even Hawking. Scientists can declare gravity was a factor at 10^{-43} seconds after the Big Bang but what brought gravity about or why gravity became or still remained, as a presence is still tightly concealed information which all are speculating on. Even in the best and most informed circles and amongst the most educated there is no one that knows what gravity is because they all ignored Kepler and for them to ignore Kepler the price they pay is not finding the principles bringing about gravity. Using Kepler even makes the method to follow and understand Einstein's discoveries shockingly simple. Gravity is the motion of space relating to time in movement.

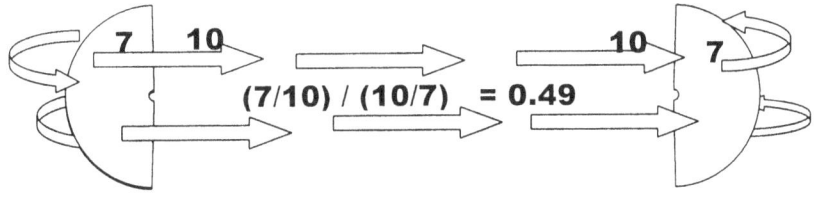

SPACE DIVIDED INTO TIME

(7/10) / (10/7) = 0.49
.7 / 1.4285 = 0.49

Taking also from both orbiting influences

THE PROCESS PARTED USING THE ROCHE PRINCIPLE

10 / 7	$(\Pi/2)^2$ The Roche influence on Titius Bode
7/10	$2.04 \times (\Pi/2)^2$ = 5.033
$(\Pi/2)^2$	$2.04 \times (\Pi/2)^2$ = 5.033
10 / 7	5.033 +5.033 = 10.066 from both objects

Crossing the singularity divide and activating the Roche principal $(\Pi^2/4)$

(10 / 7) \ (7/ 10) = 2.04

1.4285 / 0.7 = 2.04. Taking from both orbiting influences

SPACE MULTIPLIED WITH TIME

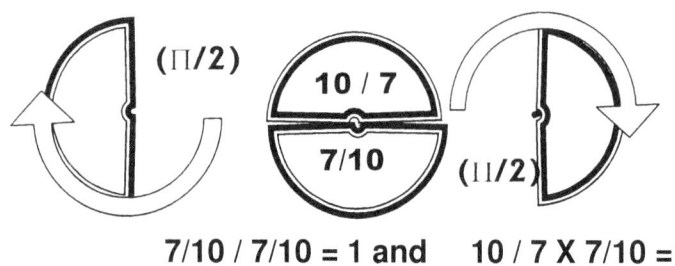

$7/10 / 7/10 = 1$ and $10 / 7 \times 7/10 = 1$

From dissecting the formula I prove that:

I prove gravity is strongest where space is least (not where the Universe goes flat as Einstein promoted).
My theory I propose is one where gravity is a relation that is based on Newton's law that for every action there is an equal but opposing reaction to the action. Gravity is not one- way traffic but is a relation of relevancies that are applying equally and without the relevancy applying between the action and the reaction in a balance there is no gravity.

In the relevancy there are opposing motion where each participant provide a motion which is contradicting the other relevant but opposing motion. My base this conclusion on what Kepler introduced when he introduced the fact that space is half the motion and the motion is the other half of space. Therefore space cannot be is if space is not moving because the movement is space repeating what was before in the instant and onto the future. Space is one part and the repeat of space from the past onto the future taking space through the present is the part forming space by repeat in motion. Space cannot be if space is not moving to provide space a past into the future bringing the one part of space that in which it is and the other part either what it was or what it is going to be. That part is time in the formula space-time. The one part of space is what there is in the present but the second part, which forms the time aspect, is what there was or what there will be. Time depends on space moving while space depends on time a position coming from or going to. In an attempt to explain my view I am prepared this one time to grossly simplify my view and relinquish accuracy in the process. There is motion expanding the confined space just like a rabbit tries to flee a dog chasing the rabbit. The rabbit is trying its earnest to escape from the dog. The dog again is trying to contain the rabbit by reducing the space that forms between the fleeing rabbit and the chasing dog. But the space between the two is what is of a major concern and not as much the running of the dog. Focussing on the running will lead to missing the process that really applies. From the view the onlooker has it may even seem as if the rabbit is pulling the dog be an invisible string in an area where both participants are remaining in an area that is repeating the continuing chase as the dog holds the rabbit in chase and the rabbit holds the dog at a distance. The rabbit has itself in a task where it wishes to never see the likeliness of the dog ever again while the dog sees dinner.

The rabbit would love to leave the dog at a distance where the rabbit find freedom from the dog as the dog's dinner opportunity. While the space between the two is a merely a common fact, it unifies their differences. Both in relevancies have to appreciate their differences by the space in the unit that is keeping them apart.

The space is the factor that has to resolve the issue being the differences in motion but cannot because different relevancies sustain equilibriums. I prove that as much as there is Newton's pulling there is Kepler's running around and the running around is equilibrium of the other factor providing the running away part.

The angle science is looking at the issue science either dismisses or cannot explain the characteristics or principals, which is there none the less. The explaining of the phenomenon is quit impossible when using the pulling rope magical attachment idea in the manner science tries to explain gravity. Therefore instead of dismissing the rope they dismiss all other factors present by gravity unleashing free motion but they would not release their idea about the rope. Gravity is motion between two particles that brings about mass. In the book I explain this in much detail but frankly there is not enough room to explain this in this web page.

Mainstream science knows about that gravity has never been defined, the Bode principal that is there in all the planets and even the fragmented planet, the Roche limit, the Coanda affect, the Lagrangian system and the sound barrier, but cannot explain any of the phenomena all though the presence of these phenomena is without dispute. It is the explanations about what causes the phenomena that is part of the dispute but in science the manner in which they defend Newton, science would rather discount the obvious phenomena than question the legality of Newton's cosmic views. I only dispute Newton as far as his cosmic principles are inclined. The phenomena being there or not becomes disputed. Science fails to give acceptable explaining of such occurrences we see in the phenomena and therefore disputes the validity of the phenomena and this failing to explain the presence becomes disputing the presence thereof. I on the other hand found a way where these explaining of the phenomena took me past the Big Bang era and introduced me to the start of all starts. Science cannot get past one specific date because they do not accept or understand the phenomena, which I prove started the Universe.

In such a light Scientists must somehow realise they are barking up the wrong tree with the information they have to use to do some explaining. They cannot refuse the phenomena and not realise they must have the cat by the tail as far as cosmology goes. Please remember that with this I am referring to cosmology and not general physics. There is an Earth versus a Universe with huge difference between the two concepts but Newtonians fail to see that because Newtonians cannot appreciate the differences thus they're not able to understand cosmic gravity; they go about blurring the understanding of gravity. If there are that many phenomena (it represents all there is in cosmology) to explain and such little ability to explain (science fails to explain even one) by using the information Mainstream science is using to explain the cosmos, then someone somewhere has to realise there is something drastically wrong in the way they present the knowledge they claim to have. One cannot be serious about science but defend your view by dismissing the validity of all unknown indicating factors presented as such. There then is some gross incorrectness in the way one reasons. The Roche limit is there and no denouncing thereof can remove it from the cosmos. They may refer to evidence received from the Hubble telescope as "the star is blowing bubbles" for the lack of explaining what is occurring but occur it does. One cannot say it is some unknown gesture presented on occasions because by not explaining the pictures present certain foolishness. It leads to tragedies in aviation and the tragedies they are incapable to understand or explain. For fifty years they lost many pilots but still has no idea what brings the sound barrier about, or find the link gravity holds in the process we call the sound barrier. Instead they try to interpret some effect established almost two centuries ago with steam trains that is travelling at the same speed that horses run. No further investigation with the science in hand brought them closer to new facts! That they should rather see as a sign telling them they are going about incorrectly because by ignoring the cosmos one produce a fantasy and not science. Nevertheless, it does not because tell them anything because Newton did not say so. When I first came upon the amount and the totality of the unknown quantities in cosmology as well as the complacency those involved has about such unknown factors being discarded, it stirred a sense of disbelief and I decided to respond.

All principles I use in the theory I introduce with the publishing of this book. All principles I apply are part of nature. I base my theory on heat stabilizing through space using motion to produce cooling. That is gravity.

I believe some of Creation remained as some particles formed by applying gravity in motion and the lack of motion in others became the lack of gravity, which inspired overheating which then formed plasma. Plasma is the result of heat where light is the epitome of heat. How light became plasma is rather obvious, which again I believe (within reason) I do prove. I believe heat is the destructed form of material and this information the atomic thermo explosions give us.

Analysing Kepler's formula without Newton interrupting Kepler's work helped me realise science has been running on an error for the past three hundred and fifty years. Please let me explain: Tycho Brahe and later Kepler made a study of outer space as never repeated afterwards. From this Kepler concluded that $a^3 = T^2 k$. We all know that a^3 is space and with the space indicated as being in the third dimension and the third dimension is unmistakably a cube that forms volume, which by definition is presenting space. We also know from the way calculations come about by using the formula of Kepler that T^2 is the duration of a specific period of time relating to a specific centre. On the one hand we have space a^3 and on the other hand in direct relation to the space Kepler introduced motion coming from a centre that forms time $T^2 k$. Kepler gave us space-time a^3 / T^2 centuries before Einstein gave the concept a name but no one ever took any notice. In the formula is space a^3. In the formula the space a^3 has direct relation to time T^2 If k is a^3 / T^2 it means that from the centre holding the gravity is space-time. Space is a^3 and the motion of space a^3 we accept as time T^2 k and such accepting is part of our understanding for the past three hundred and fifty years. Kepler gave us gravity before Newton named it as a force. Kepler gave us space-time long before Einstein named the notion. With Newton's meddling he missed Kepler introducing gravity as $k=a^3/T^2$ space / time.

I believe that I achieved an all time breakthrough success because I can now explain what gravity is. Remember that not even Newton could explain what gravity is or where it comes from, but Kepler did that without any person ever noticing. Scientists over the years paid the price of ignorance about gravity by their unwillingness to investigate the father of gravity, which coincidently is not

Newton but Kepler. From such explaining what Kepler said without Newton changing formula on Kepler's behalf, I prove the Titius Bode principal also known just as the Bode principle. I prove that the Bode principle forms gravity when using the Roche limit. These phenomena were never explained or understood by Mainstream Science although they appear more than regularly in the cosmos. In the same breath I might add that Kepler also was never investigated. My achievements came from my effort where I separated Kepler's work from the opinion that Newton formed and that he (Newton) gave his compromised views about Kepler's work to the world. For instance from Kepler's work I can explain the operation of the Black Hole, which not even Prof. Stephen Hawking understands. That is because Hawking ignores Kepler. In my opinion my explaining of gravity makes much more sense than the accepted force of Dark Age proportions…and the best part is you do not have to be a genius to realise or understand it.

Even a simple person such as myself can see it clearly! From my view a force is just motion applying and that is what Kepler said gravity is. Kepler said $a^3 = T^2 k$. I dissected **k** as a factor in the Coanda effect and found that the Coanda effect is proof of my view about gravity and the ability of establishing gravity by centralising space, which forms singularity that produces the Coanda effect. The Coanda effect is the establishing of individual space a^3 by applying motion T^2k. Where the Coanda effect is producing gravity and such producing is stronger in a small space than the gravity produced by the Earth in that spot I use that principle to show that there was some manner in which the reducing of **k** brought about a stronger T^2 just as Kepler said. This was a crucial part during the Big Bang and therefore had to play a part during the period of the Big Bang. Einstein came to this conclusion but failed to refer his view back to Kepler.

As presumptuous, as it may be on my part of trying to disprove Mainstream Physics, such a presumptions does not change the truth about Mainstream science being incorrect about gravity. After all they admit they do not know what gravity is. I am not disproving anything because they agree they do not know, which paves the way for my showing what gravity is. By admitting not knowing what gravity is they then also admit there is a chance that they can be incorrect about gravity but unfortunately mainstream physics do not see it that way (yet). The question in hand is finding what role gravity played when the Creation came about for the first time. I had to find a method that would allow me to explain why gravity played a role.

My ambition is proving the Universe not coming from nothing and therefore outer space cannot hold nothing. By taking Kepler's $k = a^3 / T^2$ and using **k** as a line I show through using the line as an example that the cosmic Universe holds everything and all concepts. However the only thing it does not hold is also the only aspect not present in the Universe at all. That is the value of nothing or zero in as much as carrying the definition of the absolute absence of any value. This means the Universe is filled to the point it is overflowing which we call the Hubble constant and not there is not room to be empty. With the line that light uses to flow the lines eliminates any such a possibility of nothing being present. Mathematics is a means of communication about matters concerning the cosmos. As an intercultural language spanning across race and ethnicity or as a principle as such mathematics cannot have zero because mathematics indicating lines, which is about not applying the numerical number or value of nothing. Everything came about from singularity and Einstein proved that. From singularity nothing ever had the chance to enter space. I challenge any person that disagrees with this statement to show mathematically where nothing as a factor ever entered the mathematics of the Universe. If there is any one there believing there is nothing in outer space I challenge that person to prove mathematically where nothing is a factor in the cosmos. Your attempt may either be before or after reading my work but my challenge will stand since mathematically nothing cannot be part of mathematics. Multiply whatever with zero or nothing and such multiplying results in nothing where nothing is then, can establish no multiplication. Kepler gave us the relation between cosmic objects as $k = a^3 / T^2$. From the formula k forms a connecting straight line filling the first dimension and not the single dimension because k in the single dimension is not zero. It is unproven how k can backtrack to become $k = 0$. I deliberately press this point and make it an issue because that removes all the theory of mainstream science from any logical base they have in support of their views that space is, holds and comprises of nothing.

In the book an open letter To Selected Academics ISBN 0-9584410-9-X the book only and exclusively deals just with the fundamental basics of my theory. I do not elaborate or explain the broader aspects or form an overall view. I found if I do that before a solid understanding of the basic concept is established no concept becomes established. The most basic to explain is that the line cannot start at zero because then there can be no line to follow zero. The cosmos has lines forming cubes and lines forming circles, which in applying 3D manifests as spheres. Between the circles and the cubes run lines, so the key to understanding the Universe is the following of a line. The Big Bang was a time when the Universe was incredibly small making the running lines small. Understanding the Universe is taking the line connecting particles through space back to its limits where such limits were during the Big Bang. But the reducing cannot go to zero because zero removes the line all together. By reducing the line to where the line will not reduce any further we will find at that point that all points land on the same spot. The spots all share one position because that is the only position there is to hold in the form singularity presents.

That is singularity being one to all but it is not zero. Finding form in that point shared by all will give a value of singularity. Extend that value received to a Universal centre and bring that value to align with Kepler's $a^3 = kT^2$ and understanding the Universe by finding the centre of the Universe makes the Universe simple as can be The Universe becomes sensible making the entire different yet unexplained phenomenon as easy as children schoolwork. There are suddenly no more mysteries in the Universe. It is only possible when we see gravity not as a grabbing force instead of seeing gravity reducing the space between particles. Gravity is not being some magic force found between particles grabbing onto everything.

The following is the mathematical proof hat through the atom time illuminates space by applying motion. Following the mathematical proof I explain how that is achieved.

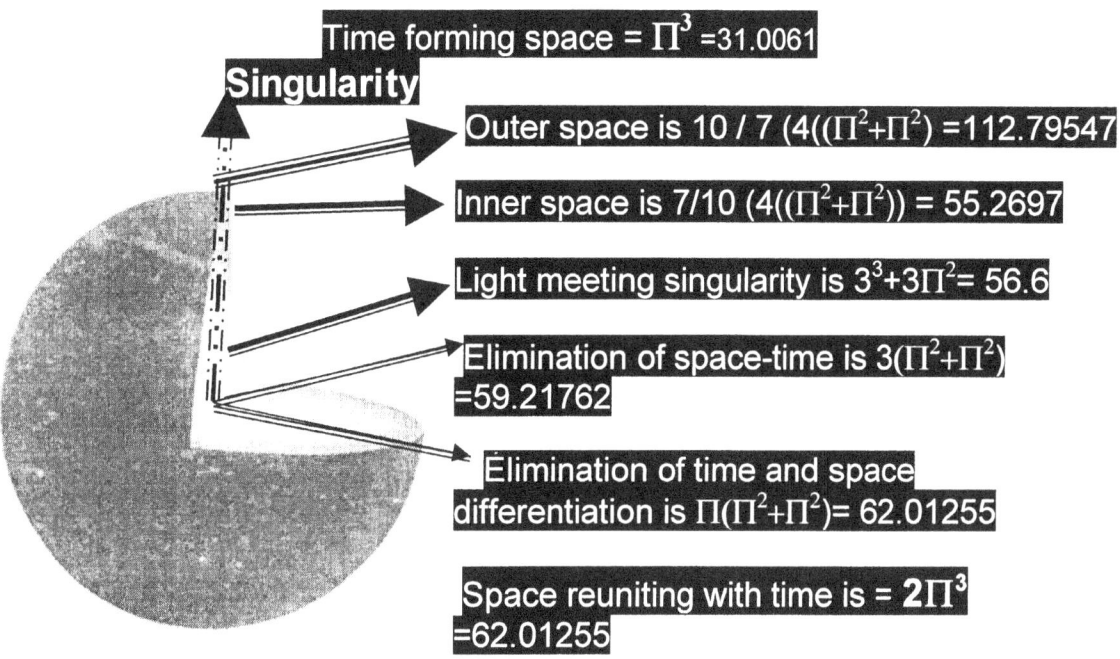

Time forming space = Π^3 =31.0061

Singularity

Outer space is 10 / 7 (4((Π^2+Π^2) =112.79547

Inner space is 7/10 (4((Π^2+Π^2)) = 55.2697

Light meeting singularity is 3^3+3Π^2= 56.6

Elimination of space-time is 3(Π^2+Π^2) =59.21762

Elimination of time and space differentiation is Π(Π^2+Π^2)= 62.01255

Space reuniting with time is = $2\Pi^3$ =62.01255

We stand on the outside 150 X 10^6 km from the spectacle and from such distance we judge the Sun . We don't even judge the Sun from what we can see but we judge the Sun from what we feel. We feel heat coming from the Sun and from that we argue that the Sun is hot. We see the Sun has heat rising from the surface as a liquid soup. That puts the hydrogen layer as the outer layer in a liquid. Hydrogen freezes on Earth at a temperature, which is the coldest amongst all other elements. Yes, the Sun is 6500 ^{0} and that is on the outside. To a human that is hot but a human has no mind judging the Sun . If the Sun squirts pure heat turned to liquid from the surface and the heat falls back into the surface the Sun is a lot colder than the Earth is. The earth requires an enormous effort to cool hydrogen down to a liquid state. We must mind the way we think of the hydrogen in liquid. The hydrogen remains a solid. The element is untouched by temperature differences. It is the heat environment surrounding the hydrogen that changes from a gas to a liquid to a solid. One removes or one amplifies the heat in which the hydrogen is and that turns to liquid or solid or gas. The hydrogen is untouched in the elements worth.

Yet we see the heat flow amongst the hydrogen as a liquid. Nevertheless we remain adamant that the liquid is a gas and the hydrogen is in a gas and the Sun is a gas bowl filled with hydrogen because to our mind hydrogen must be a gas. After all, our element table classifies hydrogen as a gas and that is the way we think of hydrogen. We do not consider hydrogen to be in a liquid state when we see the heat is flowing just like a liquid and shows all indications that it is a liquid. No the Sun is hot because the Sun feels hot.

In the Universe there are no hot or cold but a state of differentiation produced by time. The Universe parted by parting heat from cold when eternity parted from infinity, when Π^0 singularity parted from Π singularity, when 1^0 parted from 1^1. There is no hot or cold but there is a relevancy where one factor cools and another factor overheats. By retaining the Sun is the coldest space in the solar system and outer pace is the hottest there can be.

From since the time that man discovered intelligence (if he ever did) man has been with the presumption that the Sun is the hottest centre in the solar system. Later on in the present time, it came to someone's attention that the Sun also holds the solar system in gravity. The Earth by its standard and dominating its sphere of which it can control with influence is the hottest centre in the space of its domain and it holds the moon centred to the Earth. The gas planets are the hottest centres in relation with the most heat and they all hold their satellites captured by a hot centre. All space structures hold in every centre there is that is confirming their independence at that point of securing independence the centralizing of the most heat it is able to concentrate and from that centre holds all material captured or controlled in the domain of what that forms the independence of the structure. I can go on and on but heat in the centre couples gravity to space-time, just as if Kepler said before he was spoken for on his behalf and without his permission or his agreeing to it.

$a^3 = (T^2 k) = a^{3 + 2 + 1 = 6}$ with the sphere presuming the position of singularity as part of $k^0 = 1 = $ **singularity**. Einstein proved that at the point where space reduces and such reducing reaches a point where space as a factor in the third dimension disappears into the single dimension (space going flat) gravity is overwhelming. Einstein interpreted this, as the complete Universe going flat but while it may be true that the Universe is going flat, that can only be within singularity since singularity represents the Universe as flat as it can get.

The centre of any sphere has to be at the very point where space completely falls away. It is at the point where all the points of line centres meet by the crossing the centre of their individual connection coming in to contact as a group. In that way one may assume that the lines connecting the controlling points on the other end are crossing on a centre point that all that is participating in the constructing of the sphere is democratically electing such a centre. Please note this conclusion very well because this forms the heart of the Coanda principle. That will put that position where the lines cross, which in itself is centralising all space in the sphere at that point, such crossing point will become very distinct and controlling where that point forms in the single dimension and singularity is the single dimension. Kepler also solves another riddle that truly got Newtonians unstuck. This, to which I now refer, is what is referred to when they refer to the Hubble constant.

The growth we see in the Universe is an adding of space in every cycle completed by every cycle, which all the protons complete. The adding is the smallest addition that can come about in the shortest period of repeating by cycle rotation there can ever be. This growth of space-time next to singularity confirms the growth of singularity as singularity recalls the space it uses to grow in the time it grows. The margin of growth will be by the extension of **k** in the formula $k = a^3 / T^2$. Every cycle completed in the relation to space by the initial value of **k**. $k = a^3 / T^2$ leaves ultimately a^1 extending as space or as Kepler chose to indicate it as k^1. That too has to be compensated by the duration of time reducing the time aspect by the margin that the space expands. This confirms what is evident in the Hubble Constant. The further one looks at time the more time seems to race because time has the invert properties we give to space.

There is a position that is in motion that is forming the very edge of the outside. To be in motion the position must be in relation to a point from a centre. From the centre, there must be a specific allocated space ending at the object in motion and starting from a centre that has no dimensions. The object in motion determines the one limit and the centre with no sides and no space, which is standing still in singularity, determines the other limit. By that we can see there are only one way of looking at what we can observe and that is from the outside in.

The atom must be the utmost coldest and the proton is even much colder because when that cold escapes it turns to heat forming space that no one can understand. When the spin of the atom allows the cold of the atom to release the heat it had it had frozen to space the atom holds but when this heat releases from the containing form of the atom it brings about much more heat than the Human mind can cope with. One may not look at the material and judge the surroundings. The fact that hydrogen remains a gas and so does helium in outer space must serve as enough proof that outer space is hot, regardless of our interpretation of the temperature gauge telling us what we wish to hear. One must look at outer space and judge outer space from the findings only considering outer space. If helium remains a gas, it is hot. The removing of heat makes the centre of the Earth cold although we see it as being terribly hot. The only reason why it can seem to be hot is because it is cold and in such a cold environment, the heat can gather and space can collect heat because the particles find the surroundings extremely cold.

The cold in the earth centre causes the concentration of heat by space reducing, as all cold surfaces tend to do. If it was hot, the space within the Earth would expand and the space within the Earth where we think so much heat is concentrated does not expand therefore it must be cold. To gather and accumulate the space in a liquid means it became much colder being a liquid. Finding the surroundings terribly cold will allow the heat to gather and not expand but when the surroundings are hot, it will not tolerate more concentration of heat and thus will expand to rid the balance of excess heat within space. Look at the Sun and see how the Sun turned the hydrogen to a freezing cold liquid at 6500 K. Hydrogen is in a fluid state within the Sun and is colder than the hydrogen that is in a gas form in outer space. The Sun is the coldest place in the solar system. That is when the protons oversupply the removing of space to produce the cold that is so apparent. By the reducing of space, it can concentrate heat to a fluid state by producing the opposing cold that finally freezes the heat to a solid state. The expanding of space is a way of duplicating space without reducing space and by duplicating in the form of expanding it becomes just the opposite to duplicating by motion therefore reducing space by halving space in time. That is what gravity does. By motion, space duplicates and by space, halving it removes heat in space as well as by dismissing space. In all the applying of gravity, space dies. The density of the protons brings about space dense enough to harbour the heat in such quantities and visa versa applies in outer space.

The application of gravity that condenses space and bringing about heat by the compressing of space we apply in the way we go about tapping into the energy that nature provides. Internal and external combustion engines all rely on this application for harvesting motion by driving power. Compress space even today with a piston in a cylinder and then pump the compressed air into a container and such confining of space will increase the heat by the piston effort to reduce the space brought about in the container. The heat coming about inside the cylinder has no relevance to particles colliding because all compressor cylinders cool down. The walls become colder because when that cold escapes it turns to heat as the heat releases from space forming a secondary form of material forming space that no one can understand when the spin of the atom allows the cold of the atom to release into uncontrolled space. This release and unification with space that heat does is the heat it had frozen because the motion of spin to space that the atom holds, remains in a frozen state under the guard of the spinning electron. When this heat releases from the containing form of the atom frozen by the spin of the electron, it brings about much more heat than the Human mind can cope with. One may not look at the material and judge the surroundings.

The fact that hydrogen remains a gas and so does helium in outer space must serve as enough proof that outer space is hot, regardless of our interpretation of the temperature gauge telling us what we wish to hear. One must look at outer space and judge outer space from the findings only considered in the terms which outer space insists upon. If helium remains a gas, it is hot. The removing of heat from the space that contained the heat makes the centre of the Earth cold. In our universe we see it as being terribly hot because the heat then forms a separate substance but remains a form of material (8) but that is because we see the heat and not the space derived from the separating of the heat.   The only reason why the space can seem to be hot is because the space is cold and in such a cold environment the heat can gather in a much concentrated state and space can collect heat because the particles hold concentrated heat in the space separating the particles.

By removing such high concentration of heat from the space that used to be expanded heat, the space then must contradict the heat by being extremely cold. We look at the heat in the space, which by that time is another form of material and find the surrounding heat in the space hot while the space is extremely cold. The cold in the Earth centre causes the concentration of heat by space reducing, as all cold surfaces tend to do. The proton contributes to that reducing of space. If it was hot the space within the Earth would expand and explode but the space within the Earth where we think so much heat is concentrated is so much it does not expand therefore it must be cold. To gather and accumulate the space in a liquid means it became much colder when the space parted from what then is being a liquid. Finding the surroundings terribly cold will allow the heat to gather and not expand but when the surroundings are hot, it will not tolerate more concentration of heat and thus it will expand to rid the balance of excess heat within space. The concentration or release of space with heat or space from heat is a direct contribution of the singularity in control of the space-time. The regard of the singularity stipulates the conducing of heat in space or the release of heat to form space by means of bisecting the occupied space.

Look at the Sun and see how the Sun turned the hydrogen it holds captured in its atmosphere to a freezing cold liquid at 6500 K. Hydrogen is in a fluid state within the Sun and yet it is still colder than the hydrogen we find in outer space that is in a gas form in outer space. The Sun is without any doubt the coldest place in the solar system. That is when the protons oversupply the removing of space to produce the cold that is so apparent in the heat levels that the atom cannot absorb in normal growth and therefore do cannot find accommodation in the walls of the atom. By the reducing of space, it can concentrate heat to a fluid state. By producing the opposing cold that finally freezes the heat to a solid state, we find that is what matter is. The expanding of space is a way of duplicating space without reducing space and by duplicating in the form of expanding it becomes just the opposite to duplicating by motion therefore reducing space by halving space in time. That is what gravity does. By motion space duplicates and by space duplicating the material must be by dividing or bisecting - halving it removes heat in space as well as by dismissing space and in that concentrating heat. The density of the protons brings about space dense enough to harbour the heat in such quantities and visa versa applies in outer space.

The particles claim more space when heated to preserve the cold. The claim to more space produces more space and reduces more heat. Such expanding brings about cooling. When particles heat or cool motion applies in some form. Motion started at a point when the Universe was extremely hot and there was no space. By introducing motion space formed and the lack thereof produced friction that became heat that became space. It is natural, it is simple, and above all, it makes believable sense.

The application of gravity is that which condenses space by bringing about heat with the compressing of space. We apply the progress we have as a species in the way we go about with our skills to unveil ways we can tap into the energy that nature provides. Internal and external combustion engines all rely on this application for harvesting motion by driving power. Compress space even today with a piston in a cylinder and then pump the compressed air into a container and such confining of space will increase the heat by the piston effort to reduce the space brought about in the container. The heat coming about inside the cylinder has no relevance to particles colliding because all compressor cylinders cool down with time moving and not necessarily with the loss or release of particles. It is not only the discharging of air that will reduce the temperatures inside the container. The time flowing bringing motion about where the motion is not about particles escaping but heat escaping in the replacing of the heat density (not the density of the particles forming the material content within the container) but the space that compressed to heat will also bring about that the heat displaces through the container wall to the outside. This is bringing about equilibrium where heat will always flow from more dense areas to the lesser dense areas. This has no influence on the status of the particles on the inside of the cylinder but only concerns the density levels of the particles inside versus outside. After the pumping of air increased the heat in the cylinder which even can go to dangerous levels, will reduce back to room temperature when further pumping ceases and that stops further air movement into the cylinder and such surging of pumping air is what brings about heat stabilizing.

Mainstream physics ignored the clear connection completely, notwithstanding it being so very obvious. There is this far in their recognising of principles in natural physics not one single reference made to prove their appreciation of this matter. They are bent on particle colliding. When particles collide, such collision forms an atomic thermo release and that action we call an exploding atomic bomb. What principle this argument about particles colliding, ignores is that all atoms use negative charged electrons forming the atomic limit on the outside forming a definite border to the boundaries of all atoms and in both electrons from different atoms are being negative charged.

In being negatively charged, it means both will come out and totally reject the other. The closer they come the more violent the rejecting will be and such rejecting is the production of heat that will turn to space. The electrons repel other negative charged sub atomic structures, which the electrons are that form the outer borders of all atoms. With all electrons highly negatively charged (being as negatively charged as any possibility will allow to match the utter extreme) such electrons could not touch.

It is about time scientists start looking with their minds and not their eyes at the Universe and see what is truly out there to see. All the difference we find is seated in the human mind. We humans set differences because we look at the cosmos by placing humans and the life we find on Earth in a pivotal centre in the cosmos instead of placing singularity in the centre and life where it belongs; only found on Earth. Einstein proved mathematically that in the presence of a strong gravity such a strong gravity slows time down.

Surprisingly with that evidence being around this long, nobody in science since Einstein's discovery took those statements and made any further progress from that. It seems to have been left in some drawer to dry. Science still sticks to the opinion that time did not change, not even slightly, since the beginning of the time it held the same pace ever since the start of the Big Bang notwithstanding the implications this concept carries. Before the Earth took one year to circle around the Sun and even before the Sun was there a year was still the same duration of one year. How odd... don't you think ... that the only aspect in the entire Universe that is beyond change is the aspect of time? With the entire Universe including all the gravity now present and not excluding one Black Hole or dust speck pressed in such an area that was possibly the size of a lepton even then the gravity extending from that circumstances must have been beyond what words can ever describe.

When everything was that small when the Big Bang took charge, the gravity at the time was beyond light, because even today in the Black Hole the gravity is beyond the speed of light. If the gravity was that high and Einstein already proved that strong gravity slows time down, then there is one logical conclusion and that is that time was in fact at the time of the Big Bang standing still. Mathematically it is incorrect to allow gravity to compress the Universe into a spot smaller that an atom and exclude any other factors and relevancies to change.

As usual Newtonians has the relevancies mixed up. It is not the Sun that is cooking outer space to cinders but it is outer space that is boiling the Sun to steam The steam we see mainstream science promote as light coming from way in but is the Sun was hot as hell on the inside the Sun would have exploded as stars do when they overheat. Supernova are stars that overheated and if a star can overheat when everything is going wrong then the star must under heated when everything is going right. Therefore a star is a particle that is frozen into liquid and in a process of would be one day frozen into the oblivious.

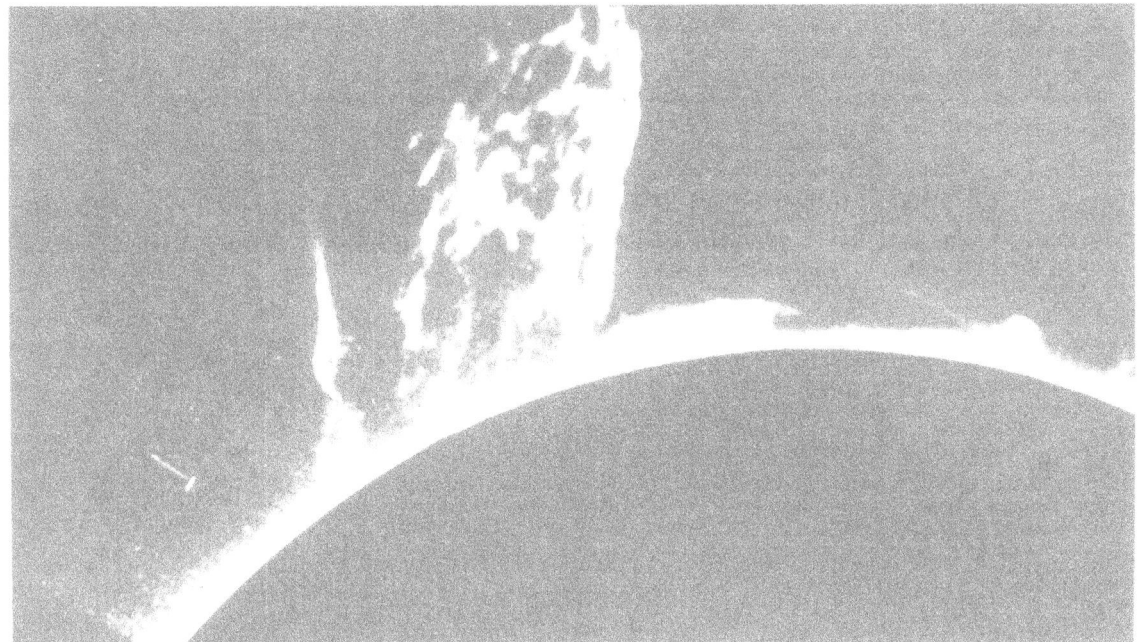

The fact that the prominence squirts out liquid in vast amounts is because there is a lot of space reducing going on in the Sun so there is less space inside the Sun . The fact that the prominence expands to outer space means that the prominence was expanding into a hotter area and does so as a liquid. However that fact that the prominence fell back into the Sun can only result from the prominence not being hot enough to return to a gas state and then through such density discrepancies had to return to a less hot area. That is science not magic. By using Newton one cannot even begin to explain any one of or the combined efforts of the above cosmic phenomena that are all over the cosmos and forms all the laws in the cosmos. Newtonian definition cannot even recognise any of the principles but only Newtonian science are taught to students. No student can have the fortune to disagree with Newton and remain a student. If the student will dare to disagree with Newton it is the end of such a students academic career. By setting this firm condition Newtonian science becomes institutionalised mind conditioning of the concepts of thought forming in physics. With my saying this I have not made one academic friend but neither have any one proved me wrong. Students are taught to accept Newton and to ignore Kepler and any student doing it the other way around will fail all examinations and other testing at Universities. Students accept Newton or they accept a ticket taking them home. Newton is an institution force fed to each following generation but saying that reserves only resentment towards me amongst Academics. According to Newtonian science space is simply nothing with no qualities but gravity separating space and space does not mingle, as one would expect if space was nothing because space does form borders.

Disasters of unprecedented magnitude arise from such borders. The Challenger disaster of February 2003 is pertinent testimony to those borders that was powerful enough to break the aircraft into pieces while the explaining contributed by Mainstream science is evidence of a shocking lack of understanding about what took place as cosmic laws were breached. I do not pretend to be of superior understanding and do not place myself on any pedestal. On the contrary the information is so simple and so easy to understand that the lack of any Academic understanding frustrates me almost witless. But academic taught culture demands all persons to miss the evidence, which is so clearly visible because academics demand researchers looking in other directions because students are forced to accept Newton's vision about Kepler's work. By the time they reach researchers status, they too have tunnel vision that can only acknowledge Newton and ignore Kepler. Our not understanding laws, provide a platform for future disasters occurring because it will lead to us ignoring more of applying principals that leads to space tragedies of magnitudes we have not thought of as yet. By not understanding the sound barrier, tragedies have and will again come about and will increase as misconceptions become more present in the future because the demand on space travel increases.

The book **an open letter To Selected Academics ISBN 0-9584410-9-X** is about that process adapted by the Big Bang, never ended and it is still bringing over, that which is in unoccupied space to material being in occupied space. Occupied space holds matter and unoccupied space is empty of solid materials. There is contraction, which we know by the name we gave as gravity. Then there is expansion, which we gave many names being the Big Bang and the Hubble Constant or better known as simply exploding or forming plasma with all the terminology accompanying that simple idea. This I show is antigravity. Apply heat and space and a balloon lift where such lifting is antigravity. There is a balance in the Universe where gravity contracts and reforms space and heat expands becoming space and produces space.

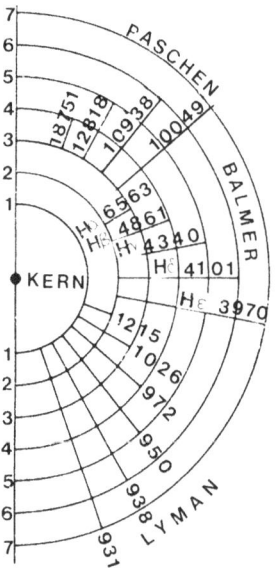

This puts my theory in line with reality. The only way anything can get bigger is when heat is added to what there already is. The Bomb at Hiroshima and Nagasaki showed how intense the heat in atoms are and how well packed the heat is which is contained in the atom. Bringing more heat to the atom brings about the atom having more of the same but in a higher measure. Only heat can make material expand and with the Universe unable to expand because the Universe is what ever can be, it must be the material inside the Universe that is expanding. In the same measure we find where space reduces heat, is removed from that space. Gravity is about contracting and that is true. In the manner that Kepler put it material that is filling space (a^3) is equal and the same as the motion ($T^2 \, k$) of the Material moving. That means to have time then time must move and the only way time can move is to move the space in the time.

However to move is either to come closer and using Kepler we can see how Kepler would reduce space in $k^{-1} = T^2 / a^3$. That means the expanding subsides to a point where contraction is and contracting is about cooling or reducing the Heat that was expanding the space. Then there is expanding of material which is as Kepler would put it $k = a^3 / T^2$. That would be when the space becomes less and the only valid way for space to become less is when what there is becomes less than what there was. That amounts to cooling where the heat is being removes from the space.

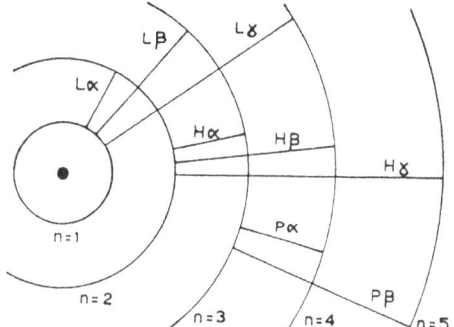

Whenever the electron jumps from a higher into the lowest (innermost) orbit, the atom gives out radiation at a wavelength corresponding to a spectral line of the Lyman series. The jumping down into the second lowest level contributes to the Blamer series. The greater the jump is the greater is the emitting of radiation to the limit of the series, which is reached when an electron enters from

the outside of the atom. Outward jumps involve the absorbing of heat and that is inclining to provide space to accommodate the increase heat levels because of the increase or rise in the absorption lines. When the heat level in the atom rises, the electron jumps to a higher band and when the heat reduces in moves down one band. The heat coming about in the surrounding of the atom produces

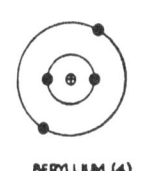

more space because the atom increases the space by applying the electron in a higher orbit ring. The moving of the electron is coupled to the giving out of radiation at a wavelength corresponding

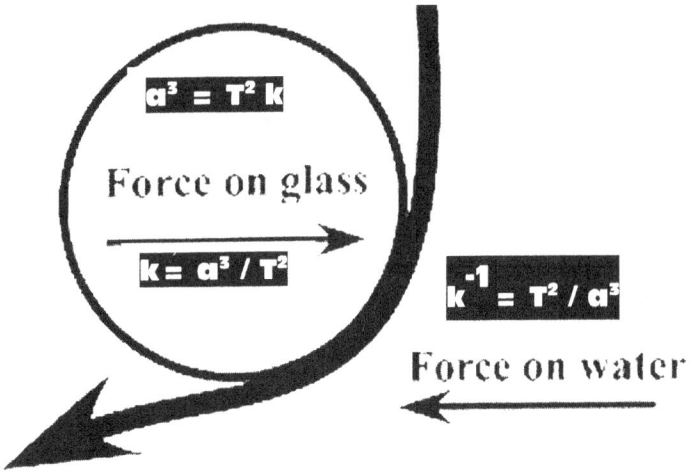

to the spectral line of When the heat level

the Lyman series. rises or lowers the space within the atom decline or increases. Every atom in every shows different heat it is in

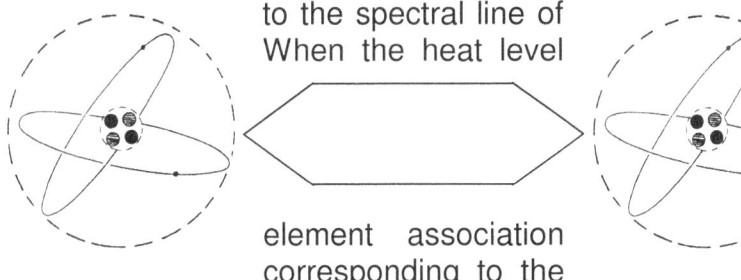

element association corresponding to the

association with. The corresponding of the atom and the reaction derives from such rises of space is a direct result of the interaction there is in the gravity contracting and the gravity in expanding depending on which of the actions of the Coanda principle is in dominance at the time. The rise in heat is a rise in the liquid part that extends the contracting by giving rise to the adding of motion. The heat is liquid because the heat is motion and the atom inside becomes the solid since the motion is conserved by the spin of the electron. Today it is the rise of levels that is in focus but this same principle had to be in use when atoms were formed that today is responsible for elements. If it was true about mass pulling mass by reducing distance the big bang was not possible and individual elements was not possible. With the cosmos down to the size of a neutron and mass confined to that space within the neutron that would be the recipe for the biggest crunch there could ever be. It the Big Bang was brought about mass confining mass by reducing the radius that implied the space within then what would ever be more applicable than that moment to bring in all the forces the hell can unleash and destroy what ever was not yet in the Universe. The entire idea of mass pulling mass to reduce space is a prehistoric thought and explicitly incompetent in explaining science. The following is a far more suitable explanation and is as true as Kepler is.

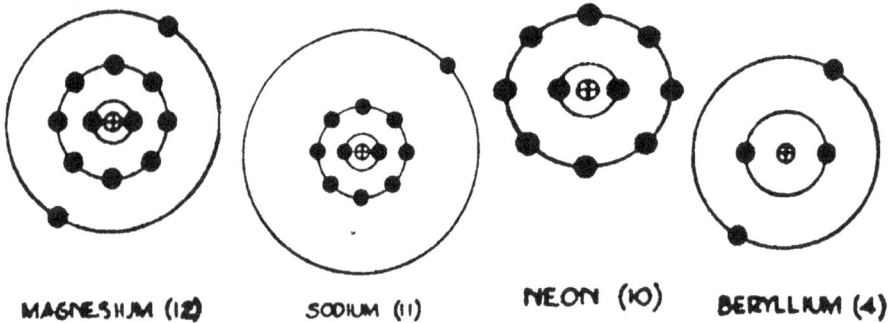

MAGNESIUM (12)   SODIUM (11)   NEON (10)   BERYLLIUM (4)

The most sensible way atoms formed is by the method of the Coanda principle. Those ones that was first were most of all very dense atoms were the first to come in place when time was eternal and dominant and space infinite and one notch off singularity.

We still find the liquid time having a vital role in the space the atom uses. At the time when T^2 was almost eternal space was infinite because T^2 will not

Element	Relative number of atoms
Hydrogen	1,000,000,000,000
Helium	90,000,000,000
Carbon	350,000,000
Nitrogen	85,000,000
Oxygen	590,000,000
Sodium	1,500,000
Magnesium	30,000,000
Aluminium	2,500,000
Silicon	35,000,000
Phosphorus	270,000
Sulphur	16,000,000
Potassium	110,000
Calcium	2,100,000
Chromium	300,000
Iron	3,200,000
Nickel	120,000

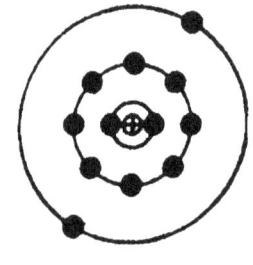

permit the space a^3 much room to be. But the opposite is   MAGNESIUM (12)
also true that if time was steady then time being so long could pack in large numbers of protons with the accompanying neutrons at the time into the time unit in space that formed.

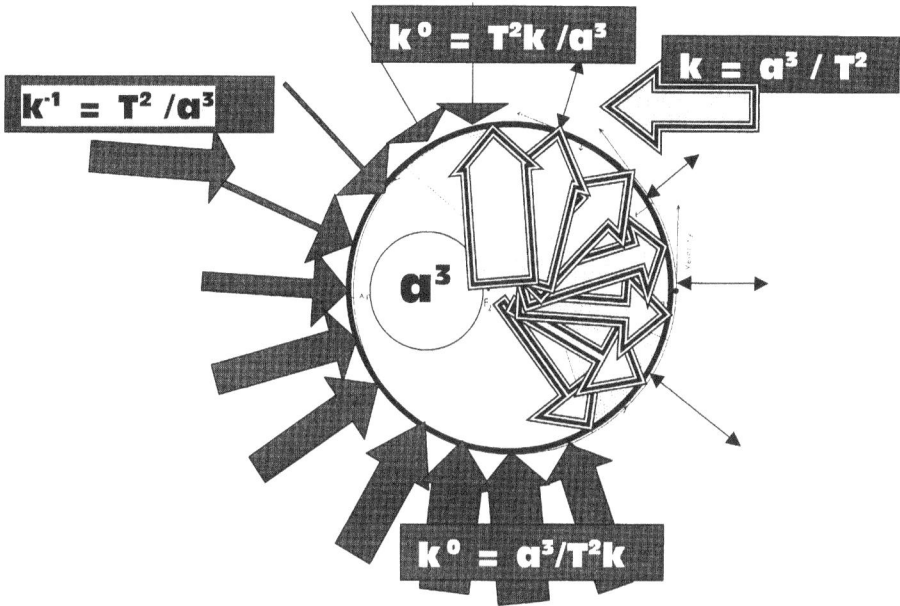

This is most accurate but this is only in concern of a unit in rotation rotating in conflict of its own spin. When time in the cosmos at large view time in progress we find that the development is not quite so simple.

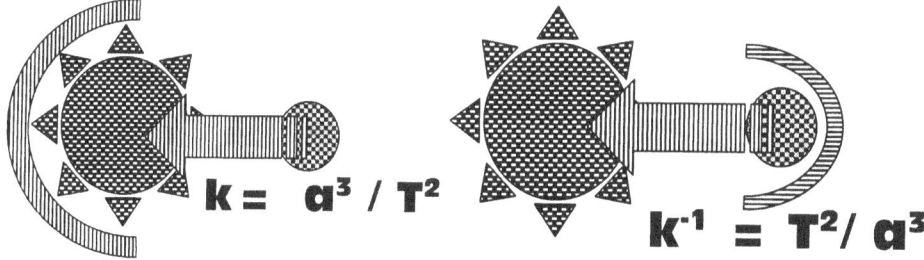

The rebound is always less than the progress and the reason for that is the flow of time. But that also forms the reason why there is a Big bang and why there is a Hubble constant in the midst of all the contracting that is shaping the Universe.

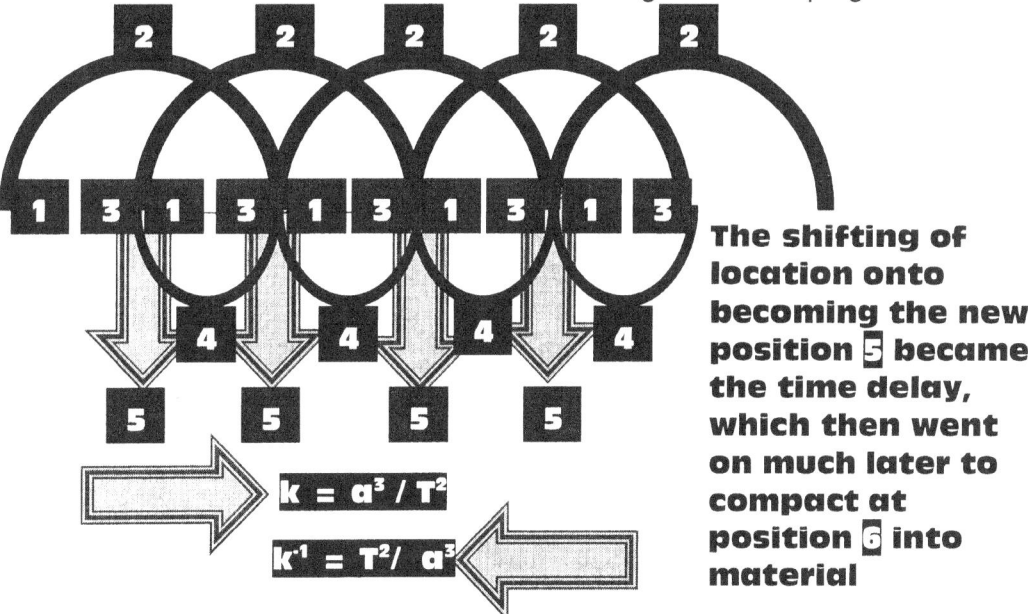

The shifting of location onto becoming the new position 5 became the time delay, which then went on much later to compact at position 6 into material

Newton presented laws on motion without realising how complex such motion is when considering what Kepler's formula proposes. We see a train standing still but while this training is standing still the Universe allotted motion to the stationary

train to the velocity of $7(3\Pi^2)\Pi^0$. This motion is the relation that the Earth allow the train to represent its entire structure in the flow of time in relation to what the Earth represents all of its entire construction in relation to the Sun representing its entire construction plus all the time lapse in between the Earth and the Sun and I can go on with this comparison until time catches the present duration. The space that forms the unit we recognise as the train is in motion with the Earth as part of the Earth and that part is what Newton recognised as mass. Every one of the smallest aspects of material refurnish singularity with a new aspect of time in relation to what it was and where it is going. The more the train moves the more train there is and the more train has to confirm with time in duplicating space.

If the train is standing still there are a number of trains following one another from the past through the present to the future. The relation is $7(3\Pi^2)\Pi^0$ and if we covert that space –time to what we humans measure by it will be kilometres (space) per hour (time). It is the number of space that repeats its relation with singularity that furnishes time with space in the position (instant) in the duration of the instant. That means while the train is motionless and getting rusted from not being in action it is moving about at a rate, which in it is duplicating and affirming the previous position to the next in the immediate. We don't see a train but we see many trains a^3 being rebuilt every time the duration T^2 is confirmed by the instant **k.** We don't see one train but a constant flow of trains that is flowing at an equal pace to our rate o flowing. If the train moves it just mean there is more trains 9in relation to time than there was before in relation to time and the more velocity the train has the more frequent will the repeat of the train be in relation to the repeat of the rest of the Universe. The faster the train moves the more train there is during the duration in relation to the number of instants breaking the duration into smaller segments.

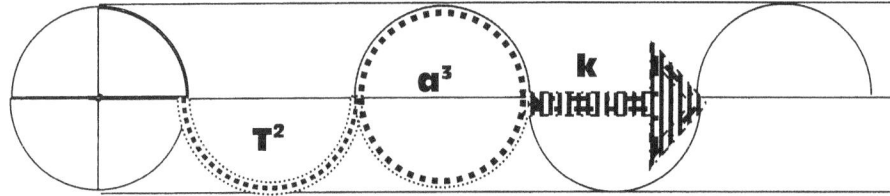

Time moves by the measure of T^2 but time in progress is by the measure of **k**. Therefore space in progress of time is $a^3 = T^2k$ where progress is T^2 but in relation to gravity we find that $a^{3\ -1}k = T^2$. In the relevancy we find the action and reaction of space-time flow is $a^3 = T^2k$ and that translates to being $T^2 = a^3 / k$ on the one side and $T^{-2} = k / a^3$. In the times we now live in we can and do produce an optical illusion of $T^{-2} = k / a^3$, but that is implementing the use of a telescope. In the true time we find as a cosmic reality the fact of $T^{-2} = k / a^3$ is rather a mathematical statement and no more than that. In reality we have $T^2 = a^3 / k$ on the one side as time expands and on the other side we find $k^{-1} = T^2 / a^3$. This we know is true because while it is possible by using an optical illusion the reality is that time can never reverse. In truth the reality about the opposing actions is that we find normal growth and that which Hubble first saw is the process of expanding space-time by the margin of $T^2 = a^3 / k$ while on the rebound we find the opposing while contracting space-time is $k^{-1} = T^2 / a^3$.

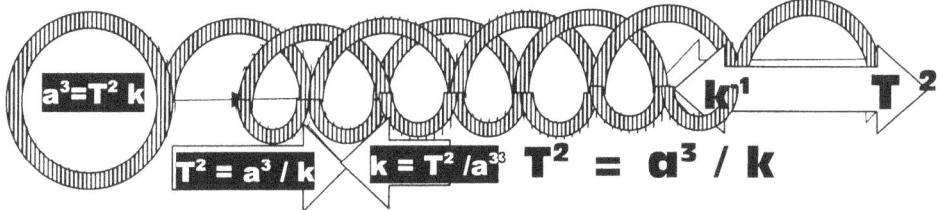

The duplication is presenting time in eternity a new motion where the duration of the spin T^2 a new position in which it affirms infinity **k.** The duplicating in relation to the confirming presents the growth of time in relation to space in relation to material. The duplicating represents a new relation to singularity in infinity where singularity in infinity is completely immovable.

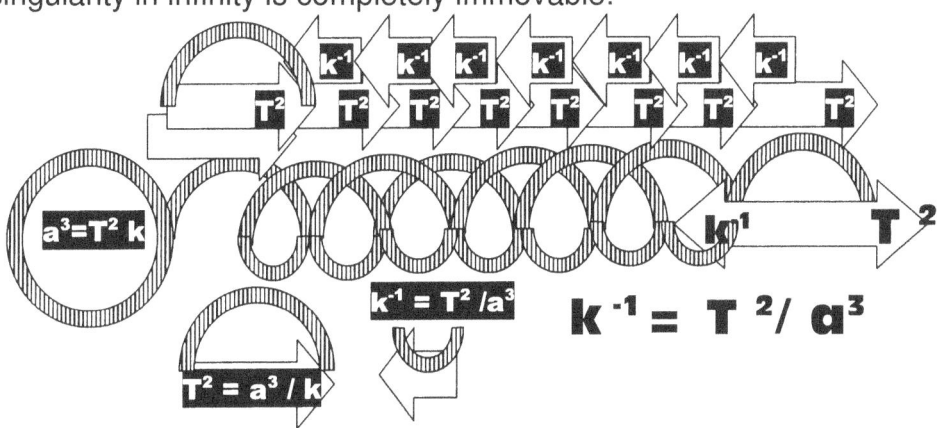

Every proton confirms what every neutron duplicates and that stands in relartion to what every electron conforms. Every proton although in one cluster with many other is a time process that takes matter through time back to what singularity affirms as time in the immediate.

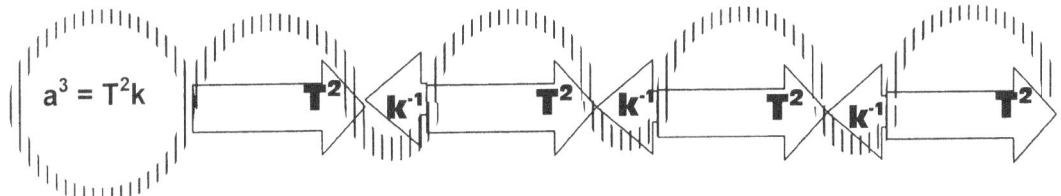

In our ability not to see the imperfect while experiencing the imperfect and at the same time see the perfect in eternity and not being part of the eternity being perfect it is little wonder we lot are so mixed up in what we see and cannot understand. Eternity lasts forever and that is why no changing is visible but we can only experience infinity because eternity is an ongoing repeat of the

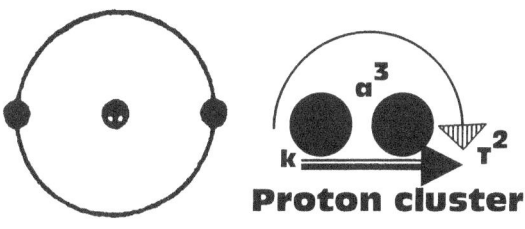

Proton cluster

same without changes. Infinity on the other hand is what interrupts eternity and therefore what we see in eternity is what infinity is interrupting. The proof of this is in the top where motion distinguishes infinity in the centre from eternity surrounding the time position of the top designating the motion from the past

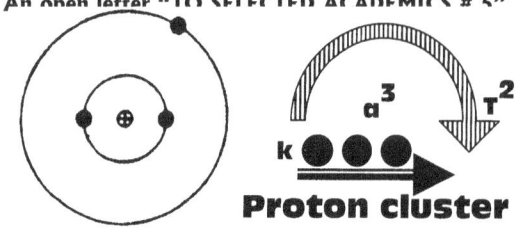

Proton cluster

infinity infinitely. The one can absorb the other just by reducing the relevancy and

through the present onto the future. Those are there for all to see and the fact of that being there goes beyond dispute.

That is the manner how stars move time back to the point of having eternity sharing increasing the flow of time. As the relative flow increase the relation is space subsides and the one become infinitely in relation to the other that then provide an eternal time. The match form and the element gain one more proton.

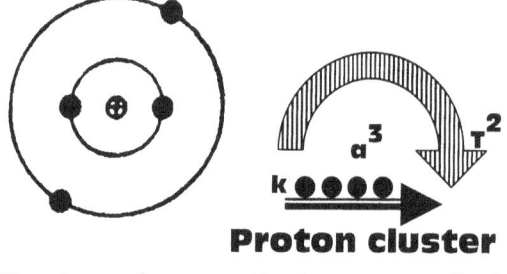

Proton cluster

The imperfect part is the part we find in infinity and with life holding part in eternity we cannot see infinity. We experience infinity while we see eternity. We see what is always there because we are unable to see what is

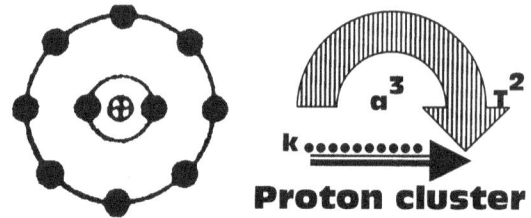

Proton cluster

changing. Look at your own fingernail growing or your hair growing or even a wound healing and you will see the nail, the hair, the wound but never the addition

by growth. You might find what I say at this point not to be physics but it is more physics than anything currently used as cosmic physics are part of true physics

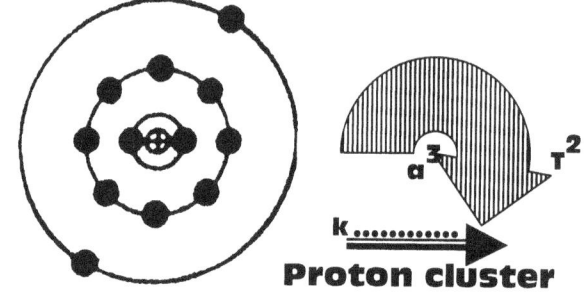

Proton cluster

When creation established space-time eternity was interrupted by infinity as much as infinity interfered

with eternity. It is not the same thing although we as humans tend o regard the matter as such. The one is having a look at it from the one side and the other is looking at what happened from another perspective.

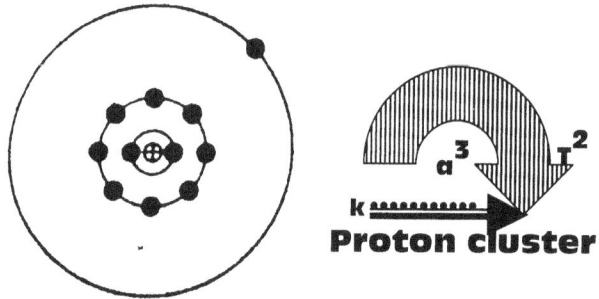

Proton cluster

Time moved on and space came about from the imperfect moving of time as well as the perfect moving of time. The roundness and the perfect6 shape is part of eternity and that in eternity is what we see with life being part of eternity (not the human body which a cosmic result of the imperfect) but life seeing it self in the position where it is occupying the centre of the Universe by studying light and night.

																	Nonmetals				

1 H																						2 He

Periodic table of the elements:

1 H	IIA			Transition metals								IIIA	IVA	VA	VIA	VIIA	2 He
3 Li	4 Be											5 B	6 C	7 N	8 O	9 F	10 Ne
11 Na	12 Mg	IIIB IVB VB VIB VIIB		VIIIB			1B	2B				13 Al	14 Si	15 P	16 S	17 Cl	18 Ar
19 K	20 Ca	21 Sc	22 Ti	23 V	24 Cr	25 Mn	26 Fe	27 Co	28 Ni	29 Cu	30 Zn	31 Ga	32 Ge	33 As	34 Se	35 Br	36 Kr
37 Rb	38 Sr	39 Y	40 Zr	41 Nb	42 Mo	43 Tc	44 Ru	45 Rh	46 Pd	47 Ag	48 Cd	49 In	50 Sn	51 Sb	52 Te	53 I	54 Xe
55 Cs	56 Ba	57-71 *	72 Hf	73 Ta	74 W	75 Re	76 Os	77 Ir	78 Pt	79 Au	80 Hg	81 Tl	82 Pb	83 Bi	84 Po	85 At	86 Rn
87 Fr	88 Ra	89-105 †	104 Rf	105 Ha	106 Sg	107 Ns	108 Hs	109 Mt	110	111	112						

That is what Darwin missed with his species being from one ancient origin. Yes that is true but the one did not develop into the other. Time did produce changes but the donkey has no family ties with the horsed. If it had the mule would have been able to multiply and be fruitful and the mule is a lot of things but that it is not. Things go along in eternity while infinity interrupts and then one-day infinity brings a change no one noticed before. The same building blocks are used built one day a new corner stone is laid and new specie arrives that has no family ties with the previous lot.

In this manner elements came about. But the placing of protons within the atom formed elements whereas the atom was there the first instant heat parted from cold. I go into that part in the **Cosmic Birth...Dismissing Nothing** and since the issue is rather complex in explaining I would prefer to leave that explaining the book mentioned.

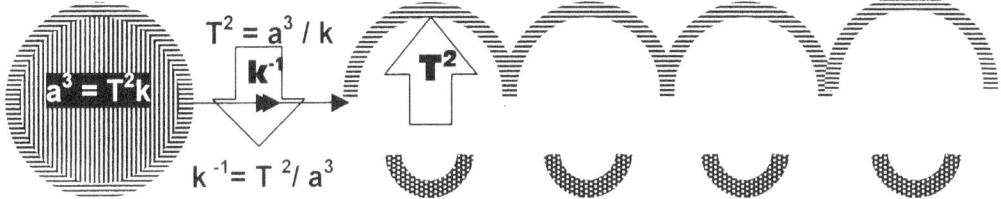

$$a^3 = T^2 k \qquad T^2 = a^3 / k \qquad k^{-1} \qquad T^2 \qquad k^{-1} = T^2 / a^3$$

Once again I have to draw the attention to what is out there in the cosmos serving as evidence. The proof we still find in the manner in which Galactica and all other orbiting objects develop. There are $T^2 = a^3 / k$ that is in favour of the promotion of one point holding singularity in the relation and to that there is another and opposing point holding singularity in prominence which is in relation the expanding contributor that holds a relevance of $k^{-1} = T^2 / a^3$ in ratio to the conserver.

The relevancy was there from moment-Alfa brought relevance from 1^0 to 1^1. We can see that there were seven in ratio of ten and we can see how the seven produced the gravity of motion relating to the ten in time. We can see when the dominance started creeping to the other side and when $k = a^3 / T^2$ got the better of $k^{-1} = T^2 / a^3$ because at a point where the sum total related to the singularity the proton ($\Pi^2+\Pi^2$) + the neutron ($\Pi^2+\Pi$) + the electron 3 = 35.89 × singularity Π =112.75 outer space.

Past such a point the expanding factor began to gain lost ground and the expanding got predominant as the containing factor started to store and preserve more than contain.

The major issue in hand is to recognise that when 1^1 overheated it parted from 1^0 because 1^0 was to cold to harbour 1^1 and by measure of 1^1 being hot, the same measure places cold on 1^0. The one cannot heat without the other establishing a border for the cold. In the Coanda principle the liquid establishes itself onto the solid by gravity as much as the solid allow the extending to lock on. The liquid is not locking onto to an unattached solid. The solid gain as much as the liquid gains but that what the solid gains are not the same in likeness to that which the liquid gains. The solid will as much prove to be colder as what the liquid proves to be hotter.

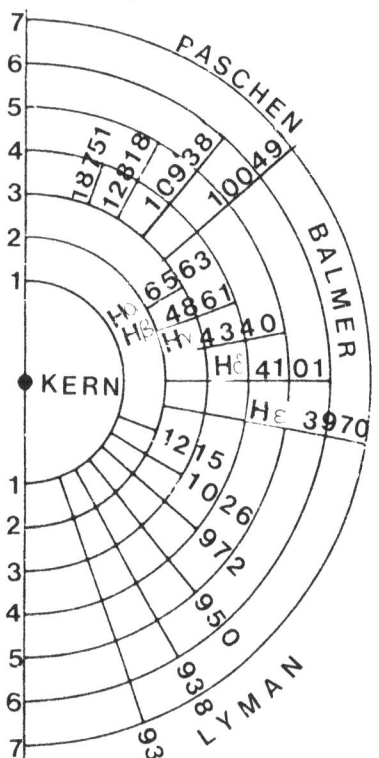

One should see the electron as the indicator where the liquid attaches and where the solid ends because where the liquid ends is where the solid start. The fact of the matter is not entirely that simple because the electron does view the neutron as a stabilizing solid while there is nothing in the cosmos more liquid than just the neutron is. The entire idea is in judging what is solid at the time in ratio to what is liquid at the time.

We see our bodies as solid while the truth is that it is life that keeps the body liquid. When life detaches from the body, the body loses its motion ability and then the body goes rigid. Then the body becomes a solid. However while life is in the body the body is used as a mobile object in relation to the Earth being solid.

When going colder the atom raise its solid level to lose some of its liquid level and the other way around. But the fact is that it is the atom going hotter that reduces the atom going hooter. Hot and cold is not by measure of temperature but it is by measure of space moving through time. If there is more space moving more rapid the atom becomes colder and the opposite is also very true.

As soon as motion commences **k** increases because although the relevancy from the Earth perspective remains the same the relevancy from the aircraft changes drastically.

From the view the aircraft holds the **k** that the Earth has becomes the T^2 that the aircraft has and the T^2 that the Earth holds becomes the **k** that the aircraft uses.

The sound barrier is just another manner in the way the Universe brings about gravity. The aircraft has to produce excessive heat and by more heat delivers the space between the Earth and the aircraft increases. The motion becomes more and that stretches the space connecting the Earth and the aircraft applying heat to produce extended motion where the extended motion leads to extending of the space the aircraft covers in the same duration of time.

The heat (formed by the release of motion in the engines) allows more motion to apply than that which the Earth generates which puts the aircraft in a higher atomic bracket where the aircraft holds more space that the Earth normally would grant the aircraft. The aircraft has more gravity (granted that it is using the motion of the Earth which is the gravity of the Earth) than it would have being stationary. With an additional source of heat the aircraft can add to the earth gravity and the Earth gravity is unrestricted motion. It has no bearing on mass whatsoever. Gravity is about motion and mass is the restricting of such motion.

It is not the motion we must be after but what causes the motion in the first place (other than being Newton's pet force). We must find what produces motion and from that then we must think further than what we can recollect from thousands of years of culture that got us this far but is getting us no further. We must see why that which moves or tend to move as we do on Earth in our gravity. It is nothing to do with mass pulling everything about but is a flow of space-time.

When one applies heat to an object it expands. That is primary school science. This states that more heat applied leads to more space acquired by the heated object. In sharp contrast to this is the growth in space when heat levels rises but freezing brings about the opposite result. When I freeze an object that object reduces its occupied space as it shrinks. Removing heat reduces space. That comes directly as nature responds to heat and I can prove that easily.

By expanding it accumulates space to increase the improving of the size of the material. The accumulating of heat is for the sake of securing singularity, which accumulates the heat in the material whereas the freezing tarnishes the overheating symptoms by the removal of material in unoccupied space using external matter and setting motion to the material until it contracts into a form which we see as visible heat. The heat is in the form of dissolved singularity that became material as material used it as growth. That is why by freezing it will diminish the space as to accumulate the heat absorbing into the heat into the material to maintain the equilibrium needed in space.

The atom is the optimum proof of the statement. The atom is the absorber of heat as well as the release valve of heat. The atom regulates heat in relation to space acquired as well as space acquired. The atom is as much about heat as controlling heat and when the atom expands space it accumulates and store heat. When it cools it reduces and absorb space. The cosmos is the atom and the atom is what the cosmos use to regulate heat.

Taking this equation of nature to outer space we seem to confuse the natural law. With outer space as expanded as anything can get we regard outer space as incredibly cold. As heat sets in the normal flow will bring about expanding of heat into the form we think of as space that limits the heat overheating. Outer space is the very edge of expanding of space where heat cannot expand into space any more. Outer space is the limit, the epitome of expanding where heat meets space at the edge of all limits once more. Therefore being the representation of the very limit of expanding outer space has to be the hottest place there is. By applying heat to a kettle holding water, the adding of heat manifests as steam and steam is hot water that traded heat as it reviewed space. By allowing the receiving of the heat to continue the container will let loose steam in order to match the contributing of space.

The manner in which heat expresses itself when confronted by overheating is to provide additional space through expanding of space. Outer space is outer space because outer space has expanded all it can it is still expanding to the speed of Hubble's $1/H_0$ which inevitably does not only affect far-off places where we cannot be, but effects us on a daily basis. As outer space is stretched to its limit, its limit will continue to stretch but while it is stretching it has to having more than it had before in that outer space holds the limit of heats expanding possibilities. Singularity has been expanding since way back when but that means singularity is still releasing heat as space-time that turns out as space in the universal time of outer space. In outer space heat cannot expand more therefore except for the continual growth that benefits all singularity throughout on a continuous bases concerning all outer space.

Every element is in relation to the heat level it uses in forming the gravity it has. One can see how the forming of the numbers of elements available in the Universe stands related to the density of the elements total numbers. More pertinent to note of that the effect of gravity is not in the mass of the element but shows a much stronger relation with the density and the density is the relation the element has with the heat that marks a boiling point or a freezing point The density factor shows what we use to classify the element in relation to being a liquid, a gas or a solid. This factor is much more prudent than the mass factor and that I show later on as the book develops.

If singularity expands when heated and there is a limit to the point it can heat, and where that point forms the maximum expanding possible, then it has been reached in the area we think of as outer space. Outer space has expanded through the unleashing of heat, where overheating is turning liquid heat into space. Any explosion is a vivid reminder of this fact and the unleashing of space is so real it destroys the space holding solids by rearranging the construction of the solids. With that in mind we can declare with great confidence that outer space as the hottest place there is. Whatever expanding there possibly is, was done to secure the cooling and all cooling that can be introduced to bring about further cooling was performed in the area we think of as outer space. Forget schoolboy culture and the temperature scales and other Newtonian scientific defects I call Xepted mistakes. Think of reality and throw out culture teachings Use the mind and not the thinking power of the past. Any place that can expand no more is the hottest place there is just because of the shear implication that it can cool no further is as hot as it gets anywhere. If that is the case then it is safe to say that galactica is freezing cold notwithstanding our concepts of heat and space and heat in space given to us by our collective culture and not by our ability to reason.

The galactica is little frozen islands in a vast see of heat. That is the reason we can see the galactica because the galactica is space concentrated by a frozen space. The galactica is slowly heating and therefore it is expanding into outer space. Outer space on the other hand has expanded to the maximum that it can yet we think it is cold when it is the extreme there is in heat that introduced the maximum expanding. What I now am saying might be deemed by the most purist as the contradiction of the century and that much I do realise. At the inner core of a star all space shrinks into the oblivious but we consider the inner core area of a star to be the hottest spot in the solar system. That just cannot be because when material shrinks it becomes cold and by shrinking into the oblivious it has to freeze into a fusing element as newly formed units. Again that is the contradiction of the century. Why will that be? The space inside the star shrunk to the minimum there can be and that tells us the space has to be cold because of the shrinking took the space to a position where no space can shrink anymore.

That shrinking of no more space can only be inside the inner star and in that region is where we locate the strongest gravity. With outer space as expanded as nature may allow the space that grew could only grow in conditions of heat because heat produces expanding and expanding is the result of heat coming about. Space shrink because it is cold: that we know and taking this law to the star centre it means regardless of our interpretation of hot and cold, that area in the star centre is as cold as it can get notwithstanding what our nature may tell us. Then obviously the same must apply to outer space for precisely the same reasons because it is so hot there it can expand no more.

At this I have to redeem myself from being human. Only in the eyes of humans are there hot and cold, but as a reality in the cosmos we will find this nowhere. We look at the hotness of space and the coldness of space but it is the relevancy to the solidity that forms the actual heat and cold limits. It is so hot no expansion can produce more space in outer space, as the outer space seems to be the epitome of what can be cold while it is truly hot and quite the opposite reveals as the true scenario inside the star in the centre of a star structure. That means the number of protons in motion has a lot to do with the cold and hot scenarios because where the protons are most dense the cold is in extreme…well in most cases. Only in the absence of space can so much heat gather in excess and the opposite is true about outer space where the least denseness found brings about the space in heat found in outer space. Our human selecting of hot and of cold and what is hot and what is not prevents us the clear vision we would have when truly understanding the applying temperature. Temperature comes about from spin and the smaller the spin density is the colder the space becomes because the more duplication produces the most cold. We think of outer space as 0^0 Kelvin but in fact it is as hot as no other place can be in the Universe. The coldest is where material is freezing solid as material does when frozen solid and the hottest is when by boiling the material is going into a gas with liquid being the intermediate position where heat acquires the space to perform as a flexible substance.

When we look at particles in outer space we see the particles being frozen. It is because there is such a severe contrast between the particles and the environment surrounding the particles and not the particles that is so frozen. The particles are in a gas state because the particles do not form a part that is part of the space unit. Hydrogen clouds of hundred of light years in diameter are a common sight in outer space. The heat we find filling space is not part of the space but like the particles the heat is a separate issue. That heat filling the space is another form of material that could conduce by diverting from space or marry the union of space by becoming more space. If it were that cold which we think it is, it would not have expanded into such a massive cloud but would have contracted forming a cube of frozen hydrogen. But as we can see the cloud expanded the gas as far as the gas can expand.

That expanding is indicative of heat and has extremely little to do with gravity or is it just a matter what we think of as gravity. If you are of the opinion that those hydrogen clouds will contract one day into forming a star, well then think again as there is just no such a chance that that will ever happen because that is not the manner that form of gravity functions. Because outer space is completely overheating the condition it has in support of the particles makes the particles appear to be in a state of freezing but the particles is counteracting the heat limit it meets. However the particles do not contract, as the heat is immense. The space in outer space has absorbed all the heat by means of expanding and will appreciate still further as it will never depreciate. That is not because outer space is freezing the particles but it is because in contrast to the heat of outer space the particles seems to be frozen.

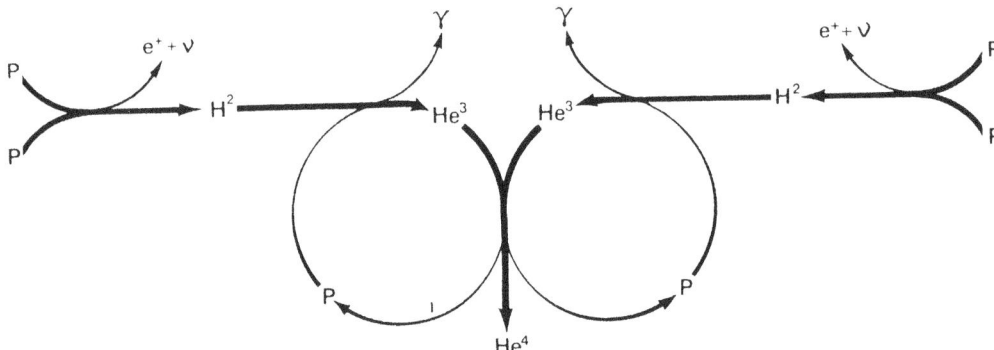

The atom must be the utmost coldest because the proton is even much colder than whet the electron can freeze. In fact the proton is 1836 times colder than that which the electron is able to freeze. We find that when cold escapes it turns to heat and the heat relieves by forming space, however it seems that that no one can understand. Motion brings about cooling. When the spin of the atom allow the cold of the atom to release the heat it had, which it had frozen the heat returns to space. This is what the atom shows in the electron bands or rings the atom holds. This must not be confused with uncontrolled release of heat. When the motion of the electron is interrupted such motion reducing results into the utmost expanding there possibly can be. When this heat releases from the containing form of the atom it brings about much more heat than the Human mind can cope with because no human mind can comprehend the total devastation a nuclear release of space may bring forth. In this I am not referring to the normal way material relates to heat. That is a totally different matter altogether.

One may not look at the material and judge the surroundings. The fact that hydrogen remains a gas and so does helium in outer space must serve as enough proof that outer space is hot, regardless of our interpretation of the temperature gauge telling us what we wish to hear. In the vent of outer space truly being the coldest we have, then hydrogen and helium should be frozen crystals clotted in balls of material. One must look at outer space and judge outer space from the findings only by considering outer space without the prejudgement of teachings about ideas when persons were still held in prison for being suspected werewolves. If helium remains a gas it is hot. However we can witness hydrogen being a liquid in the Sun. That squirting from the Sun is liquid heat that is frozen as a form of material in the hydrogen layers and holding the hydrogen in form in the hydrogen layer.

We might think it is hot in the centre of the Earth but that type of thinking is as Newtonian as thinking of big stars as mighty gravity pools. The removing of heat into a liquid makes the material in the centre of the Earth cold although we see it as being terribly hot. The only reason why it can seem to be hot is because it is cold and in such a cold environment the heat can gather and space can collect heat because the particles find the surroundings extremely cold. Then again we confuse heat and time altogether and completely but more about that later on…

The cold in the Earth centre causes the concentration of heat by space reducing, as all cold surfaces tend to do. When material reduces space, it parts the material from the heat within and places that heat within the electron bands on the outside of the electron bands. By removing the heat the atom contracts and by contracting the atom reduces space. That heat forming space has to go somewhere. If it was hot the space within the Earth would expand and the space within the Earth where we think so much heat is concentrated does not expand therefore it must be cold. To gather and accumulate the space in a liquid means it became much colder being a liquid. Finding the surroundings terribly cold will allow the heat to gather and not expand but when the surroundings are hot it will not tolerate more concentration of heat and thus will expand, to rid the balance of excess heat within space. That is the terms in which to think in when thinking in terms of cosmology.

Look at the Sun and see how the Sun turned the hydrogen to a freezing cold liquid at 6500 K. Hydrogen is in a fluid state within the Sun and is colder than the hydrogen that is in a gas form in outer space. The Sun is the coldest place in the solar system. That is when the protons oversupply the removing of space to produce the cold that is so apparent. By the reducing of space it can concentrate heat to a fluid state by producing the opposing cold that finally freezes the heat to a solid state. The expanding of space is a way of duplicating space without reducing space and by duplicating in the form of expanding it becomes just the opposite to duplicating by motion therefore reducing space by halving space in time. That is what gravity does. By motion, space duplicates and by space halving it removes heat in space as well as by dismissing space. In all the applying of gravity space bites the dust. The density of the protons brings about space dense enough to harbour the heat in such quantities and visa versa applies in outer space. However it is not purely the density of the protons that produce such cold but the exquisite motion forming a rapid duplication of material and such duplication brings the contraction by removing space. Removing space is also removing heat that is separating material.

We have to accept that the coldest place in the solar system is in the very centre of the Sun because there the most number of protons sharing the least amount of space producing the coldest area that can allow therefore the hottest density of heat within the cold environment. Later I will show why the star is so extremely cold it freezes material together and outer space is over boiling with heat expanding into more space. We have to see what forms space and why space can be the absolute basic container through which gravity can relay the influence it carries.

We have to realise that whatever forms space has to be that same ingredient which also is the basic component that forms the lot of everything in the entire Universe. It is than which becomes more making everything seem more and it is also by removing that which reduces every aspect of the Universe. That which becomes more is what the Universe is built with and it is that which the Universe uses to form its entirety. When particles heat up the particles expand the space the particles hold to limit which the rising heat demands in relation to the heat rising. The particles claim more space when heated to preserve the cold that the material is protecting. The claim to more space produces more space but that in turn reduces more heat exaggeration. Such expanding brings about cooling. When particles heat or cool motion applies in some form. Regarding this fact we can claim that motion started at a point when the Universe was extremely hot and there was no space. However I have indicated that hot and cold are only factors with little specific or formal value in the Universe. By introducing motion space formed and the lack thereof produced friction that became heat that became space. That must be the way the Universe then started.

The application of gravity is the same as the condensing of space and bringing about heat by the compressing of space we apply in the way we go about tapping into the energy that nature provide. Internal and external combustion engines all rely on this application for harvesting motion by driving power. Compress space even today with a piston in a cylinder and then pump the compressed air into a container and such confining of space will increase the heat by the piston effort to reduce the space brought about in the container. The heat coming about inside the cylinder when being compressed has no relevance to particles colliding because all compressor cylinders cool down or become colder when that cold escape through the walls of the cylinder. As soon as the pumping stops the heat releases from the inside space. There is an immediate stopping of the increase of heat as soon as the pumping stops. The material inside the container forms a secondary form of material that comes about since the space reduces and the forming of space is in a turnabout. The compressing of the space inside brings about a rise in the heat levels within the container but apparently that no one in Newtonian circles can understand. By compressing the spin of the atom increase and the motion of the material removes additional heat from the ranks of the inside of the atom. Thus, when the spin of the atom increases it allows the cold within the atom to release the heat the atom holds into uncontrolled space. This releasing of heat and unifying the released heat once again with space increases the levels of heat in the atmosphere of the containing cylinder. What that heat does is the heat that the material absorbed as material within the atom was captured as frozen heat because of the motion of spin to space that the atom holds remains in a frozen state under the guard of the spinning electron. But when this heat releases from the containing form that is the atom in being the biggest cosmic heat container the heat becomes in a frozen state through the motion within the atom. Forming a frozen substance by producing motion that is faster than the speed of light the heat is frozen by the spin of the electron. The spin of the electron brings motion and such motion reduces the heat to a frozen state which is the frozen state of heat we named material. Therefore one may not look at the material and judge the element state of form by its surrounding which is heat it surrounds its electron to the outer side of the containing spin.

Again we must look at the state of material in outer space and realize that the fact that hydrogen remains a gas and so does helium in outer space must serve as enough proof that outer space is hot, regardless of our interpretation of the temperature gauge telling us what we wish to hear. One must look at outer space and judge outer space from the findings only considering in the terms which outer space insists upon. If helium remains a gas it is hot. The removing of heat from the space that contained the heat makes the centre of the Earth cold. In our Universe we see it as being terribly hot because the heat then forms a separate substance but remains a form of material but that is because we see the heat and not the space derived from the separating of the heat.

The only reason why the space can seem to be hot is because the space is cold and in such a cold environment that rejects the heat within the atom. There the heat then must gather in a more concentrated state and space can collect heat because the particles hold concentrated heat in the space separating the particles. By removing such high concentration of heat from the space that use to be expanded heat, the space then must contradict the heat by being extremely cold. We look at the heat in the space, which by being in a liquid state should be by our standards considered as another form of material and find the surrounding heat in the space hot while the atomic material in space is extremely cold. The cold in the Earth centre causes the concentration of heat by space reducing, as all cold surfaces tend to do. But the numbers of protons contributes that reducing of space and the removing of heat captured by the material. If it was hot the space within the Earth would expand and explode but the space within the Earth where we think so much heat is concentrated is so much it does not expand therefore it must be cold within the solid parts. It is the motion of so many protons in such a little space that allow the heat to be contained as a liquid and the extravagant motion by the many protons in such a reduces area forms the ability to contain the heat as a liquid substance without allowing the expanding of the heat into gas or space. To gather and accumulate the space in a liquid means it became much colder when the space parted from what then is being a liquid. Finding the surroundings terribly cold will allow the heat to gather and not expand but when the surroundings is hot it will not tolerate more concentration of heat and thus it will expand to rid the balance of excess heat within space. The concentration or release of space with heat or space from heat is a direct contribution of the motion controlled by the space-time. The regard of the space-time providing the motion, which provides the cooling of the space, stipulates the conducting of heat in space or the release of heat to form space by means of seizing the occupied space.

Look at the Sun and see how the Sun turned the hydrogen it holds and which is captured in its atmosphere to a freezing cold liquid at 6500 K. Hydrogen is in a fluid state within the Sun and yet it is still colder than the hydrogen we find in outer space that is in a gas form in outer space. That must be because of the enormous motion of the particles within the confinement of the Sun. The Sun is without any doubt the coldest place in the solar system and that is because as the ferocious motion within the Sun. That is when the protons oversupply the removing of space to produce the cold that is so apparent in the heat levels that do not join outer space. By the reducing of space it can concentrate heat to a fluid state. By producing the opposing cold that finally freezes the heat to a solid state we find that it is what matter is. The expanding of space is a way of duplicating space without reducing space and by duplicating in the form of expanding it becomes just the opposite to duplicating by motion therefore reducing space by halving space in time. That is what gravity does. By motion space duplicates and by space duplicating the material must be by dividing or halving. Halving the material, which is heat, is at the same time doubling the space, which is bringing about cooling. By doubling the space as the duplicating of material removes half the heat from a single space and distribute that same quantity of heat over a double amount of space it removes heat in space as well as by dismissing space and in that concentrating heat. Again it is apparent that in all the applying of gravity it is space that bites the dust. The density of the protons brings about space dense enough to harbour the heat in such quantities and visa versa applies in outer space.

We have to accept that the coldest place in the solar system is in the very centre of the Sun because there the most number of protons sharing the least amount of space producing the coldest area that such intense motion can allow therefore the excessive motion brings about the hottest density of heat within the cold environment. It is the duty of scientist to look far beyond the ordinary and find why the inner star will be so cold and as to why outer space will be so hot while being seemingly so utterly cold or hot in humanly applied standards. It is the duty of the professionals to find matters as they are and not as they would seem to look from a human vantage point. Later I will show in much better detail why the star is so extremely cold and outer space is over boiling with heat expanding into more space. We have to see what forms space and why space can be the absolute basic container through which gravity can relay the influence that it carries. We must come to realise that whatever it takes to form space it has to contain something that is the same ingredient, which also is the basic component that forms the lot of everything else in the entire Universe. When particles heat up the particles expand the space the particles hold to limit the heat rising.

The particles claim more space when heated to preserve the cold. The claim to more space produces more space and reduces more heat. Such expanding brings about cooling. When particles heat or cool motion applies in some form. Motion started at a point when the Universe was extremely hot and there was no space. By introducing motion space formed and the lack thereof produced friction that became heat that became space. It is natural and it is simple and above all it makes believable sense.

The application of gravity is that which condenses space by bringing about heat with the compressing of space. Compress space even today with a piston cylinder wall in an engine cylinder and then from that action pump the compressed air into a container and such confining of space will increase the heat by the piston effort to reduce the space brought about in the container. The heat coming about inside the cylinder has no relevance to particles colliding because all compressor cylinders cool down with time moving and not necessarily with the loss or release of particles. It is not only the discharging of air that will reduce the temperatures inside the container but the time flowing bringing motion about where the motion is not about particles escaping but heat escaping in the replacing of the heat density (not the density of the particles forming the material content within the container) but the space that compressed to heat will also bring about that the heat displaces through the container wall to the outside. After the pumping of air increased the heat in the cylinder which even can go to dangerous levels, the heat will reduce back to room temperature when further pumping seizes and that stops further air movement into the cylinder and such surging of pumping air is what brings about heat stabilizing.

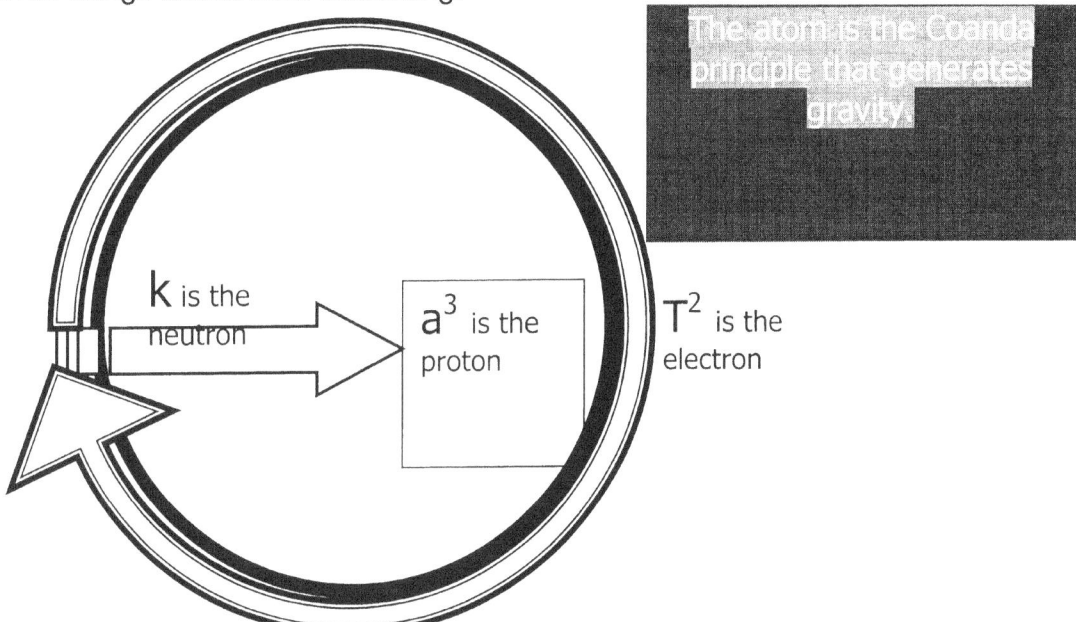

Mainstream physics ignored the clear connection completely, notwithstanding it being so very obvious. There is this far in their recognising of principles in natural physics not one single reference made to prove their appreciation of this matter. They are bent on particle colliding notwithstanding the much nonsense such an idea promotes. Atoms cannot touch simply because electrons are all negatively charges and will therefore repel one another long before there is any possibility of touching coming about. However in the case when particles do collide such collision forms an atomic thermo release and that action we call an exploding atomic bomb. What principle this argument about particles colliding ignores is that all atoms use negative charged electron forming the atomic limit on the outside forming a definite border to the boundaries of all atoms and in both electrons from different atoms are being negative charged. In being negatively charged it means both will come out and one totally reject the other as much repel the other or cast the other away. The closer they come the more violent the rejecting will be and such rejecting is the production of heat that will turn to space. However that rejecting will increase the motion and the increased motion will reduce the space occupied. The electrons repel other negative charged sub atomic structures, which the electrons are that form the outer borders of all atoms. With all electrons highly negatively charged (being as negatively charged as any possibility will allow to match the utter extreme) such electrons couldn't touch. When the pumping of the air container commences the balance at first favours the forming of heat from the space coming in and being reduced in the containing size they are squeezed into is reducing the space from what it was on the outside. The space distribution inside then changes considerably and reduces a great deal compared to conditions outside the cylinder wall and with the decrease of the space distribution inside compared to conditions outside that space then becomes reduced and charges with excess heat on the inside.

The electrons will disallow any contact directly between atoms. No force can be big enough to enforce such touching. It is because of that contact rejection electrons bring about that science has to use an overload of neutral neutrons putting them in the atom nucleus to fake a complying of charges that will eventually lead to atom touching each other but that is through enticing a neutral stance which is enticing a positive overload for a short while. When the touching of electrons does take place the event is called a thermo nuclear reaction where heat is released in unmatchable quantities and the atoms in reaction dissolves into a liquid heat. The increase of heat by the distribution of particles in the space that is forming the connecting space still keeps particles separate. The heat rising is a separate issue that has nothing to do with contained particles colliding because why does it stop when pumping is seized. This ratio of heat reduction is time connected as much as it is motion dependent. Motion reduces space by expansion as much as time contributes to space distribution by allowing the flow of heat. When the pumping stops the heat immediately starts the reducing thereof. Most important is the realising that every atom constitutes of two parts. In fact the entire Universe constitutes of the two parts, which I go about mentioning in this entire book. On the inside of the atom there is a circle formed by a rotating electron that contains the outer wall of the atom forming the sphere and holds material in contact with the protons. On the outside there is heat surrounding the inner material part within the sphere and distance the inner material from the space between it and the next atom. The electron forms the division between heat uncontained and heat contained. This is why the Roche factor is so very important. There can be friction between particles in reduced space under controlled circumstances where such particles are grouped together in a unit and as a unit elects a group singularity forming the centre of the chosen form of the unit.

The Universe separated heat from material by covering the exterior of material with heat that forms space. Some material became softer by uncontrolled overheating while others remained more solid by containing form through controlling the overheating. On the outside of all elements there are a layer that is the heat the element uses in relation to place relevancies between such an element and the rest of the cosmos. In the case where many atoms form a unit such as an aircraft coming in from outer space the space surrounding the craft becomes liquid heat as the space becomes more intense within the atoms combining as the structure in concentrated space that forms heat. In an aircraft coming in from outer space at altitudes that high there can be no particle in friction and even more so way up there in the atmosphere at the altitude where the cosmos meets the atmosphere just because the particles up there are so sparsely distributed in that part of the atmosphere. Above and beyond this lies the fact that all the so called air particles are very volatile and excitable by nature and they are known to turn the slightest heat into rapid motion thus establishing a scene where the particle that supposedly are in contact with the aircraft sheeting will move away from the hot incoming aircraft. The gasses will become more gasses when the heat levels surge. If then and not for any other reason why there can be no friction then it is because the particles are highly volatile and exceptionally sensitive to heat. Airborne particles are prone to motion just because it is the airborne element nature to change heat into motion and the motion comes about from their sensitivity to duplicate. No particle in the air being part of the space we call air, which is in a free floating in that air can produce friction because of the volatile nature that those elements have. The craft's coming into the atmosphere produces a point where $a^3 = T^2k$ changes to $k^{-1} = T^2 / a^3$ (the explanation is forthcoming a little later on) The distance separating the incoming object from the Earth centre reduces rapidly therefore the object start to descend towards the centre of the Earth. We must also acknowledge the fact that there is one specific point of specific entry where this will occur more than before.

That point will rapidly increase the time factor where the incoming object crossed such a very visible border. By the reducing of distance k space a^3 will have to compromise in the relation of all the factors forming the equation since T^2 will very suddenly grow more acute. What happens is that the applying gravity reduces the space a^3 and the compromising factor comes about since the time factor T^2 moves back to a time where outer space was as dense back then as the density we now have within the atmosphere that then became the Earth atmosphere. It is outer space that remained denser than what the outer space currently is. I am now referring to a process that I introduce as this book unfolds which is by nature completely different to what is accepted by mainstream science (as you might have noticed in this short space of reading). That which I refer to as outer space back then was the same density as that which the Earth now supports but outer space in the meantime expanded while the motion that the material that forms the Earth structure provided, came about at a point just before the Earth established an atmosphere that grew through gravity and by the measure of the Earth gravity became separated from the atmosphere. While the gravity of the Earth contained the space surrounding the Earth in a much denser packed envelope the area not under the direct influence of the Earth governing gravity became more spacious.

The Earth contained its atmosphere and it relatively grew as much denser as the solar system developed into what it is today and outer space reduces its density. It is a matter of the kettle not being able to call the pot black. As the atmosphere released from what we think of as outer space that releasing from outer space made the atmosphere much denser in the space just above the Earth, which is using a reducing time factor. It is there that the applying gravity makes the Earth atmosphere more compact. That established the T^2 factor to be that more condensed when one compare in ratio the density with outer space. The density that was there at the time when the separation came about in outer space when such parting between the limits of the atmosphere and the limit of outer space separated and such separation allowed outer space as a separate object to move away. This parting brought a barrier that is in place between the Earth and the outer space and any object coming from outer space into the Earth's atmosphere will have to negotiate its entry by passing through that division. The incoming object then would have to reduce the measure of the space the craft holds as the containing singularity set new standards applying to the incoming object with which the craft then needs to confirms its form and its status within the contained space of the Earth. The reducing will then suddenly no longer use space as the compatible factor but the focus will shift to the time factor that dictates to the space what the space can be. Such reducing comes from the switch there is in space – time where it was in outer space performing as being $k = a^3 / T^2$ to what it has to be within the Earth atmosphere $k^{-1} = T^2/a^3$. When the atmosphere grew apart from the outer space there are two ways of looking at the event. One can think that outer space expanded by the implication of the Hubble constant or that gravity withdrew the atmospheric space of the Earth at the time that the parting of space came about. But however you look at it there was a time when both outer space and the Earth's atmosphere shared equal density as we find it still applies on the moon and on Pluto. Then the Earth became dynamic and now they do not share any density at all. Things were overall more compact back then than at the present time and that included all things in the Universe. The space component is reducing the time component by compacting space to alter the space – time ratio.

This is portrayed by Kepler's formula $a^3 = T^2k$ It shows space as the density of space decreases.   The Earth still compact space by reducing the volumetric confinement of space $T^{-2} = k / a^3$. This we call the atmosphere, as the atmosphere becomes denser towards the Soil of the Earth. There is a change in the time component. Most evident of this is when studying the pendulum. Just as we can see in the pendulum swinging, we can see that the swing reduces. Such reduction is because as the space diminishes every time the arm rocks from side to side. With this there is proof that in the developing atmospheric space of the Earth the ratios change from outer space. This is proved by the pendulum arms that Galileo's experiment used to show that the swinging pendulum indicates $k^{-1} = T^2/a^3$. Further more it proves that Galileo was correct after all and unnoticed by science Kepler helped Galileo prove Galileo's point. In this the net outcome establish Kepler as being correct and the Newtonian argument of friction brought on by gasses fall apart which is at that altitude where such friction supposedly should take place, the material in friction is not even present in the atmosphere.

 Nevertheless science will stubbornly cling to the old theory with persistency that would warm any warring Field Commander's heart. In retrospect the following information is established in the past few pages: Every element stands in different

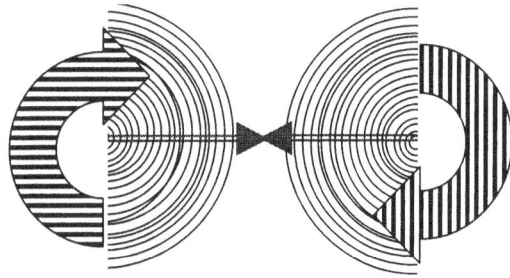

regard to the heat surrounding the material, which makes us consider the material to be either a gas or a liquid or a solid. The material in every element there is as such is all three forms and not one of the forms in particular. It is the way under which the circumstances is presented that the element allows the heat to gather and accumulate as the surrounding heat occupying he surrounding space. Every particle is unique in the way it regards the heat to material ratio and how much heat it uses to form either the gas liquid or solid state. The fact of being a gas or liquid or solid is so much more complex but in time we will get to that explaining. If space a^3 declines then so must motion in relevance will have to compensate by reducing **k** and limiting T^2 because space a^3 must always be equal to motion T^2k

Space shifts as heat releases space and converts the Universe in one direction bringing about expanding into more space but less dense space. Remember how the heat came down from 10^{34} to 0 K at present? The density of heat in space surely diminished considerably since then to now. Gravity on the other hand is exchanging heat through the concentration by removing space bringing about space loss with increased density of particles and therefore heat concentration. In the centre of all spheres, which all stars are it is hot. In fact the heat in the centre of the star is the product of the space it concentrates to form heat and in that we can read the gravity the star can produce. The ability to secure heat by reducing space becomes the measure of the star. Momentum is the second form of gravity symbolised by Kepler, as **k.** The Big Bang is the result of heat expanding into the forming of space. Gravity, on the other hand is about concentrating space back to heat, and take recouped heat through to material, acting out a balance of expanding while contracting. This way gravity is applying the onset of the Big Crunch by destroying space while space is converting heat to material occupying space. The Big Crunch is coming about because the Universe is expanding where the two processes are one principle.

The relevancy there is between the aircraft and the Earth is precisely the relevancy we find between the proton and the electron in the atom. When heat released provides more space between the aircraft and the Earth the distance between the aircraft gravity relevancy and that which the Earth allocated to the aircraft by only providing Earth gravity without the adding of heat by the aircraft allows the aircraft to respond exactly as the electron does in the case of the Atom. The aircraft falls into the role of the electron, the atmosphere takes up the role the neutron has and the Earth retains the aircraft therefore the Earth has the role of the proton. When the aircraft has more heat than that which the Earth provide through the atmosphere the neutron position has to expand in order to facilitates the new dimensions which the additional heat that drives the aircraft provides. The ratio that the Earth initially holds becomes stretched as the aircraft suddenly finds more heat and therefore more motion that becomes more space between the allocated position and the position the aircraft claims by individual motion in addition to the motion the Earth has provided. It is all about heat released that generates motion and motion provides space differentiation.

Throughout the entire cosmos is leaning on the four pillars which is the phenomena and the four culminates in one which accommodates all the others an it is the Coanda principle that establish space which provide the gravity which allocates the motion a position within the space that forms. This very principle of electron / proton is in gravity. Gravity I shall prove is motion and not mass inspired. In fact of mass being a factor corrupts gravity by restricting motion. Gravity is anti mass and mass is anti gravity because the neutron is all motion with no mass. The gravity of motion is heat driven because it is heat that drives gravity. When an atom is in outer space it is surrounded by an atmosphere of 0 K. That puts a limit on the atom as far as structural differentiation goes.

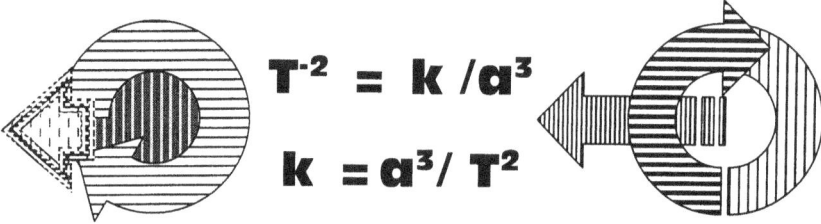

$$T^{-2} = k / a^3$$

$$k = a^3 / T^2$$

By looking at the construction of the Coanda effect we find that the space takes the liquid as an extending of the space that increase the domain the space claims. The space is always the solid acting factor that holds the space a^3 in relation k to the liquid T^2 and the Coanda effect is the personifying of Kepler's formula stating that the space holds a direct value to the motion connected to the space $a^3 = k\ T^2$. When putting Kepler's formula into the correct connotation the Coanda principle is the materialising of Kepler's formula. The motion of the liquid limits the space by adding the motion to the claimed space.

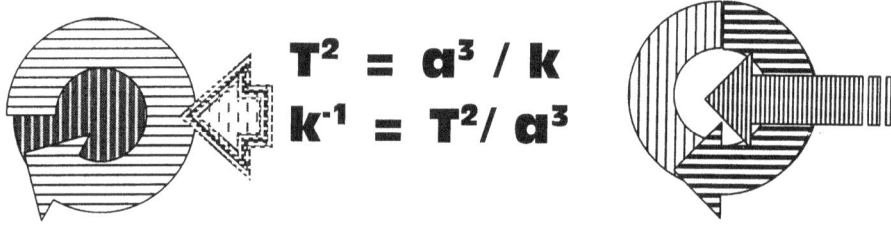

$$T^2 = a^3 / k$$

$$k^{-1} = T^2 / a^3$$

Take this formula into context from the liquids point and we find that there is quite another and opposing connotation to the same formula. From the liquid perspective we find that the liquid T^2 attaches k to the space a^3 by adding one more layer to the unit $k^{-1} = T^2 / a^3$ and the motion T^2 is an addition k to the space $= a^3$ by measure of $T^2 = a^3 / k$. By removing the extending that the motion T^2 of the liquid offers the space this reduces the space by the value of k
$T^2 = k / a^3$. **This means the liquid extends the boundary of the space while the space includes the liquid as the motion attaches to the space.**

The inside as well as the outside must be zero Kelvin because outer space has no other scale that being zero Kelvin. When the atom is on the Earth the relevancy goes that the atom is 40^0C, which is 313 K.

If the outside is 40^0 of the atom is hot then the inside of the atom must be 40^0 cold. The heat on the outside must generate a condition on the inside, which opposes the condition on the outside. The inside is in relevancy or in division of the outside because there is a mass differentiation of 1836 times.

It is true that when concerning the Earth and outer space where there is little to choose from when comparing what changes is occurring in the atmosphere of a star. The Earth is as close to outer space as common civilized decency will allow.

When an atom finds a location in a minor star such as the Sun we are filled with surprise. It seems to us that Sun is very hot and with the Sun that hot the atom has little validity to stay intact. The atom should explode being in such a hot environment, and yet it is there and very it is much undeterred. Any atom we would heat to a temperature of 6500 ^{0} C as the Sun temperature is, will destruct with an enormous bang. Well it is good and well to say gravity keeps it from destroying but when saying that we should use that as a clue and not as an answer. It puts what is in the Sun in another class of structure and confinement.

If the temperature on the outside of the atom rises to 6500 ^{0} C, then the temperature on the inside should respond to what applies on the outside.

It is quite true that when temperatures rise the electron jumps a band. The electron moves apart from the proton as the circle widens. It is not the amount that the circle widens that should be of any interest to us but the total response. On Earth the electron ring would enlarge but at the same time the proton should equally respond by reducing. Place the atom in the circumstances we find in the Sun where the atom heats to 6500^0 C. On earth the atom would explode but in the Sun the atom remains well formed and very intact.

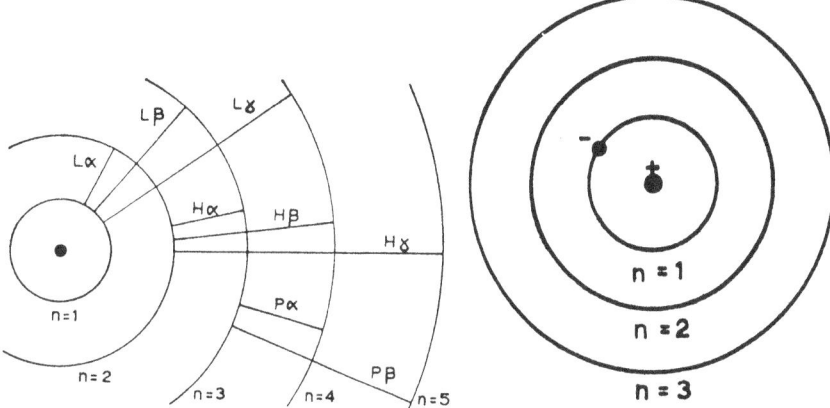

The atom does not explode because the atom does not get bigger and extends to outside its proportions. In such an event where the heat rose enormously and the atom remained as it is in outer space it would mean that the atom therefore must have gotten smaller because the enormous atmosphere kept the atom in tact. Yet with such temperature rising there has to be a change to the form the atom has and that means the atom shrunk in size. The proton became smaller when the temperature rose because the atom had to respond in some way to the rising of the temperature. Putting all this down to gravity is tiresomely attributed to laziness on the part of the human thinking capability because it proves how far we will go to restrain our ability to think. If gravity controls size by heat contribution then gravity has more to do with temperature than it has to do with what mass contributes.

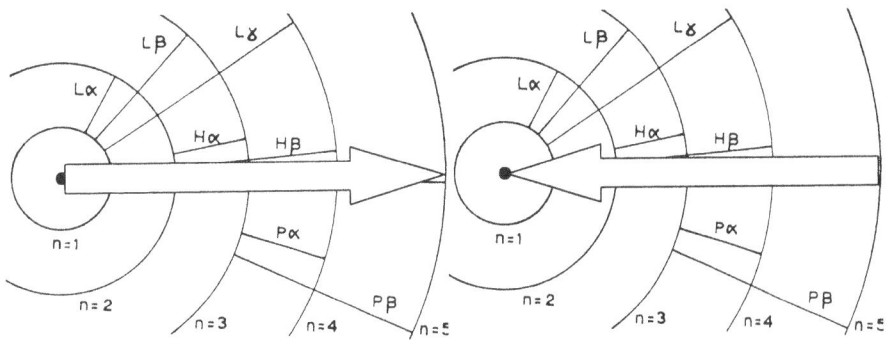

What we see as heat is relevancies because as the relevancy within the Sun changes the atom adapts to the changes. The atmosphere of the Sun becomes denser, which we see as being hotter and the containing becomes stronger. The atom has to reinvent it by adapting to the changes or different surroundings. In this manner the motion that the star provide which is so much more than what is the motion is we find in outer space that the hot / cold dynamic changes all together.

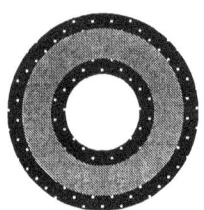

When an object is in outer space that object, encounters a specific relation with what we presume is space. This comes about by motion and through material volumetric size. The space the object encounter by moving through outer space puts a value of a ratio between the space it moved through and the space moving through which Kepler introduced as $a^3 = T^2/k$. That means there is a contact ratio between space containing and space contained by.

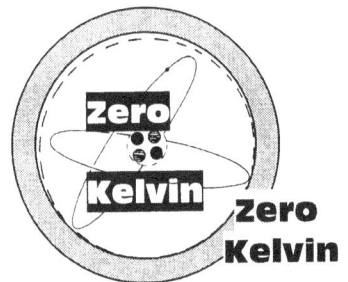

When the atom is in outer space the atom is surrounded by a temperature of zero Kelvin and that is because zero Kelvin is the presumably the coldest any temperature can get. Being zero Kelvin on the outside and with zero Kelvin being the coldest temperature there can, it would make the atom also zero Kelvin on the inside since there can be no colder than that. That would mean the entire atom is then zero Kelvin.

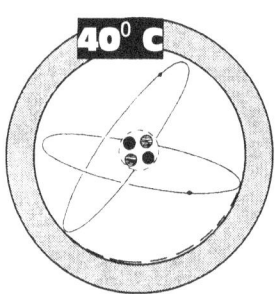

However applying motion reduces temperature and there is much motion going on inside the atom. That means the fact that zero Kelvin produces the coldest there can be makes a little nonsense of such a statement. When the atom is 40^0 C the outside of the atom must affect the inside of the atom because from the fact of what the Balmer and the Lyman series would represent and that proves that the outside temperature of the atom does influence the inside temperature of the atom. The normal summer's day temperature on my farm is 40^0 C normally in the shade because at that temperature little in loony enough to venture outside the shade. We consider that the atom must be 40^0 C because that is what the daily temperature is outside the atom. We feel and experience the 40^0 and we presume that all around is suffering from the heat of 40^0 C.

We know that the action brings about a reaction and the actions leads to a

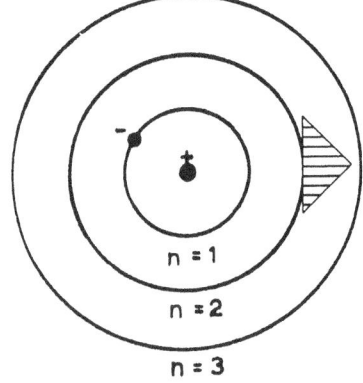

response. If the atom heats on the outside by measure that it finds a need to reposition the electron by one band, then also the inside got smaller in relation to the growth by one band. We associate such repositioning

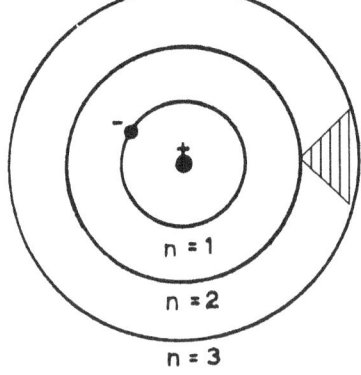

with the heat on the outside that amplify or reduce. However the adding of heat brings on a faster flow of liquid, which results in higher motion and it is in the motion that we find the answer to the cosmic principle applying. In the cosmos there is no hot or cold. There is higher or less motion.

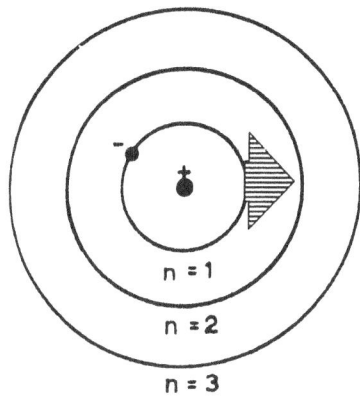

The relocating of the electron into a new position where the electron jumps a band is done by implication of the Coanda effect. From the Coanda effect we know that the liquid attach to the solid using the formula $T^2 = a^3 / k$ where as space

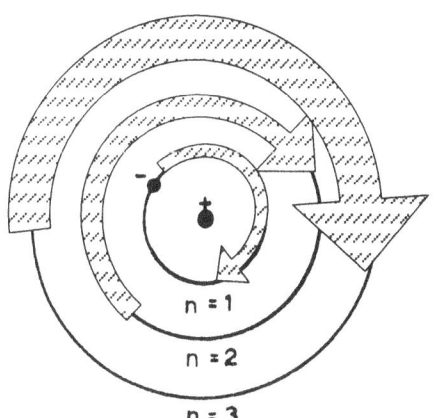

identify new boundaries by identifying the allocated boundary set by the liquid as $k^{-1} = T^2 / a^3$ where the space then forms the limit at $k = a^3 / T^2$. Every time the motion of the liquid intensifies the motion will attach to the solid by applying a new relation, which alters the relation of the solid by extending the space the solid has differently.

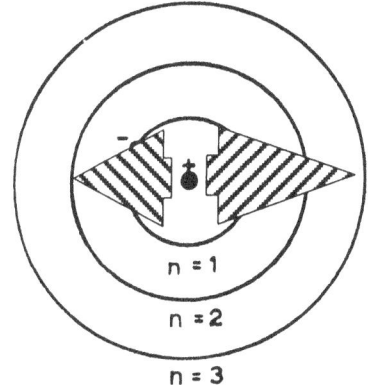

However we must not lock our focus on the heat but we must refocus on the motion that intensify or weakens. It is the motion that produces the new electron allocation and the motion produces a heat that establishes a cold. The focus is on the motion

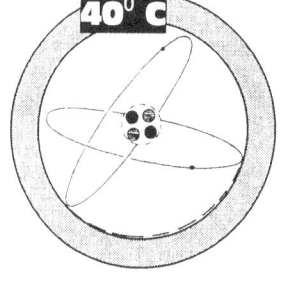

because the motion brings on accelerated duplication and accelerated duplication produces cooling that brings on a relevant cold within the atom.

If the temperature on the outside of the atom changes from zero Kelvin to 400 C it is not the temperature that changes but the atom is responding to higher motion. With the atom in outer space the atom is subject to lesser motion since the atom is only in distinct and personal orbital motion in relation to the Sun . That is why the atom can be subjected to zero Kelvin. When the atom is within the boundaries of the Earth and circling around the Sun in a location set by the singularity of the Earth, the motion is distinctly more than what it would be if the atom were located in outer space.

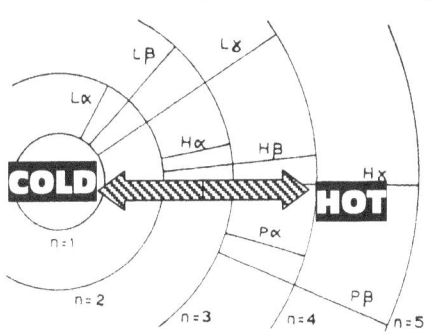

The outside of the atom calls for a direct response to condition inside the atom since the outside can change very little if the inside does not respond in an opposing manner to what the outside produce. In such a relevancy there are always three factors performing as gravity and in that is the

Coanda effect in charge of committing the standards by applying the gravity or the motion in relation to the solid.

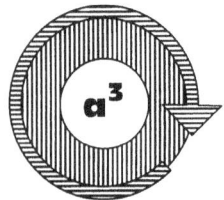

The material revolving through the space holding the material and allowing the material the privilege of motion is in the amount of material per time frame that makes contact with the space which serves time and that it encounters as the space duplicates its position it holds coming from the past through the present into the future

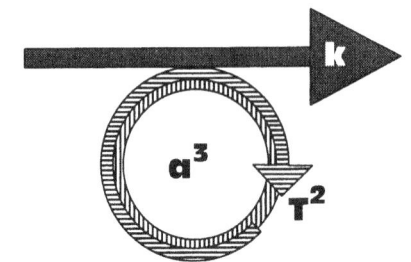

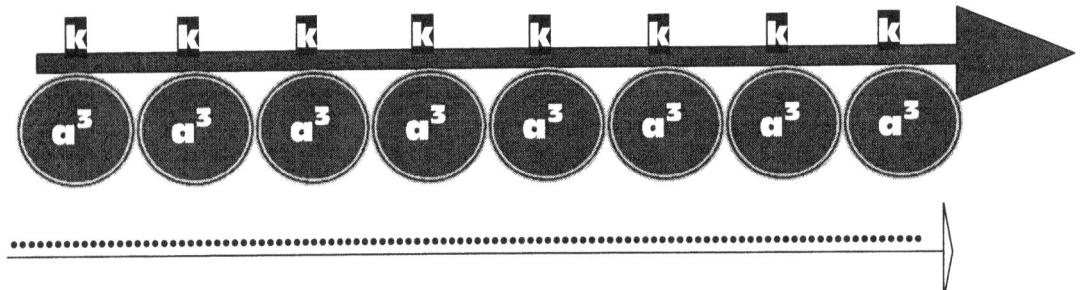

The movement reduces the size the material occupy by duplicating such vat amounts that the duplicating freezes the material into the oblivious.

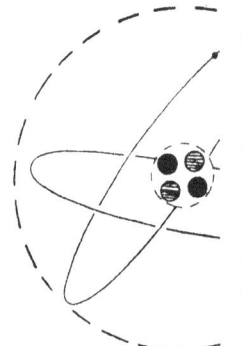

There is a definitive relevancy between the electron and the proton and that factor is what the neutron fills. The neutron is unrestricted gravity or liquid motion whereas the proton as well as the electron is very much restriction of motion of space-time flow, hence the mass. It is proposed that when the atom becomes hotter the electron jumps a band but that statement is not altogether the truth. The proton shrinks as much as the electron jumps a band just as much as the neutron fills the vacant space.

By jumping a band the space within the electron becomes more and the neutron fills that relevancy therefore the neutron becomes more. But if the neutron becomes more the neutron is there to bridge the gap between the electron and the proton and that will have it that the proton needs to respond just as much by becoming colder in the presence of the electron facing more heat. The heat is not the factor but the motion contributed by the heat is what brings about the larger jump in spin.

The neutron facing off the electron as well as the proton will respond on both sides of that which it influences because the response is that of bringing over more motion from the electron to the proton. One cannot gauge the electron behaviour without extending such behaviour to the reaction that the proton would have since the neutron fill the gap and also provide the response on both sides and the changes is what the neutron contributes by suffering the greater discrepancy in changes. However in the ratio or relevancy there will never be any change. The changes come in the form of an amplifying of the motion, which is a relation the space has with time.

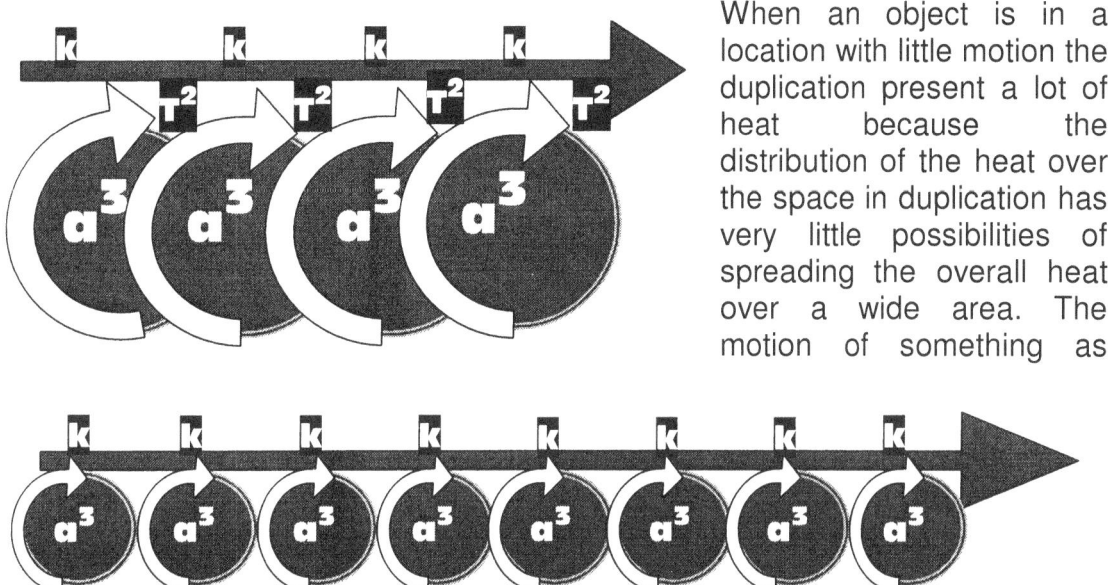

When an object is in a location with little motion the duplication present a lot of heat because the distribution of the heat over the space in duplication has very little possibilities of spreading the overall heat over a wide area. The motion of something as

small as the earth will confine the atoms into a relative hot area since the space in duplication does not reduce the extent of the heat by distributing the heat over much space.

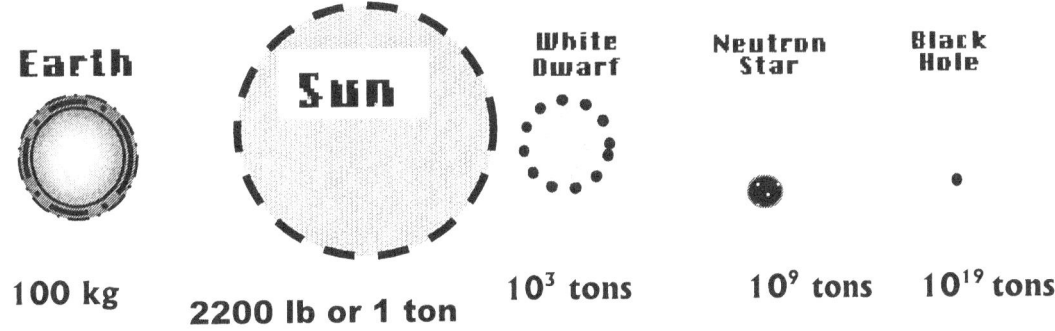

In a structure with the size of the Sun the motion of space is enormous by the sure quantity of space in need of duplication. Shifting that volume of space needs duplication that is millions if not billions of times more extensive than what the earth may produce. By duplicating such a vast area in a period reduces the individual atom to a fraction of what the situation on earth would allow. The more the spin of the liquid is in relation to the solid state of space is reduces the space and extends the material in quantifiable measure many billion times over to what smaller stars are. It is not the space that holds the matter but it is the spin in relation to what the matter holds that puts the relevancy of hot and cold within the star. The more cold there is because of the mot\re liquid heat bringing about motion, the colder would the atomic material be and the higher the relative contracting gravity that the star produces. This we see in the admitting of Mainstream science confessing that the reducing of space produces an increase in mass and because mass is the frustration of material unable to move, it admits to the fact that mass in volumetric size has no influence on gravity.

 The physics we encounter on Earth allow us to use a common and a constant, a fit all and an all-purpose because we find us captured by the Earth singularity. The Earth provides the space we may claim as well as the time in which such material duplication will take place. The earth does not provide conditions found in outer space and neither does the conditions found in the Sun be remotely compatible with the conditions the Earth prescribes. On Earth we find conditions little different from the conditions applying in outer space.

There is a certain ratio of heat to space that allow the material the motion to reduce heat to the extent that will grant the material a certain volumetric size in space-time. The material is hot because it is holding large quantities of heat in the structure of the atom. By moving lowly more space is duplicated less giving less heat being distributed over a smaller area covered by material. By the motion the conditions give a specific ratio of time that allow space to duplicate to that specific required ratio. Since outer space is as hot as time can be, there is no more expanding of singularity possible in outer space.

We call this dynamic speed or velocity, which is just another name for a motion in ratio with time. There is a volume (meters moving) in time Seconds flowing and that ratio produces the size of the object in relation to the time the object allow the ratio to be in contact with the time the object moves a distance. We also know by blowing over a body the "air" cools the body. That means the more "air" that the body is in contact with, the colder the body will get. To this argument there is a lot more and later in this book I return to the matter. However
It is a ratio that is coming about.

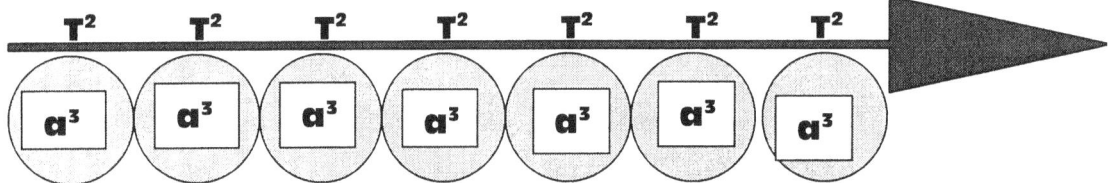

If the motion of the material is more it is more in contact with "air" which is not "air" or even "space" but it is time, the time (or space or air) effects more of the material since the same volume of material moves through more time. The time is a constant and therefore the material cannot increase the time but the ratio can produce more material (in contact with time) than moving at a slower speed. The material reduces in relation to the time it moves through and therefore the material shrinks allowing heat to flow from the material to the time aspect. This is the same, which we find when compressing air into a container used for storing compressed air.

That means the relevance between "space" which the material moves through or is in contact with, reduces the size of the material in ratio to the space it encounters. But we know that this effects the heat balance more than the size because the material moves through more time therefore more heat is transferred from the material to the space surrounding the material It is for this reason that we blow or radiators with fans. By the excessive motion of the massive Sun the material reduces allowing the material to become so cold that the space outside the atom, become 6500^0 C in a normal day on the Sun.

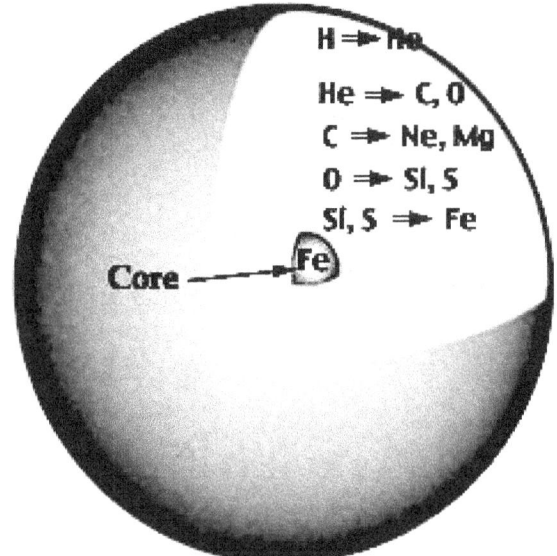

With the enormous container that the Sun is there is nowhere remotely the duplicating by motion going on anywhere else in the solar system. The outside layers are formed by elements known for their volatility, which is another term for duplication. Hydrogen and helium has very high ratio of interaction with heat and in that they have very high freezing temperatures. It is not by coincidence that the most mobility is on the outside and as the layers reduce space towards the inside we find the containing elements preserving space on the very inside.

The Sun is as enormous as it is because it freezes material in ratio as the motion shrinks the material to a fraction the size it holds on the lesser solar structures. By the massive duplicating of space-time it contain heat in vast quantities because it freezes hydrogen at 6500^0 C to a liquid. Deep inside the Sun it gets so cold that the restriction ion motion freezes hydrogen to other elements and this process is called fusion.

It is a case of the motion cooling the material and the cooling is shrinking the material while it is excelling the heat in response to the material cooling. That way gravity is all about motion and heat that contracts material as motion cools material in relation to the outside of the atom that has to rise because of the lowering of the coldness and size of the material. Gravity has very little to do with mass and has so much to do with motion and it is gravity in motion cooling down material that shrinks material to accommodate more dense heat on the outside of material which becomes prevalent within stars.

If the Sun is 18×10^6 on the inside of the Sun one have to take into account that the inside of the atom therefore would presumably be zero Kelvin. If that is not the case then the exercise is fruitless because the 18×10^6 will be meaningless. If there is a limit to the side being hot, then the side being hot must have another side being cold in order to give the being hot side any validity. In that case we also have to mention that when the hot side is zero, the inside of the atom that is zero Kelvin from the other end must be minus 18×10^6 when gauged on the reaction side because for all actions here has to be equal reactions. Everything is in a relevancy and only nothing can be unattached. There is always a relevancy coming about as one part in the relevancy does thee expanding factor and the other part is producing the containing factor

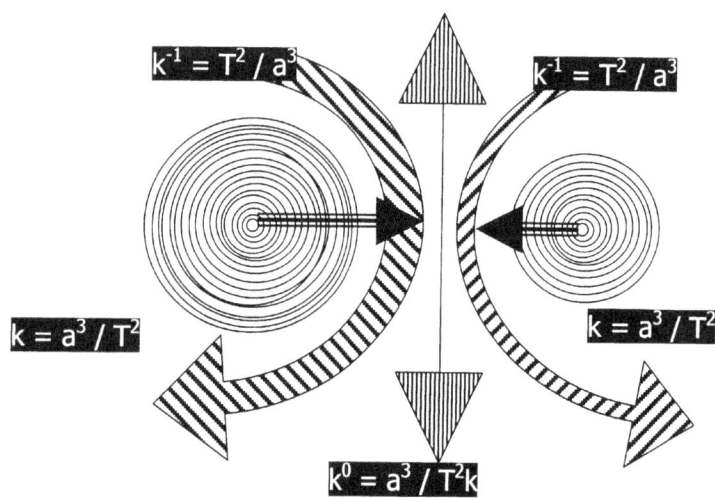

As the rotation commences around a centre point there are changing relevancies as the one cosmic object orbits the other cosmic object ands there are forever cosmic objects orbiting another cosmic object. Even in the case of the Black Hole eternity is orbiting infinity as eternity melts back into infinity. Light is the only factor that responds directly to time by joining time as light is in the space sector $a^3 = 3^3 = 27$ and the motion part is $\Pi^2 3 = 29.6$ giving a total

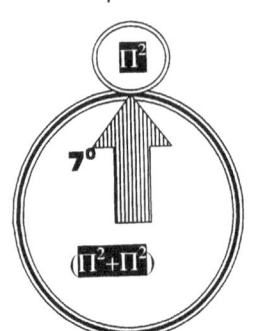

displacement relevancy of 56 .6 within the star. Yet even in this case where light is overheated singularity there still is a cyclic flow of time T^2 through the four quarters, which is in the orbiting

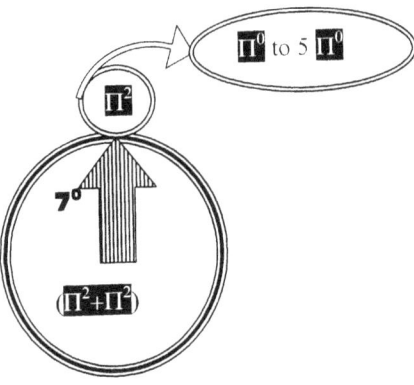

contexts of cosmic structures forming seasons. It is where singularity changes the dynamics in the relation k has with T^2 and a^3. The locations of positions and the allocations of positions of material in motion places the dynamics within motion in different concepts in the ratio they have to each other.

In the cosmos every aspect there is indicates an atom's behavior pattern. Even the behavior witnessed when objects move shows expanding relating to contracting and the one forms the electron or expanding factor while the other form the proton or contracting factor and the two factors are joined by a liquid that holds pure gravitational and unrestricted motion. The part that connects the two factors form a neutron dynamic that is free flowing within the unit.
$$\mathbf{a^3 = [T^2 = 7(3\Pi^2)] [k = \Pi^0]}.$$

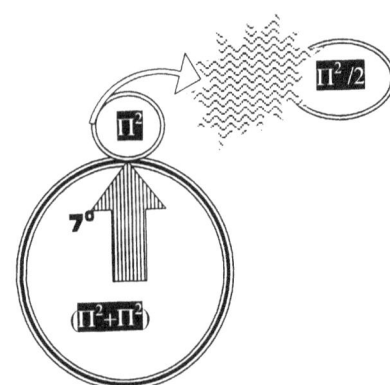

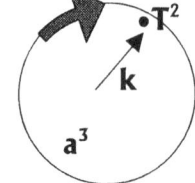

The Earth holds a specific size in relation to the motion of the individual object being the Aircraft $T^2 = 7(3\Pi^2)$. Since the craft is stationary the distance between the craft and the object is $k = \Pi^0$

$$\mathbf{a^3 = [T^2 = 7(3\Pi^2)] [k = \Pi^0]}.$$

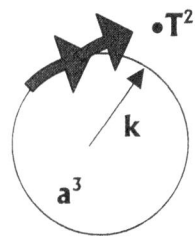

$\cdot T^2$ As the motion of the aircraft accelerate the Earth still holds a specific size in relation to the motion of the individual object being the Aircraft $T^2 = 7(3\Pi^2)$. However the connecting flexible link being the atmosphere which plays the role of the neutron has to extend in order to compromise to being $k = \Pi^0$

$a^3 = [T^2 = 7(3\Pi^2)] [k = 1 - 5\Pi^0].$

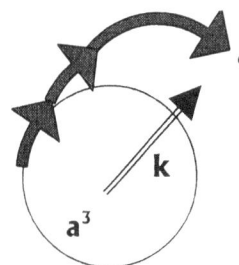

$\cdot T^2$ The Earth holds a specific size in relation to the motion of the individual object being the Aircraft $T^2 = 7(3\Pi^2)$. Since the craft is now in motion the first beacon to arrive at in relation to singularity Π^2 extends the distance between the craft and the object to $k = 2\Pi^0$ then $k = 3\Pi^0$ and so on.

$a^3 = [T^2 = 7(3\Pi^2)] [k = \text{exceeding } 5\Pi^0].$

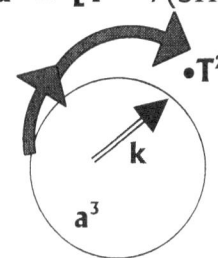

$\cdot T^2$ As the motion of the aircraft further increases the Earth still holds a specific size in relation to the motion of the individual object being the aircraft $T^2 = 7(3\Pi^2)$. However the connecting flexible link being the atmosphere now has to extend to being the most furthers that the neutron can possibly extend and such extending can compromise up to being $k = 5\Pi^0$

$a^3 = [T^2 = 7(3\Pi^2)] [k = 2\Pi^0].$

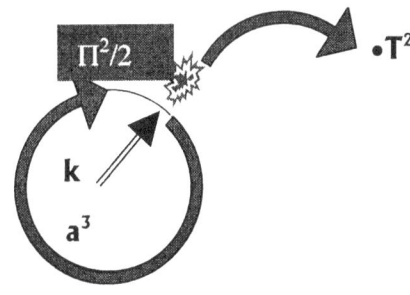

$\cdot T^2$ When the motion of the aircraft further increases the Earth expands beyond what the limits will allow being the Roche factor of $\Pi^2 / 2$. This puts a cap on the specific size in relation to the motion of the individual object being the aircraft $T^2 = 7(3\Pi^2)$. In this the neutron of the aircraft parts in a dimensional time value from the neutron the earth holds and the connecting flexible link being the atmosphere then cannot extend beyond what the neutron can possibly extend and such extending breaks down at $k = \Pi^2 / 2$. The entire issue is about the atom of whatever proportions containing heat in relation to a specific center.

Stars can and stars do **overheat**, sometimes and the **Polar Regions** where **the Titius Bode matter-to-matter applies** holding the square matter (7+7) in relation to the square of space (10) and **other times** in a double relation to the **square of space** 10 to that of matter in a half square (7 /10 or 7/ 10). The fact that stars overheat should tell Newtonians something but Newtonians are not told because Newtonians tell the cosmos something. Stars going out of tune show every sign possible there are of overheating taking place. Saying that one has to differentiate between heat and overheating because if a star can overheat in the face of outer space then the star and outer space are polarised where on is the hottest and the other is the coldest. The star can get hotter but there is no evidence what happens and when it happens when stars go cold. Forget pressure because the star cannot have pressure simply because if the star has pressure then what happens to the star when the star overheat.

If the star expands when overheating a star represents the coldest space in the Universe and not the hottest space. Pressure is as man-made concept as mass is and as temperature is. It is standards set by life according to the practises life apply. It has no validity as a measure in the cosmos, just as mass has no measure of validity or as hot and cold has any validity as measurements. When a star overheat it means the star must be "under cold". The star expands and the star explodes which is all part of undeniable tell tale signs of overheating. That means the star cannot be as hot as the limit would allow otherwise the star being at an ultimate temperature could go no higher. Only when the star is far under a limit can there be any possibility of overheating. If it is gravity that contains the star then the gravity has a direct link with temperatures because when temperatures go array and increase the gravity becomes dysfunctional in containing the star as a container. **Heat and cold are relevant dynamics** forming **in appreciation of singularity. Heat and cold had to part when the first moment of cosmic development came about. Then at that moment eternity became the hottest since motion is with eternity and motion apply as heat keeping infinity the coldest which it then still is. Eternity expanded because the time provides motion and infinity remained frozen because infinity is still incapable of producing space.**

The Sun is the coldest place in the solar system and that is fact. Looking at evidence the Sun provides contradict everything science wishes to believe about cold and hot. Science wish to see the cosmos through the eyes of what fits the needs sustaining life on Earth and what benefits maintaining surroundings in support of life as one find on Earth whereas life has no part in the cosmos except for the speck of dust we call Earth. Looking at the cosmos impartial to life the evidence support another view. Every aspect in **the cosmos is the very opposite of what**

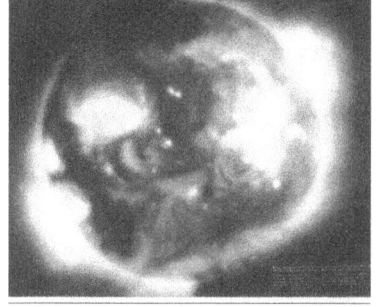

science believe it is. The Sun is **not a ball of gas but** a **giant sea of liquid,** frozen **without any** form of **gas or air** in the interior.

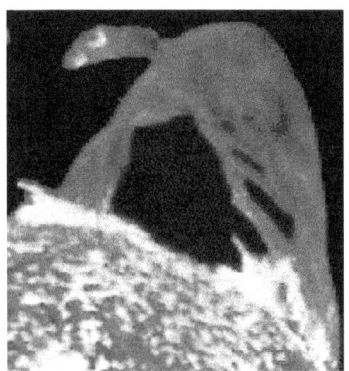

Having a liquid interior **the Sun has no pressure** but has the **very opposite of pressure** to which there is yet no name given. **The liquid comes from singularity freezing** space-time within the atmosphere of **the Sun,** and such is the case with

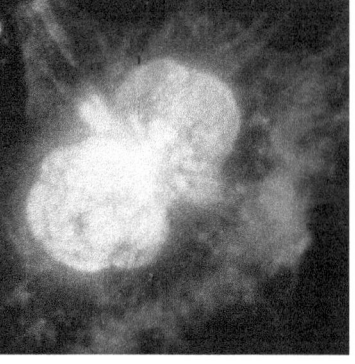

all stars still in the shining phase. **Stars more developed than the Sun is frozen solid causing fusion.**

In **the picture to the left** we find not withstanding whatever name we attach to the **red liquid substance flowing from the Sun into** space and back to the Sun, **that liquid is heat** in a very direct form. **If outer space was the coldest place in the solar system** the heat **should** immediately **escape to outer space** and **not return to the Sun as** it clearly does. If **outer space were colder the heat would not return to the Sun.** If the liquid were heat the result would be a burst of space as one can see in the next picture. The expansion of the star going supernova shows an outpour of uncontrolled heat while the picture of the sin indicates very well controlled heat coming from a liquid that heat at the surface while the rest of the structure remained very cold. We tell the Sun that the Sun is hot but the Sun tells us back that the Sun is cold. If the Sun were hot the Sun would have exploded the way the picture to the right shows the star to go. It is a typical portraying of heat surging above controlling limits. If the star was under gravity control and gravity contained the heat levels of the star by preventing it to surge beyond control then gravity is all about heat management.

Where the ultimate destroying of stars come about one can clearly see the liquid bursting out from the centre. The story of stars exploding shows clear signs of heat management that went beyond the control of the gravity. It shows that balances in heat management was not that well managed and what was suppose to be cold became uncontrolled heat. The liquid we presume has to a result of the heat exploding but the heat coming from the Sun draws a picture of heat being in a liquid form within the Sun. We associate the fact that it is a liquid with immense heat but in worst-case scenarios the liquid clearly turns to gas as the liquid expands. The gas is a higher ratio of heat than is liquid and by going to a gas from a liquid the gas takes the liquid into a higher state of heat. That is science and not picture fiction we draw on because we are culture programmed to believe what we wish to see.

It is exactly the same as the case with mass. We are retained on Earth and out retaining comes with a measure whereby we are retained. The retaining is in measure by a value we prescribe and with it we prescribe mass as a value of such retaining. The fact that mass has no validity when we fall is top the Human

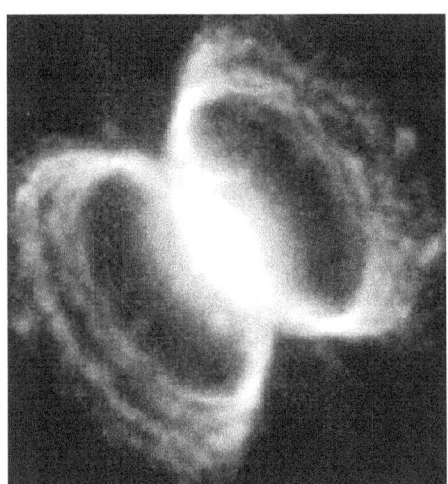

mind of no significance because we can only appreciate the retaining when we can connect the retaining with a measure we associate such retaining with. We feel it is a hot day and we feel it is cold. When the Sun shines it is hot and when the Sun is absent it is cold. With the Sun we are hot and without the Sun we are cols. The Sun must be hot and outer space being the other pole must therefore be cold. We do not look at the cosmos and find direction in what the cosmos present but we find our needs and apply such needs to what the cosmos would need.

All elements forming matter in as much as the heat forming **an atom is** as much a **liquid as it is a gas and a solid**. **There is**

no hot as there is no cold. It's about storing energy in space or in heat, which is another Cosmic equal being opposing similarities. Hot and cold are **relevancies brought about by singularity valuating space-time** and during **the Big Bang** the universe was **freezing cold** at **three billion degrees C**. It is the relation matter has with heat that provides the form the particle has at that moment.

The increasing or decreasing the heat will alter the form of the element. Therefore all elements forming **matter is as much a liquid or not than it is a solid or a gas. It is the space surrounding the atom which provide the form the atom find its relativity to the rest of the atoms it share space with. Hydrogen is as much a solid as tungsten is a gas depending on the heat in relation to the space matter is within.** Should **you reply** that it is **the gravity pulling the heat back to the Sun** , then that **confirms** my theory that **gravity is all about collecting heat onto matter** with outer space being the hottest place**. It is the concentration of heat in space being relevant to form. When** 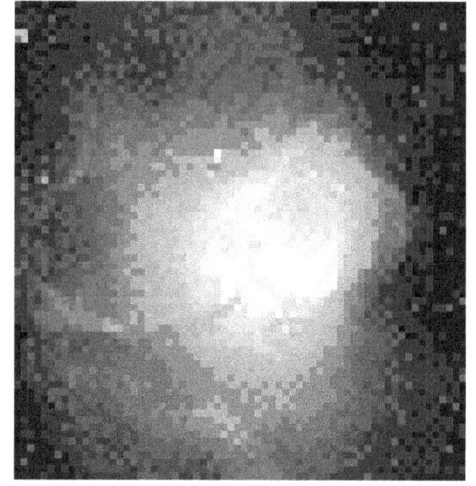 **overheating a star turns its liquid to gas whereby it merely transforms it's interior to a relevancy it has from pre- to post- Big Bang.**

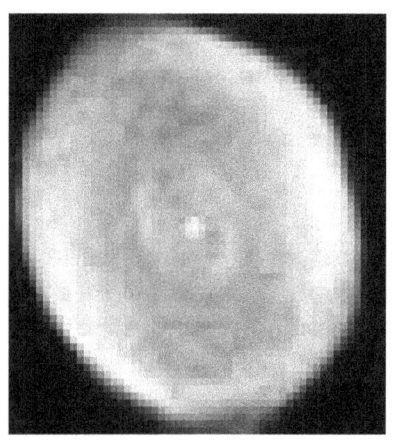

As star going supernova paints a picture of liquid going onto form gas. When liquid goes into a gas form it is a sign of heat rising. When gravity goes wrong and heart levels raise the indication resulting from that is that gravity is about containing heat. It puts heat in containing while it keeps cold inside the star by keeping heat out. When the heat comes in the star explode and that is what the picture of the supernova shouts out, if only Newtonian arrogance will learn from the picture and not tell the picture what Newtonians wishes to teach the picture to tell.

One thing we humans have to realise and that is that heat expands and expanding is produced where more material hold space and therefore heat is present while claiming more space. It is having what there is just more of it proportionally than the rest has or more than what was. If it contracts it is colder notwithstanding our human mentality and when it expands it is hotter notwithstanding our human perceptions. When it moves it is hot and when it does not move it is cold. The fact that the Sun contracts serve as proof that the Sun is cold. The Sun expands at the borders because the Sun heats at the edges but heating at the edges is little proof of massive heating going on inside. In order to have space reduce space has to go cold. Some space has to release some of what it has contained to heat or it has to contain heat which is released but that would be a cooling of heat.

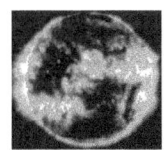 We humans on Earth think that hydrogen is a liquid at – 259⁰ C but that only apply to the Earth. We say according to standards we established that -273⁰ C is cold and we say 6500⁰ C is hot and we stipulate what we intend to confirm. This we do while ignoring all the signals from the Universe where all indications are that the very opposite applies in accordance with cosmic standards. When any object is in a form and heats the levels of space required to hols such heat will rise. The demand on more space to be filled by more material is a natural result from heat rising. When a star expands Newtonians draws the conclusion that gravity got mad. The fact that more space becomes filled because of a demand that developed is a sure indication of heat becoming more. If the heat becomes more when things go array the heat was less when things were normal. When the star heat it shoots the content within the star in the direction of where that which fills 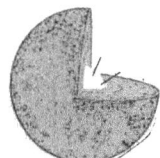 the container at that point can expand no more. In outer space no heat levels can bring about any increase in space except the controlled expanding of the Big Bang exploding. The picture clearly shows the **heat in a liquid** flowing **from the Sun and back to the Sun.** In the **Sun the hydrogen holds enormous quantities of heat in a liquid at a temperature of 6500⁰ C.** When a star has its singularity secured the star is bitterly cold because it has heat in a liquid form flowing back to the point of singularity although we may regard the star to be rather on the hot side. The Sun   (fore instance) freeze hydrogen in a liquid form at 6500 ⁰ C. If hydrogen remains a liquid at 6500 ⁰ C, just think how cold it must be as the star's interior approaches the point of singularity.

Therefore fusing protons comes from cold and not from heat or pressure. By allowing the singularity to overheat the star overheats and heat within the star flows from singularity to outer space freely. In such an event outer space is then colder than the star because the heat releases to outer space with no intention of returning whereas in the Sun it returns as soon as it leaves. There are two ways to reduce heat; one is to bring about expanding space, as the photographs clearly show. The second one is where heat will reduce when in motion by spin. When withholding or retarding motion matter will overheat.

Gravity is the motion of unoccupied space through the dimensional transformation to occupied space. Motion and space therefore is the anti-, the opposite the negative to heat being the positive. With singularity overheating the expansion of the singularity drives heat into space, creating space to compensate overheating **That is a natural phenomenon.** The only reason why **heat will** rather **flow back** to the star than **escape to outer** space once the star released it into outer space is **if outer space presents more heat than does the star,** because **heat always flows from hot to cold** no matter what influences may arise. **Outer space must hold more heat than does the star but the accumulation of space in relation to heat makes it seem colder bringing expanding of heat to become space. Space and heat directly relates being the one form of the other**.

The cosmos is all about **converting space to heat** which we see **as gravity** and **returning heat to space** as a **control mechanism** always **keeping** a very delicate **balance** which we see as **a star shining or being normal.** The purpose of the converting of space to heat is to supply the core where singularity is with heat. **It turns space to heat** sustaining matter but sometimes singularity overheats and then matter converts to heat allowing heat to convert to space. That we call many names amongst others exploding into super nova. Whatever the names used is less important because the **process rests on space and heat interacting to form energy**.

That was what **the Big Bang** was and **the Hubble Constant** is all about where **matter converts heat to space.** I show that **space and heat is the very same thing** and there **is no such a thing as pressure** but releasing **heat produces space** and **concentrating heat reduces space** with the two interacting on singularity demand setting time to space with time being the spin or motion of heat in space. **Heat and space form the second singularity** caused by the **fragmenting of singularity to compensate overheating during the pre-** Big Bang matter forming era**.**

That is what we see as **light and space,** which again is the **same thing and is fragmented singularity forming radiation and heat, where the star re-transfers heat back to space due to an overload.** This comes about through the overheating of singularity (7+7)/10 (top) or layer overheating 10 / 7 (bottom) The fact that stars overheat is evident throughout the universe and as such should not be surprising. The reason why they overeat is very simple and very surprising. When I make controversial comments nobody finds any reason to listen to me. Everyone finds an incoherent novice trying to make sense of some incompetent view that strays from the accepted. Nobody takes the accepted and compare that with the in views I have. In all this they do not believe me but moreover is it 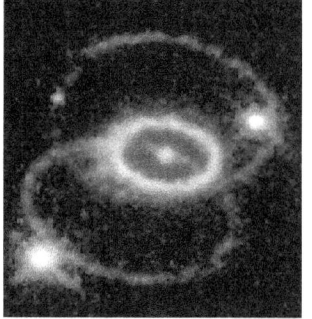 that they do not believe their eyes. What I say they can see but what they see is not what they believe.

When gravity can no longer contain heat by motion that reduce space, which will cool heat, the motion that should contain the heat is unable to do so and the heat releases due to a lack of sufficient motion to contain the heat. When the material in the star is spinning the star too slow to keep the heat in check the heat will release from the inside of the atom and when on the outside the heat then is comparable with outer space.

Being part of outer space the heat will expand to what outer space would allow. The motion that singularity demand will keep the flow of time acting in a manner to sustain the law of singularity by following Π as the indicator. That is the curvature of space-time after all. But the motion will apply to the measure of Π while heat is released from the cold star to the hot outer space. The motion inside the star condenses the space by retaining the heat. Motion cools the star and when there are insufficient motion to contain the material as cooled heat in the space the star reserves the space will expand and increase the space. This when the star heats. Gravity cools the star by preserving the heat through motion and motion comes by measure of duplication as well as containing.

That is the reason why so much "mass" would fit into so littler "space" by such a lot of material, and even moreover is the fact that the smaller the star gets the more material can fit into so much less space while the temperatures get so much higher. Gravity is motion that reduces heat which brings on cold which reduces material size which compacts space into more dense heat that multiply the gravity or motion of the relation there is to "space" or one actually should call it time and space which we actually call matter.

Once upon a time a very long time ago everything we see sprang from one point. Since then all points are a precise duplication of such a point where one forma the expanding factor and the other forms the contracting factor. Both factors are equal since both is the same but one appreciate space to the benefit of the unit by expanding while the other part is conserving the space by contraction also to the benefit of the unit. Since the unit is equal in the value notwithstanding that the unit is served by different factors, the factors hold the equal value is in all aspects of the rotational gravity. By contradiction of the two sides falls to one the factors that form rotation apply to both. Each one of the factors is covering one side of the Universe they form since both have it in goal to preserve and maintain singularity. So often Newtonians talk about pressure within stars and heat pressuring to commit to fusion. A Star that is under pressure is a star that is destructed. For that they have a fancy name. They call it a new star. Can you believe it that after the star has come to pass and blew up like a cherry cracker on New Years Eve, they call that star new? In the star there is supposedly pressure and with the enormous pressure the star pushes elements into fusion …and best of all is that they walk around with the doctorates in op physics! Let's have a close up in the process that applies when the air or pneumatic compressor is pumped with air.

$$R^3 / T^2 = R^3 / T^2$$

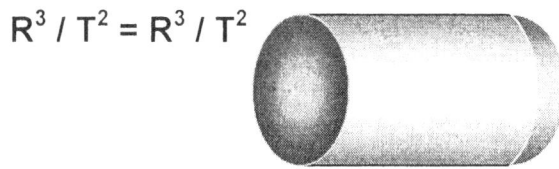

When pumping a cylinder with air the air inside heats up. The scientific explanation is that the molecules bumping each other to the extent that friction must occur because if not how does the heat come in place cause this. It is hydrogen, oxygen, nitrogen and a bit of helium that is pumped. It is not copper vapour tinted with iron and tungsten. The so-called gasses which is extremely volatile and very much a gas, finds the air inside the cylinder so cramped they collide.

This is rubbish, as the next day the container is cold. What made the molecules calm down because through all evidence, the air is still there and the container

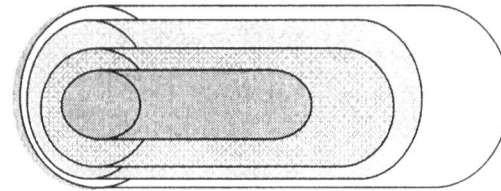

wall is cold. Pumping the container will increase the heat levels inside the container. The inside gets hotter but we are taught at school level that heat will flow from hot to colder areas. There is another way of thinking about this issue, which might seem more accurate in the final analyses. When any object is heated

it expands and when the object is cooled it shrinks or that is what those carrying the flame of knowledge tell us. When we pump air into the compressor the air gets more inside the cylinder. The compressor gets hot while the air gets more. The air gets more while the size of the container remains the same. Seen in another way we can think of the air remaining even while the compressor is getting smaller. The compressor is containing more while the air level is at a constant.

$$R^3 / T^2 \simeq R^3 / T^2$$

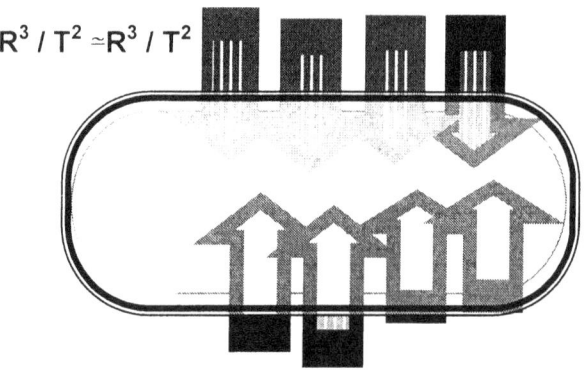

The heat on the outside of the cylinder is at first the same value as the heat on the inside of the cylinder. Then by pumping the air into the cylinder, the molecules take with the heat (unoccupied-space time) they contain. Inside the container the relation to heat gets more because the volume of the container remains the same except that the container walls get hotter that holds the air in place. Because the walls get hot we may assume the walls try to stretch because by heating the walls should expand as it gets hot inside and therefore it will force the walls to expand. By this token it is clear that as much as the container is filling the container at the same time is also shrinking. Because it is also the size of the compressor that shrinks as much as it is t6he content growing more the space outside the cylinder has to accommodate the increase in flow of heat coming through the compressor walls because the overall practise of science is that nature rules by equilibrium.

As the air becomes more the walls of the cylinder will reduce by the same token. There is more air connecting with the cylinder wall and therefore there is less

$$R^3 / T^2 < R^3 / T^2$$

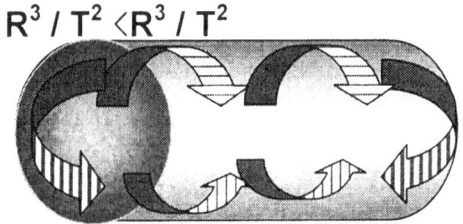

cylinder wall with which the air can connect. The flow of air inside the container encounters less of the cylinder wall and

$$R^3 / T^2 = R^3 / T^2$$

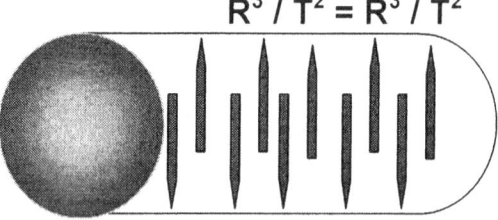

more space and in that the truth is about cosmology. The air that came in brought with it the same volume as heat as what it had related to per volumetric ratio as was applying in the atmospheric space when the air was outside.

The volume of air expanded but when anything expands it gets hotter. There is no evidence of anything expanding without increasing heat and even in the spectroscopy we have evidence of just that. It seems the bouncing has the increase in heat except that when gasses increase in volumetric capacity the gasses become volatile. If that was the case then there was more heat within them container that the air brought in and since the container got smaller the space held more heat per measure of atoms than was the case when the air was outside the container.

With the air increasing, by the very same ratio will the size of the cylinder keep reducing. That is why the compressor walls get hot. It cannot stand the reducing and ties to grow and expand, as it should while the air remains the same volume. From the container side there is no growth in the sir volume but there is a decline in the wall size of the container and that is why the container tries to expand the shrinking walls by allowing heat to try and expand the heating walls. The molecules are then more to the inside than the outside, the heat containing them, is also more on the inside than the outside (bigger ratio inside than outside).

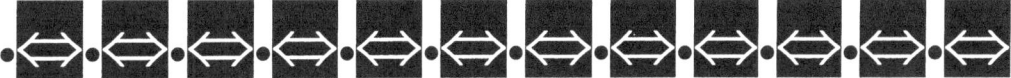

What we find taking place in the wall of the container that is shrinking is the same that is taking place when concerning the position of the space not filled by material The space is becoming less that is holding the material that is becoming more. If the material per volumetric molecule is taking up more space then the space holding the volumetric molecule per unit is getting less.

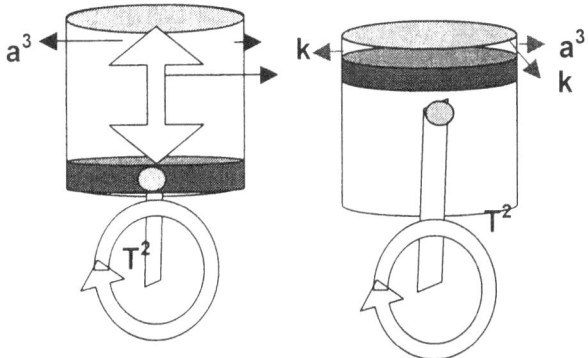

The more the molecules are the less the space must be that the molecules claim and the more the molecules has to reduce volumetric space to compensate for becoming more. Then the same applies in relation to the space parting the molecules as the space in ratio also have to reduce in order to accommodate more space claimed by the molecules as well as space pumped in that was accompanying the increasing number of molecules entering.

However we find the same process in the internal combustion engine with the only exception in that the process is put in reverse. Notwithstanding the different application the end result remains the same. In the case of the Diesel oil engine we find spontaneous combustion occurring when the process becomes at its peak of compression. However there is not a pumping of air but normal airflow into the container. At a point an intake valve ends all further airflow. Then the piston moves up and the piston reduces the space.

This time it is the space of the container that becomes less by motion reducing and the volumetric reduction of the space. In this the heat level rises to a point 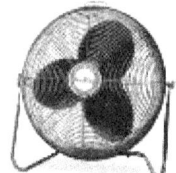 where the air gets so hot it makes oil combustible. The volumetric space reduced and in that the particles became more. With the increase in the number of particles per space available the space available between the particles also became more in ratio. What becomes very clear is that the reducing of space brings about an increase in temperature.

The very opposite is also true and we use that principle for cooling in everyday life. By blowing with a fan over a surface reduces the temperature of that surface. By making the space available more the space between the particles also becomes more. Then the space being more reduces the heat level surrounding the object, which is cooled. There is air blowing over a surface normally not moving and therefore by blowing over a surface that is not moving one gives the surface not moving the opportunity to be in a position that it can move in. In that way we enlarge the surface that is not moving by duplicating the surface not moving as the surface finds a location where it enjoys a larger ratio with the space it does no occupy in the same period of time.

Past going onto present and becoming the future
We have two persons standing still. The one is a thinker and the other is an accomplished and distinguished but sincere Newtonian scientist and which one of the two is which, that is for you to decide… The problem we investigate is how does both come from the past move through the current and leave for the future. The defining characteristic about time is that as time moves on the position of every object changes in relation to future and past positions.

We find that in any given area there is a ratio of Matter filling space and
In this ratio is built in another ratio of Matter holding time.
Matter determines THE RELEVANCY OF Matter to space during time.
Space holds heat in A RELATION OF SPACE-TIME unoccupied- occupied-, densified and singularity. Motion or moving by time or otherwise is the most

complex issue there ever can be. Time relocates the structure by breakind down the entire structure as to relocate the entire structure and re assemble the entire structure to the previous spacifications and by perfect duplication.

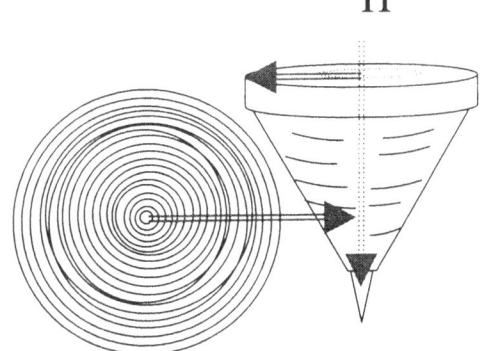

The position of the following instant neutralizes the previous position as it takes the place of the previous position.

In order to understand this concept it will be best to return and see how space and time started. The location where it all started is still present in the entire

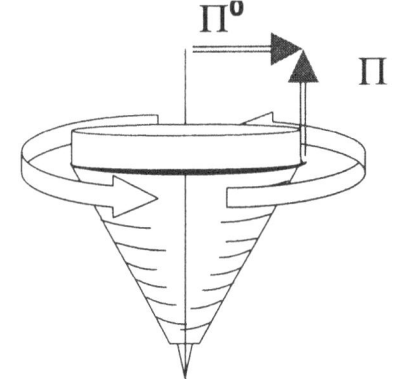

Universe in use today. Fortunately we do not have to move back that far but investigate how the ordinary top is enabled by motion of rotation to stand erect. In the centre is a point that was there before time began. Time evoked the point back then as time still evokes the point in the present.

The point is so small it holds all points in one position. All for points are there and are rotating but by rotating from point 1 to point the point number 2, as it leaves 1 it lands on three because from there it moves to for which is also 3. All the points rotating are on the very same point. The point was eternally rotating and the rotation was there but the points became undefined and blurred because they were allocated to the same position. In such a simple concept as motion there are so many relevancies that has to establish new relevancies before relocation my motion can take place.

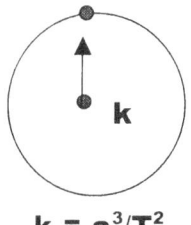 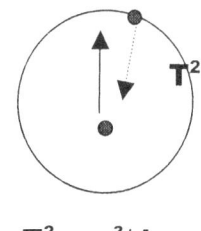 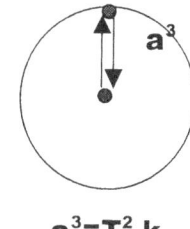 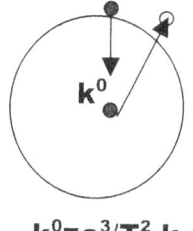

$$k = a^3/T^2$$ $$T^2 = a^3/k$$ $$a^3 = T^2 k$$ $$k^0 = a^3/T^2 k$$

$k = a^3 / T^2$ We have the fact that in the moving of space-time brings a new identifiable location for space to centre.

$T^2 = a^3 / k$ The motion will establish such a centre

$a^3 = T^2 k$ The space provides the motion to continue into the future while the space fills the one side of the Universe holding a position in eternity.

$k^0 = a^3 / T^2 k$ Singularity establishes and relocates space-time successfully by completing the motion.

Simple wasn't it? Let's run through the process once more and find the simple matter of motion in time.

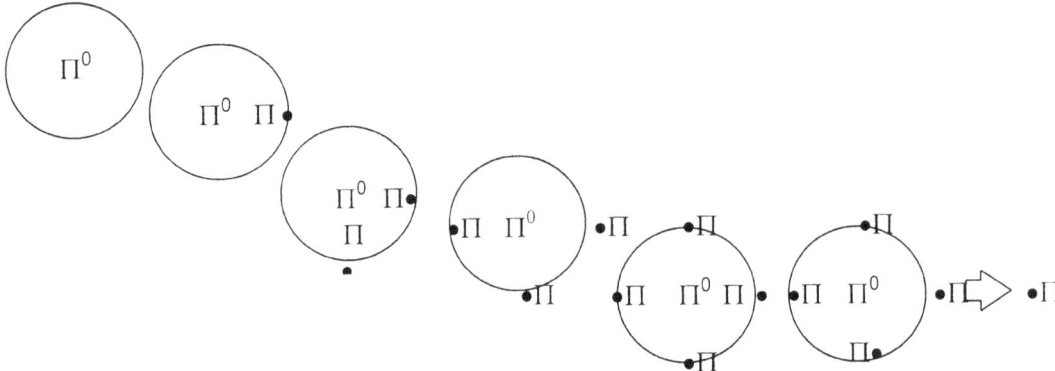

Singularity shifts from Π^0 to Π, which is a spot forming a dot. In our simplistic Newtonian way of thinking is that a spontaneous sphere formed as the spot expanded into a dot. Beware, all is not that simple because we have then small matter of $k = a^3 / T^2$ to deal with. Every spot has to find a position in accordance with rotation as well as a position with relocation. The same dot that was 1 became 2 because it was relocated and not reinvented. Then the dot was allocated a position in position three by reinventing 3 as well as relocating 2. This became $T^2 = a^3 / k$. At the very same instant $T^2 = a^3 / k$ did not disappear because what once is in the Universe is always in the Universe. As the motion took the dot to 4 then 4 became the new 1 because motion took singularity from Π^0 to Π where Π^0 was placed into a new allocated position by establishing a point as point five and relocated 4 as 1. Simple is it not. Try do that to every point that has a possibility of holding 1, 2, 3, and 4 in one position where all share the same position.

All this is rue because $k^0 = a^3 / T^2 k$ singularity positions pace in time by circling the straight line and repositioning the allocated spot.

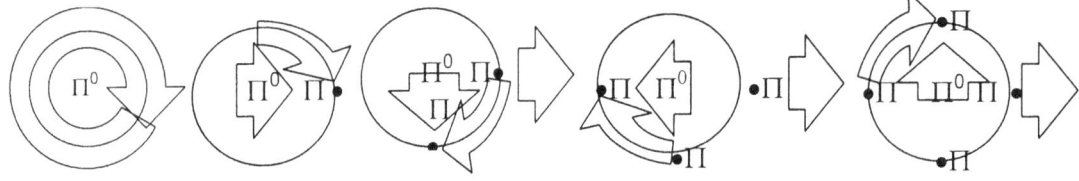

This is the prominence we find in the Lagrangian system using 5 points in the system where singularity forms four plus one. The motion in time in eternity is in a

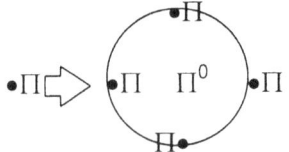

direction, which we might call progressive but also there is another relocation of the dot taking k from k_1 to k_2 that will form T^2. In all of this it is vital to see that there is the rotation as well as the linear and that forms the allocation of space where material is the time delay caused by locating the position of distribution and not being able to remove the previous allocated positions quick enough. Material is the time delay of heat distributed.

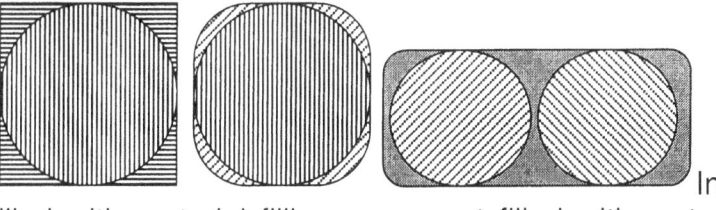

In the cosmos we have space filled with material filling space not filled with material while both are filled with heat. By moving the material filled heat faster than the cosmos relate the filled 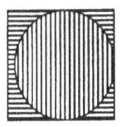 heat with unfilled heat a specific such action will surpass limits of a specific ratio and that ratio we call time. By blowing air over the hot surface that is not moving we are relating that area with a larger unfilled space and therefore without moving the filled space becomes larger in relation to the increase in unfilled space. With the filled space becoming larger the heat within the filled space becomes distributed through a larger area because the relevancy of the filled space has increased the filled space in size by matching the filled space to a greater ratio in unfilled space. That means by moving the air we are increasing the size of the material and by increasing the size of the material the material has to duplicate more often and by duplicating more often the material is shrinking in size.

With this information fresh in mind let's return to our compressor cylinder filled with air.

The cylinder had an initial size to begin with. While the air was expanding through the pumping the inside of the cylinder became smaller as a result of the pumping of air into the cylinder. The Newtonians say the molecules are colliding and bumping and that friction brings about the heat. Then why is the cylinder cooling with time because the particles doing the bumping is still doing the bumping if they were doing the bumping in the first place. Yet the heat does subside and no air has to leave the cylinder to get the heat to subside.

As the space reduced the air got more and as the space reduced the material became smaller and with the material becoming smaller the material had to dump heat from the inside to the outside. Therefore not only did the air not filled with space compress and heated but the heat inside the atom had to disperse some of the heat to decline and dispense of some filling because it had to reduce the initial size it had.

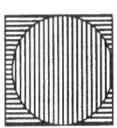 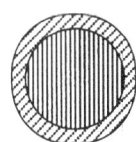 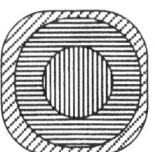 

The process just described relies on pumping, on pressure, on retaining by an outside wall, by confining through deliberate replacing of material, which is confined into a cylinder that offers more confining.

Most important is that the entire process relies on life and if not for life intervening in cosmos affaires none of this would be possible. So how does this comply with conditions in side a star? Well it does not comply even by a stretch of the imagination and only a Newtonian that is prepared to forsake logic in favour of madness and forces of unknown origins can see any connection between the star and the cylinder having pressure.

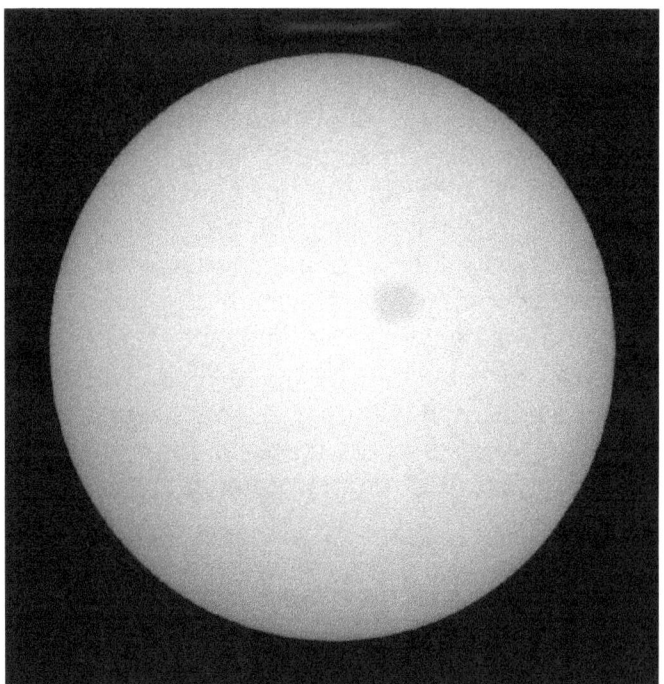

Looking at the Sun as a cosmic object I do not see any retaining cylinder wall and therefore there is no material seeking to find a way out. There is no pump putting material against the flow of nature into the container. There is no forceful relocation of material from one side to another side and there is no escaping from what is unnatural circumstance. All there factors contribute to what makes me not accept the view of pressure inside a star.

There is no bursting of what is inside to what wishes to be outside. There is no evidence of retaining what is inside. It is a round structure and therefore it holds what is inside in accordance to singularity applying. There is no possibility of life intervening in any way or life interfering with the process. The scope of cosmos affairs just is limitlessly beyond what life has as possibilities.

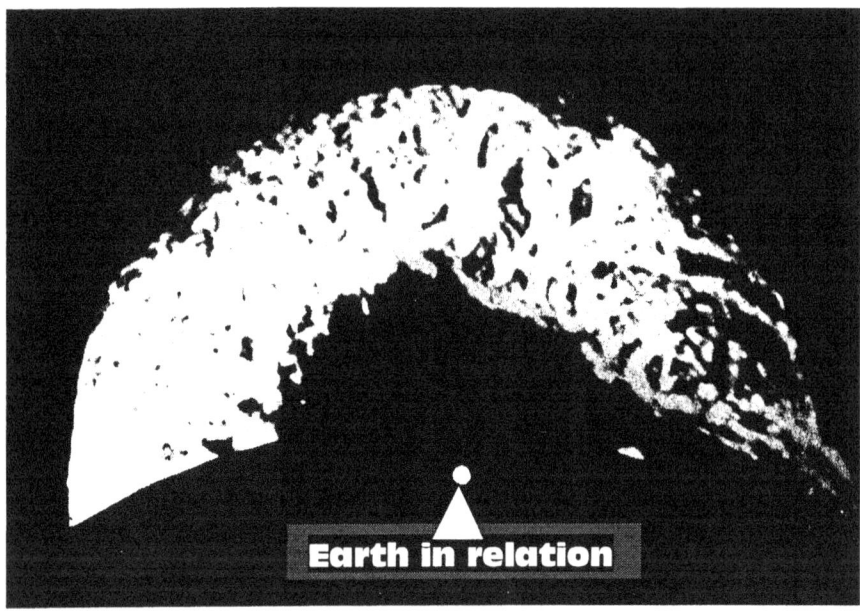

Earth in relation

Yes we do see what is inside trying to spill to the outside but it is far more evident that the spilling out is the forceful behaviour and the retaining is what comes naturally. There is no deliberate escaping from the pressures within but when released that which was inside flows back as a natural reaction and defies the whole idea of unnatural pressures building up inside of the retainer. There is no comparing the cylinder of pneumatic principles to the star that holds liquid and not gasses inside that star. There is no evidence of any gas although the flow of photons emitting light rays is by some imagination some part of a gas.

Inside the star the movement of all atoms combine in motion that establishes a centre governing as a principal all conditions applying in the star. The rotation of the atoms forms a synchronised motion that establishes the line, which parts infinity from eternity. The motion confine the material to the star but it is because the material elects a principal to confine the conditions to a status which all material inside the star agree on and benefits by the conditions of space-time we find in the star. As the conditions serves the star gravity will come about and gravity sets freezing conditions within the star.

 What happens inside the cylinder is the part that is compatible with what is applying in a star when we exclude the pumping, the pressure and the container idea. Lets go back to the fan blowing wind over an area in need of cooling.

When heating the object increases its initial size from normal to becoming larger. The heat increases the size the object has to a larger ratio than what applied before. To cool the object an increase in the ratio is needed on the airside to keep equilibrium and to bring in cooling even bigger ratio is required. By increasing the air we are decreasing the material and by increasing the heat we reduce the material by progressively anticipating more duplication as the ratio of space to material is increased by more space duplicating more material. The heat has to increase the size of the object within in order to match the ratio set by time on the outside. At this point I think it worthwhile to remind the reader that during the Big Bang the lot outside seemed hot and today the lot seems a lot cooler. I say it seems because it is not truthful. The heat has to increase the size of the object in relation to the match it has to find in the space it is within. With the heat coming into the object the relation that the object has with the heat or air outside makes the object that many times bigger, because the ratio in the heat balance is disturbed. If we blow air over the object we increase the size of the object by allowing the surface of the object to make contact with much more air in the same period of time, which will bring the size of the object back to the normal ratio it was before, because in relation and considering the contact with air the object expanded by the motion that increases the amount of air being in contact with the object. In the normal flow of time the object has a heat to space relation set by the time the dictates. Then we go and increase the heat of the object and in that event we actually increase the size the body has in relation to the heat in the air. By blowing air over the body we increase the air and therefore we increase the size of the body during the same period of time. There is now a dispensation of many times more air where the body is carrying more heat and making more contact with the surface whereby it is contacting heat or air which brings the equilibrium back to normal what ever normal then is. There was a body size ratio and by applying heat the balance shifted to the reducing of the body size in relation to the heat. The body then had to expand in heat because the body was to small to incorporate that larger heat. Then by blowing the air over the body it increases the size of the body and heating the body decreases the size of the body in comparison with the air it comes in contact with. The body is either expanding or the body is reducing and the balance in heat places the body in relation to either gravity cooling by contraction or by expanding by overheating. The very same principle applies in the sound barrier.

The gravity motion of the Earth is 7 ($3\Pi^2$) which is the distance of space in relation to the time it takes to displace that space and any motion above that is an extension of the atmosphere where the atmosphere accepts the role it has as being the neutron of the Earth. The extending can go from Π^0 to $5\Pi^0$, which will then be the moving, object extending its neutron part while still attaching to the Earth atmosphere. Above that Lagrangian limit of 5 the Roche limit in sharing neutron status sets in at $\Pi^2/2$ and the attachment there is between the atmosphere linking the aircraft and the Earth is severed.

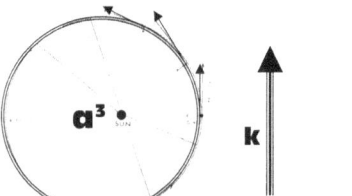 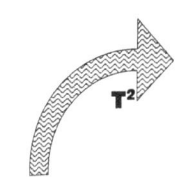

In physics there are always two relevancies at work, which has nothing to do with mass but is solely directed on singularity achieving motion.

The current notion of mass pulling mass has no comparing with reality because as I explain mass has only a counter effect of gravity. In the Universe there are a flow of space-time and a relevancy bringing about expanding as well as contracting gravity. This has no bearing on mass and is a control mechanism we find in the cosmos. It is a product of the cosmos we named the

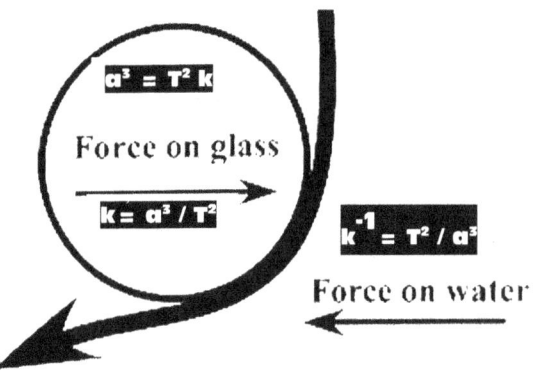

Coanda effect. We have to see that there are two parts in gravity where the one is expanding while the other is contracting and center to this is the control we call gravity.

Many moons ago long before I dreamed of becoming an author of any of the books including this letter I started my search on the basis of a certain remark that Einstein once made on a realisation or a conclusion that Einstein came to in his younger days while still being a clerk at the patent office. Apparently the idea Einstein came to was concerning the subject of gravity. This happened while Einstein was still being a patent clerk in his younger days. Apparently Einstein was looking out a window of the multi story patent office, when Einstein suddenly realised that had he, Einstein fall out of the window from the roof to

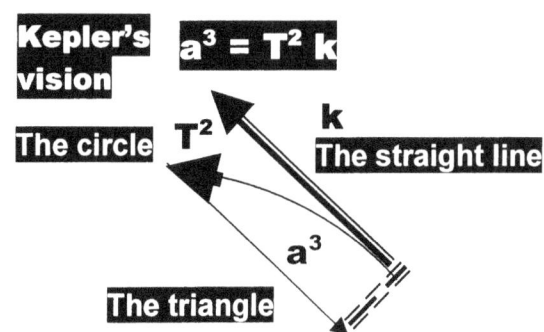

the ground of the patent office where he was working at the time, then he (Einstein) would feel as if he was weightless during the time of his fall. Not only that but also so would all the articles in his office that surrounded him at the time being his office chair, his desk and a pen. By falling with him those articles would feel equally weightless should they accompany his fall down as being part of the falling process in his imagination. As the objects were travelling alongside Einstein down the building to the ground the lot would travel at the same speed from the top to the bottom of the building. That is what Galileo concluded about five hundred years ago. Then I

▲ **The pulling away of the smaller space. a^3**

▶ **The double counter-acting referee. T^2**

▼ **The pulling towards within the larger space k**

went one step further by supposing the Einstein group's falling was real and no imaginary thoughts were set in the fall, then what was the imaginary factor then? Let's pretend Einstein did fall with his pen, his chair and his desk and Einstein was not imagining his fall. Einstein as a human being can imagine but his falling companions can't. Then during a true fall Einstein may have had an imagination that could tell him about his feeling and in particular about the condition of his weightlessness, but the pen, the chair and the desk had no such imagination and they were travelling at the same speed as he did downwards and therefore had the same weightlessness as he (Einstein) had while they all were being in a downwards fall. If Einstein was imagining his weightlessness, it might be psychological, but in the case of the other travelling companions it was not possible to imagine anything. The falling companions had no such a luxury as having an imagination, however they too had to be weightless as they travelled next to Einstein all the way. There is an immense difference in size between the falling companions and that notwithstanding they travelled the same speed while descending. If they travelled the same speed as Galileo proved and they all hit the Earth the same time, which then indicated that their weight and mass, that which gravity used to drive and what propelled them downwards and that which was causing the drawing of what the mass was instigating to allow the motion of fall to commence, was equal. Size changed nothing to the equality there was in speed. Einstein should only have thought a little further than he did at the time because that would have made him realise what gravity exactly was and what Kepler found gravity to be. Kepler found space a^3 being equal to the motion thereof T^2 in relevancy to a centre point k. Kepler found space had to move.

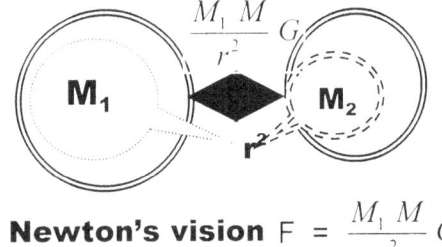

Newton's vision $F = \dfrac{M_1 M}{r^2} G$

Realising this part made me doubt the correctness of comic science. In the cosmos all things are moving therefore all things are falling never to get there. Galileo said that all things falling fall equal. If there were Newton's mass discrepancy when falling this would not be possible because part of the driving force of such a fall is the mass. Newton even put the product of the mass in relation to the destruction of the radius. The mass forming the driving force then has to allow for heavier objects falling faster by measure of mass discrepancy. But that would sideline Galileo and Galileo could not be sidelined. It is either mass driven with Galileo being incorrect or it is as Galileo said that all things fall equal rendering mass equal during everything falling.

When reading this that evening so many years ago, I came to realise that Einstein could only feel weightless if it was true that he (Einstein) was weightless. He could not feel as if when the as if was part of his imagination because he was truly falling, and in truly falling the falling was then without his imagination doing the pretending. Einstein had to feel his weightlessness as a cosmic fact in the true sense because if he was truly falling, then the part, which was the falling experience, was what he was experiencing in reality by three dimensions with one dimension in time. Then he (Einstein) was feeling weightless through falling and that feeling came as a result of what was happening to him as a cosmic interpretation of reality. He was not pretending to fall whereby he then would feel as if…he was really falling and with that there is no "as ifs". What he then would have experienced came by means of what he was experiencing in reality because of his cosmic state in relation to his relevancy with gravity. If Einstein was experiencing weightless ness, it would be because he was weightless while falling, then Einstein would not imagine the weightless ness because Einstein was truly falling, thus carrying out his cosmic state he was in. His body being in motion ($a^3 = T^2k$) was at that moment truly weightless while experiencing unrestricted gravitational motion. Einstein, the pen, and the chair had the same weight since they were all weighing the same in falling. If there were any mass differences there had to be speed differentiation for the force of the one would generate more motion than the force of the other onto the different mass components but since there is not mass discrepancy amongst the falling while falling the lot is having the same state of weightless ness, they adopt the same speed in the fall. After all it supposedly is the mass that is doing the pulling and more mass does more pulling…except if the mass is not doing the pulling in the first place. With more force applying to different masses there had to be more speed involved and an increase in mass in some participants has to generate more force. All four items including Einstein, would be equally weightless during the falling…that was what Galileo found because objects of different size and different mass travel at an equal pace (distance over time or space moving divided by time flowing while the object changes position in relation to the Earth ($a^3 = T^2k$)) while descending. The bigger objects do not fall quicker than a smaller object and that can only be attributed to one fact; it can only be true if the four weighed the same while falling and no one weighed anything while falling. That means the gravity applied while time flow in relation to the space that was applying the motion, which was what gravity is $k = a^3/ T^2$ according to Kepler. The single line falling is represented by the factor **k** being the relevance of space a^3 that was relocating its cosmic position while all that was happening in relation to the motion of the Earth T^2, which was in relation to the Earth spinning around the Sun and that rotation gives us our time T^2. While in motion the four different objects weighed the same since they travelled at equal speed downwards. However, when they stopped moving and came to a standstill, they then weighed different, which then indicated a difference in mass factors amongst them. By standing still the objects had mass differences and when they were in motion they weighed the same. When the motion became frustrated by being blocked by another space that was also filled with material and that was holding the spot too where the motion was directed, they then had different weight. The two had different levels of frustration with the larger party being more frustrated in the inability to move. The pushing resulted from the bodies striving to remain independent. It is the independence of the two bodies and the desire the bodies have to remain independent and not to share space that bring about the mass or weight. The two objects were in a fight to claim the position each desired, and that was to fill the centre of the Universe. Being $(a^3 = T^2k)$ was being in the centre of the Universe because the centre of the Universe was $k^0 = a^3/T^2k$. $a^3 = T^2k$ $k^0 = a^3/T^2k$

From this one can deduct that gravity is motion or the intent to commit motion and mass is when the motion of gravity is frustrated by some solid structure blocking or preventing the continuing of the motion. Then one may conclude that gravity is motion of space and mass is the restricting of the motion of space. Having mass does not bring about gravity but it does restrict gravity's motion, which is what brings about the mass and weight. Gravity produces mass but mass does not produce gravity or in fact mass produce weight but mass is not responsible for the intended motion. Gravity on the other hand is the intention that the body has to move the very instant the blocking is removed.

The intent on moving while being blocked by another object is frustrating the motion of gravity in both cases and the higher the frustration on motion is the more mass there is co0ming the way of the bigger object who then has the greater desire to move. The reason why it has the desire to move and why space is equal to the moving in time of the space in relevance to the centre of the Universe (which at that point might be the Earth or be the Sun) is what I am trying to explain. Mass is the restraining of motion and gravity is material moving about by committing gravity. Mass only comes into the application thereof when two objects filled with space moves into a position where both want to claim the very position in space the other occupy. It is the motion and the independence they show to hold onto their individuality as independent cosmic structures that prevent them the sharing of space which in turn prevent further motion that causes mass. Gravity is in essence where mass is present, still in a tendency to commit motion but is then in the frustration of motion and gravity at such a point is the commitment to move once the blocking of space is relinquished. Because the one object that has more "mass" would put in a more assertive effort to move in relation to a smaller object and the effort to move will constitute to a greater resisting effort by the blocking object in a fight not to relinquish its position on the space both object claim that the tendency to move and the tendency to block the movement will bring the effect of greater or smaller mass being present during the effort and in line of resisting the effort. However while any space is in motion, the gravity of motion is equal to all and puts everything on an equal basis. Therefore there is no big and small and the big Sun does not pull the small Earth closer. The big Sun allows the small Earth to glide past in a circle year after year without interfering because the two does not claim the space each other has. Mass is when the motion is prevented that a differentiation in motion effort becomes part of the picture.

Do not be fooled by the seemingly innocent explanation that space is the motion thereof which is what gravity produces because of all things the cosmos creates, motion of space through time is the utmost complex manoeuvre and without bringing a restraining of mathematics into science, it is so complex there is no viable explaining in physics about how the cosmos produce the act of motion of space in time. To get every atom to spin as every atom follow the lead of the atom in front and gives direction to follow to the atom just behind while giving coherency to the structure the lot of atoms are holding as an individual unit times the units there are going around in the entire Universe is beyond what the human mind can absorb. While the atom in front is vacating space to fill the space of the atom in front that is vacating at that instant, the atom behind is filling the space that the atom in front has vacated in order to vacate and relinquish the previous position in favour of the following position to honour the direction gravity is insisting upon. Times that with every atom there is in the Universe and one may grasp the significance of the calculation. The coordinating of moving one atom from one point to a next point requires the skills that the human mind may never conquer.

We may see the moving of an object through space being as simple as merely accepting it as a given fact, as science has done in the past, or we may reason about the complexity as civil person's should do, and come to realise that the complexity of motion of matter is beyond the scope of human understanding. Removing material from space by filling material into a position of new space sounds simple because the complexity has never been realised. This was all a result of understanding the dynamics of Einstein's arguing about gravity and mass. Then with this information I further realised gravity is motion differentiation between objects. It is the independent motion providing a different speed while sharing a common centre of attracting that allows a discrepancy to establish mass under specific conditions applying between the two in relevancy. While falling the gravity applies as moving of space that is putting time in relation to the distance travelled. That means there is a speed relevancy between particles in motion and synchronised motion would bring about equal orbit around a shared centre. That is the result of gravity functioning. While the object falls the motion confirms gravity. When motion ends mass sets in and becomes the constraining of the object preventing further motion. The motion is still there but now it is reduced to a tendency to move thus establishing the object mass as the limiting of further motion. Preventing the motion by implementing mass is the resting of objects against each other by resisting the motion to continue, which then is where the mass takes the place of the motion. Where a confronting of objects restricts gravity the action then implements an introducing of the mass as a substituting factor for motion that then replaces motion as substitute for the motion that would be and the mass is providing the tendency of gravity being the motion of space. However mass then restricts motion and becomes motion in a tendency to apply motion. While falling gravity applies and motion neutralizes size, mass or weight. Mass counters motion being when the Earth restrains further motion of the falling object and the moving object is stopped from further movement where mass is then preventing or hindering gravity. This is the result of objects claiming an individual and personal claim to space occupied in a dual or in fighting for their individuality and independence of each other while wanting to be in the **centre of the Universe**. While falling or moving there is no opposition to the body being independent. When the motion seizes the falling object remains individual and still tends to move while Earth individuality resists further movement of the falling body's movement. Further movement is disallowed as other material fill space that falling body wants to lay claim to. The only manner to remain independent by the falling object will be to relinquish to motion in the securing of mass as a substitute to motion where it then finally comes to rest. Mass then sets in not causing the motion but substituting the motion and from that motion restriction becomes resistance that becomes mass. While falling the object is experiencing gravity because the object is in gravity but when on the soil the object experience mass which is the restricting of gravity or motion by other space filled with material. It is a fight of objects to secure and retain the position they have of being in the **centre of the Universe**.

Moreover, I then came to another conclusion of equal importance. When any person is standing on any place anywhere, while viewing the Universe, that person is filling the **centre of the Universe**. Let's get more personal. When you, the person that is reading this, are standing at night and are looking at the Universe you are seeing the Universe from the position that one only can have if that person is filling the specific spot in the **centre of the Universe**. All the light, every single beam that ever left any destiny at any time acknowledges this fact. You are the most important person in the Universe because you are holding the most important position in the Universe. All the light that come across and travelled all of the vacant space from any and all possible positions in space runs directly towards your position using a straight line towards you where you are filling the **centre of the Universe**. Not excluding the effort of one photon, all light is heading to meet you where you are in that centre spot and not one photon will pass you by. Not one photon dare miss you because if they do they miss the effort that all light has to accomplish and that is to locate you as the person filling the **centre of the Universe**.

Should you decide to shift your position to any other place in the Universe, you will shift the **centre of the Universe** to that location as well. If you install a camera on Mars, the light is obliged to acknowledge your relocating the **centre of the Universe** at your will to reposition you're being that **centre of the Universe**. All the light that ever left its destination crossing the vast spaces of the Universe, excluding no particular light, travelled all the way just to find you filling the **centre of the Universe**, right where you are. By you're standing anywhere, you fill the **centre of the Universe**, and the entire Universe admits to that because all the light comes to meet you there. If you shift from the North Pole to the South Pole you will shift the **centre of the Universe** because all the light travelling throughout the Universe will find you where you then moved the **centre of the Universe**.

The light left its destination billion years ago as it travelled through space at the speed of light anxious to acknowledge you're being in the very **centre of the Universe**. No photon will be able to pass you by where you are in the **centre of the Universe** because all light is heading your way from their starting positions. No wonder every person born has the idea they were born to fill **centre of the Universe**, which we do fill. The Universe is spinning around you or I, which is filling a centre where all motion is connected. That is the Coanda effect on the utter-most grandest scale imaginable; nevertheless it is only a manifestation of the Coanda effect. It implicates gravity as wide as can be... Some things mathematics is able to explain but other explaining goes beyond mathematics. Try to explain mathematically the colour of the sky being blue in a clear Sunny day and changing to black when nighttime falls. Do the explaining in mathematics to a blind person that had no vision since birth in such perfect mathematical detail that would allow the person afterwards be able to explain the difference between blue and black to other blind persons by using only mathematics. Some aspects of the Universe go beyond mathematics and some even go beyond words. It is our task to find space, to find time and moreover it is our optimal task to find the Universe. We have to see what is solid, what is liquid and what causes gravity. It is therefore very important to see what is a solid and what is a liquid. Again we must put culture in the background and value the cosmos by using cosmic standards. Everything that moves, do so in relation to another that is relevant stationary is a liquid notwithstanding that it may or may not contain material. It is a liquid nevertheless. Everything that is relatively stationary is a solid in relation to the liquid that moves about the solid that anchors the liquid by gravity. Gravity **is to move or apply the intension to move** space a^3 **at the** distance or relevancy of **k** while T^2 is the time it is going to take to **apply gravity** or move the space filled with material space a^3 at the distance of **k** in the time period of T^2. That confirms Kepler's attribution to gravity where according to Kepler space a^3 is equal to the movement T^2 (time it takes to move) at the distance **k** from the centre specific.

Do not frown on this being in the **centre of the Universe** or regard it too lightly because from that I can prove life being eternal and life being part of the other side of the Universe. That is not part of this letter since that I do prove in another book with another title under the article heading **Man – in- Motion**.

I then subsequently reviewed my vision I received from the vision Einstein received and applied such a vision on the findings Kepler received from the Cosmos. It puts all aspects of gravity in the Universe in new dimensions. But the visions formed the beginning because the visions unleashed many new questions. If gravity is motion, what causes motion? What stops motion? That answer is in the Black Hole. In truth the explaining of the Black Hole is as complicated as the Universe may represent and as simple as the cosmos truly is. If a star is about fusing atoms and with such fusing of atoms is thereby growing, what happen when all the atoms fused into one all collective atom in one already all—atom-accumulated star? What is the gravity if the star has melted all atoms it had into one all-inclusive atom and this all-inclusive atom is providing all the gravity that the star had when the star still had massive volumetric space? If all that space that once filled an entire giant star fused into one specific space less centre holding singularity 1^0 then the enormous gravity is applying to the centre of such a non existing space-less atom and that entire enormous force has been secured in the space less than that which one atom holds. In that case the atom would then show a force that would pull the surrounding Universe flat. The purpose of fusion is to reduce space and magnify space less ness inside the sphere. Where does the gravity of the star end when all the atoms in the star became one giant atom by fusing all atoms into one nucleus? Gravity is smallest where space is least. Where space of an entire massive star is left in the size of one atom the gravity coming from that will pull the Universe flat at that point. However fusing means freezing together because only by reducing the heat can the removing space be accomplished and by reducing space to the point of freezing material permanently together is getting material frozen permanently. That means the Sun and all stars are as cold as they can get and not hot!

The motion is a product of heat and the motion produces a cold that sets in as the motion comes about. When the object moves it moved because the heat became excessive but by moving it is doubling the area it holds by halving the area as it divides the space between where it goes and from where it came. As soon as the motion halves the space used to move the halving of the space halves the heat which produces the cold which brings about the containing or the reducing of the space. Then with the motion completed the contraction retains heat where the heat increases to bring more heat rising that leads to more expanding coming about and the cycle once more repeats.

I am not getting into that argument now, but because of the size the Sun has and the size moving through such distance the Sun in its very centre is the coldest place in the solar system and outer space is so hot it is over boiling. That is why outer space is expanding. It is because the heat rises as much as the stars reduce the heat by containing the heat as material inside the atoms. Nevertheless, gravity is motion and motion comes from overheating which the motion then produce the cooling that contains the overheating by accumulating the contained heat inside the atom. To do that the Coanda principal is employed and the Coanda principle sets the Titius Bode law in operation and the Titius Bode law produces gravity. By spinning liquid heat around solid space a relation between 10 / 7 and 7 / 10 produces a relevancy that contacts heat.

It is about. Gravity is all in Kepler's $a^3 = T^2k$ and $k^0 = a^3/T^2k$ where then relevancy $k = a^3/T^2$ and in response to keep equilibrium applying $k^{-1} = T^2/a^3$

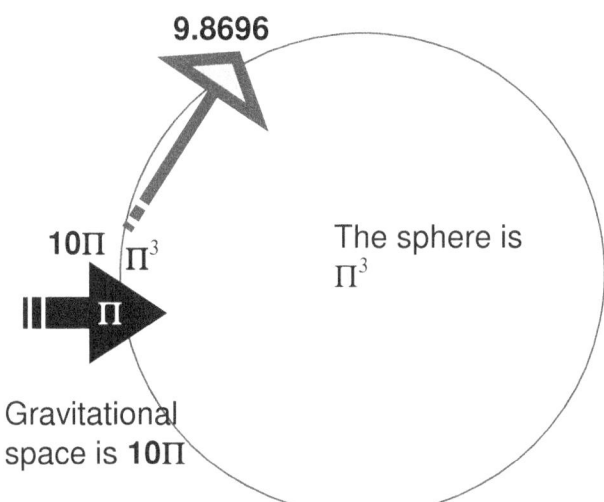

9.8696

10Π Π^3

Π

The sphere is Π^3

Gravitational space is **10Π**

This proves the reality of the Titius Bode law, which too I have to add, the Newtonians put down to a coincidence. This proves that the Titius Bode law is part of the chain that brings about gravity. Most of all this disqualifies mass as having any importance what so ever in the producing of or the conducting of gravity.

This proves that gravity is the motion where space interacts with time to give singularity the significant control it has as not being part of the Universe and yet being responsible for all action in motion taking place in the entire Universe. This is what keeps the top erect while spinning and it keeps the Earth in gravity as much as it built the solar system to a mould that built the entire Universe. The fact that motion brings about gravity in line with singularity must be proof to all Newtonians that their perception on mass has no grounds. Even where those Newtonians are unable to show what brings about gravity even after so many centuries of investigative research while trying and getting no results does this simple arithmetic proves more than all the multitude calculations on their part that proves zero about the manner they promote gravity as being a pulling force.

Matter in relation (part of) to the total dimension of space.
(10 / 7) \ (7/ 10) = 2.04

1.4285 / 0.7 = 2.04 Taking from both orbiting influences
SPACE DIVIDED INTO TIME

(7/10) / (10/7) = 0.49
.7 / 1.4285 = 0.49 Taking from both orbiting influences
SPACE MULTIPLIED WITH TIME

7/10 / 7/10 = 1 and   10 / 7 X 7/10 =1 Therefore not influencing change
THE PROCESS PARTED USING THE ROCHE PRINCIPLE

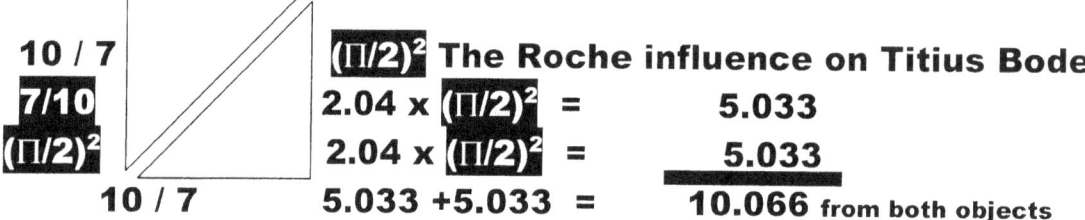

10 / 7
7/10
$(\Pi/2)^2$ $(\Pi/2)^2$ **The Roche influence on Titius Bode**
          10 / 7   2.04 x $(\Pi/2)^2$ =   5.033
                        2.04 x $(\Pi/2)^2$ =   5.033
                        5.033 +5.033 =   10.066 from both objects

SPACE DIVIDE INTO TIME

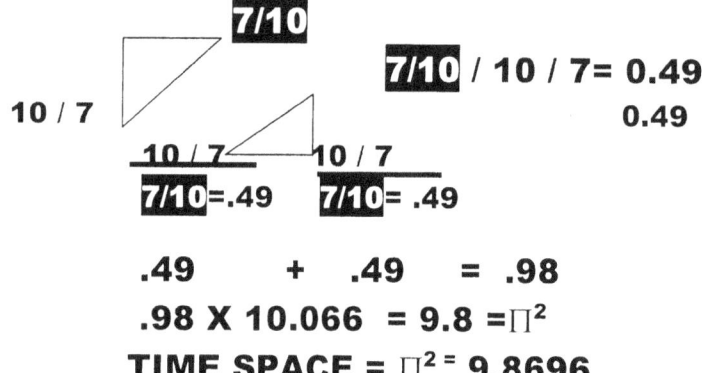

7/10
          7/10 / 10 / 7= 0.49
10 / 7   0.49
      10 / 7   10 / 7
      7/10=.49   7/10= .49

          .49   +   .49   = .98
          .98 X 10.066 = 9.8 =Π^2
          TIME SPACE = $\Pi^2 =$ 9.8696

TIME SPACE =Π^2=9.8696= Space and time in a dimensional implication.

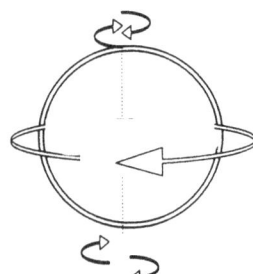

On the inside, there are the seven markers of which singularity is the focus point in the centre of the centre. The markers are representing one aspect of space, which for argument's sake let us call it cold. Then there are three more markers on either side being part of the space but not captured in the space. It is space in motion by the influence of the motion of the Earth.

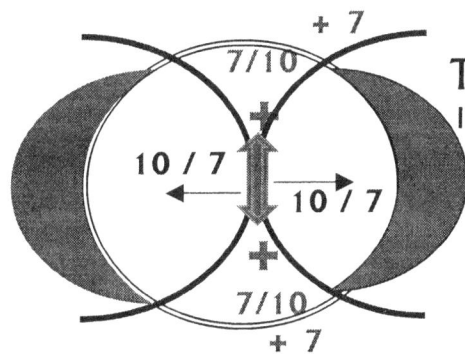

TITIUS BODE LAYER CONNECTING
Inside the cosmic sphere

$7/10 + 7/10 = 1.4$

Singularity in the square of matter

$10 / 7 = 1.42$

Singularity in the square of space

$1.4 / 1.42 \times 10 = \Pi^2$

MATTER HOLDING THE SECOND PROTON COUPLING THAT TO THE NEUTRON TO COMPLETE THE NEUTRON. Due to the influence of the matter dimension on the space dimension, the curvature of space-time comes into affect by dominating outer space.

The Titius Bode Principle is equal to gravity @ $= \Pi^2 = 9.8696$

Proving that the Titius Bode Principle is a product flowing Directly from the growth of singularity forming space-time The Titius Bode principle directly valuating TIME to SPACE $= \Pi^2 = 9.8696 =$ MATTER HOLDING THE SECOND PROTON COUPLING

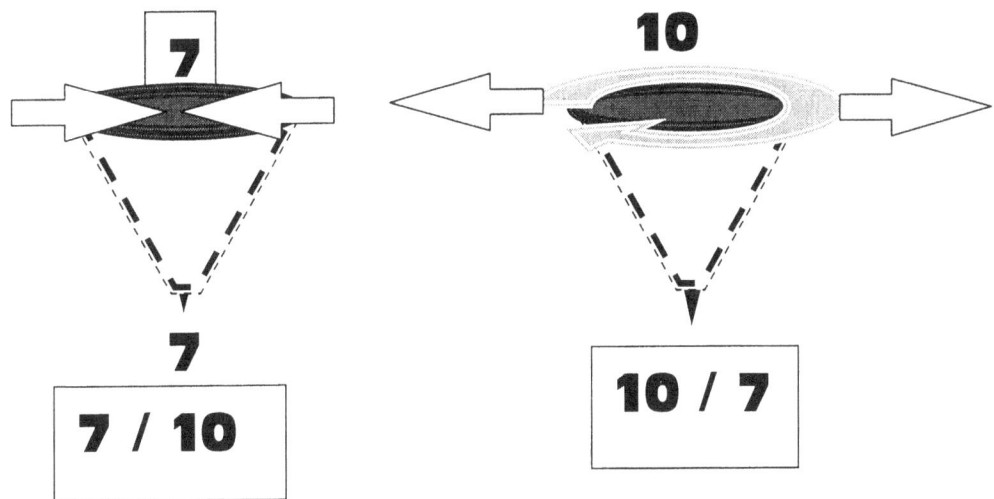

In this maintaining of cross referencing of singularity located in individual atoms providing spin to the governing singularity that maintain structural form in solids, many factors of singularity all form a close knit network and being inseparable as one unit, by the same margin it also is strictly individual to a point of destructing. From the inner or governing singularity outward all is concerned as space-heat. The relevancy in the material sector always includes the governing singularity and the very next one to the inside. All the others do not form any part of such a relevancy to the object forming the relevancy. On the time issue it is only the relevancy forming a connection with the one in question and the governing singularity. All other objects in the line are merely space-time with no value to the object that holds the relation. The fifth object will link in the material sector to the fourth and then directly to the governing singularity skipping or excluding all others from one to there. In the case of say three, it will connect to two and skip one while all points holding singularity to the outside is no consequence to the rotating object.

The spherical positioning layout forming the Titius Bode Principle

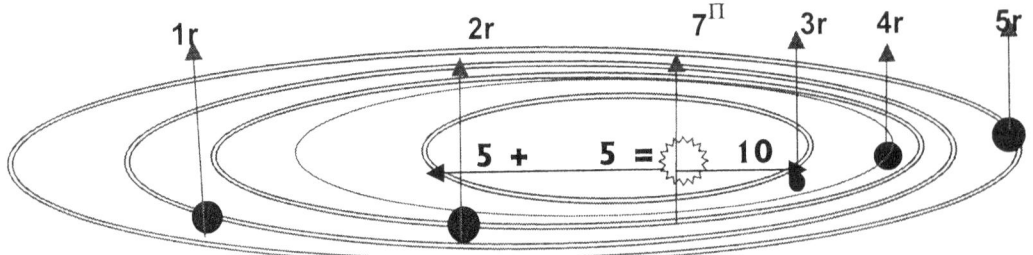

From the matter-to-matter relation in the Titius Bode configuration there are 7 / 10 + 7 / 10 = .7 + .7 = 1.4
From the space-to-matter relation in the Titius Bode configuration there is 10 / 7 = 1.42

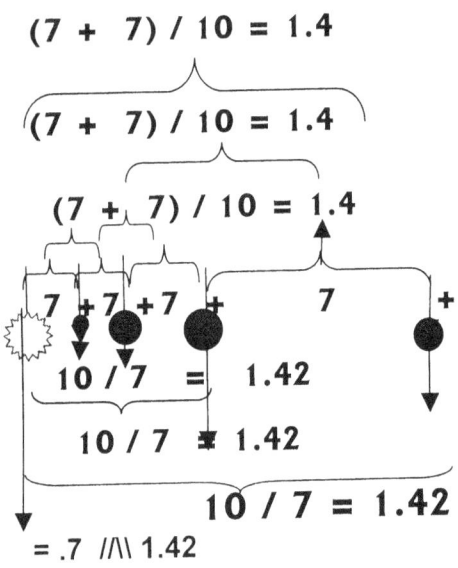

The $5 + 5 = 10$ is a position of dimensions as space loses value to singularity. The 7 that matter diverts in points from singularity may seem as coincidental but is valid. Still in accordance to our perception valuing the number in degrees, it seems coincidental but if it is coincidental, it is nevertheless a figure of diverting proven as accountable in all other calculations and plays a most dynamic role.

The Lagrangian 5 point system results as much from the Curvature of space-time as does the form the Black Hole holds. The Galactica is the opposing equivalent of the Black Hole and has identical but opposing similarities being the five points positioned to singularity. The galactica is generating space and the Black hole is degenerating space.

= 1.4 /Λ\ 1.42. **Because the space-to-matter is in the square at 10 placing the matter-to-matter at a square of .7 + .7 = 1.4 the space-to-matter forces the matter-to-matter to double the distance by number as structures are place father from the mainΠ^0 maintaining singularity.**

1 3 6 12   24   48

Reasons why this does not fully apply to the solar system I give in book # 7.

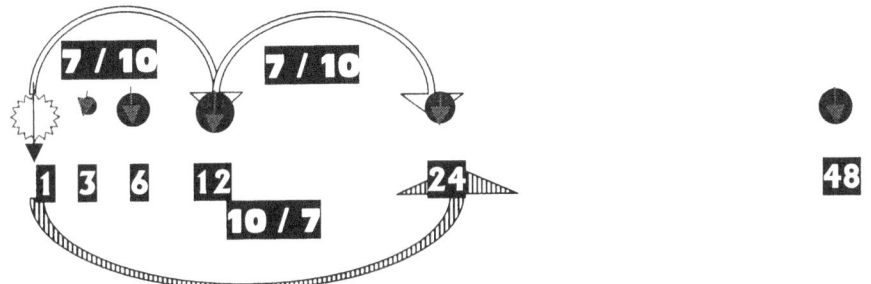

SINGULARITY BY DIVIDING SPACE INTO MATTER AND MATTER INTO SPACE, ANG ALL OF THIS ACCORDING TO THE TITIUS BODE LAW OF 10 / 7 AND 7 / 10 IN CONJUNCTION WITH THE ROCHE PRINCIPLE OF $(\Pi/2)^2$

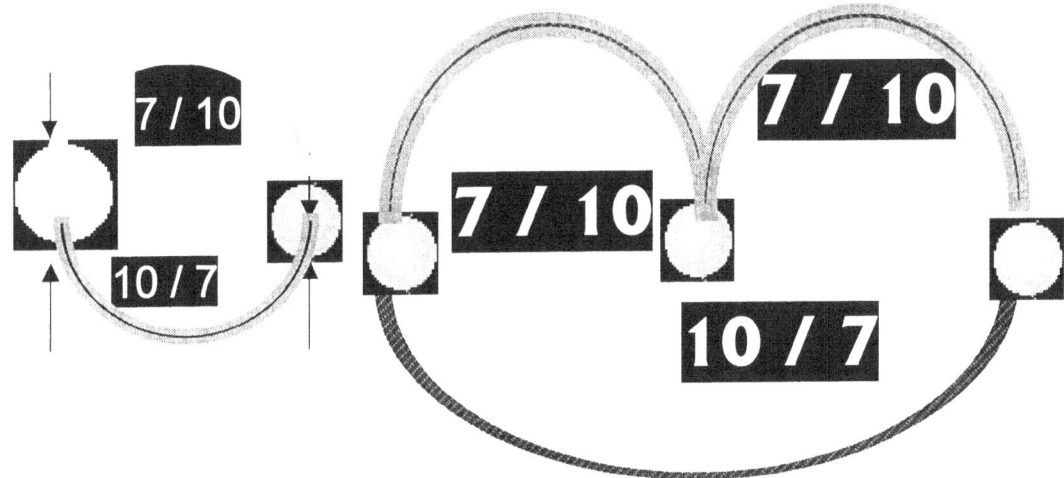

Time started at zero, eternity, whatever you wish to say, as long as you say time did not move at all. Then the command came and time overheated for the first Π^2 in time. That brought space into play.

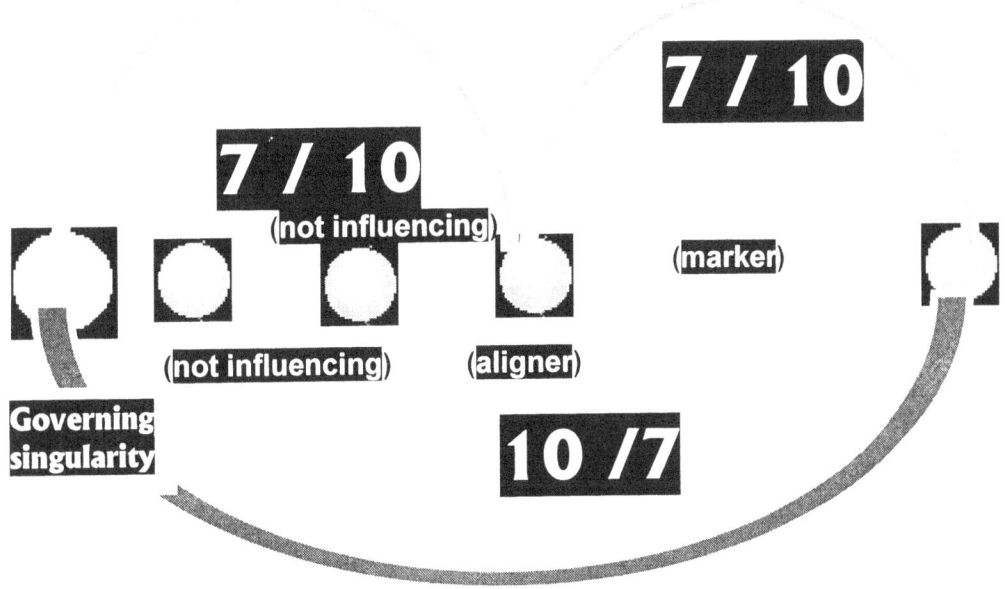

From the dividing singularity only one reference holds a matter value forming the position next to the governing singularity and therefore 7+7 becomes a factor and not all the dividing singularity between the point of reference and the governing singularity. That way the star to the outside takes a position doubling the distance every time. In balance everything in space to the outside of the governing singularity is space be it space or matter that makes no difference therefore that is 10.

The extension of Π is well received as a dimensional implication to matter holding seven positions from singularity and space having four quarters through out the rotation of singularity forming the centre to the five dimensions (one side lost to the cube's six sides connecting to the five remaining sides) making the total sides facing space from the point holding singularity at any given instant at a value of twenty (4 X 5 = 20). Then adding the singularity cross of Π being (1+1) = 2 the relation becomes 22/7. This is crude because in more precise calculations it becomes .91 + 1 = 21.91/7 = Π

The sectors provide individual singularity as a means in sustaining governing singularity by which provision comes through maintaining governing singularity the required spin in maintaining cooling. If this process did not apply, there would be no connecting individual singularity to major singularity. The sectors provide individual singularity a means in sustaining governing singularity by which provision comes through maintaining governing singularity the required spin in maintaining cooling. If this process did not apply, there would be no connecting individual singularity to major singularity

SINGULARITY BY DIVIDING SPACE INTO MATTER AND MATTER INTO SPACE, ANG ALL OF THIS ACCORDING TO THE TITIUS BODE LAW OF 10 / 7 AND 7 / 10 IN CONJUNCTION WITH THE ROCHE PRINCIPLE OF $(\Pi/2)^2$

This ratio there is between the governing singularity and the marker (innermost planet and planet marking a position according to the Titius Bode law. There are reasons why some diverting stems from this but in other books I explain that using much better detail

From the orbiting structure (planet) aligning singularity only one structure the very inside singularity applies as a position of reference and that is reference to the distance applied between the points in governing singularity. From the Sun (governing singularity) the matter marker is 7/10 = 0.7 with the only one other forming a marker 7/10 = 0.7. The two form 1.4. From the Sun (governing singularity) the outer planet forming the marker in search of position holds space in the square 10 / 7 = 1.42 in aligning with the 7 forming material of the Sun . Therefore there are two sevens relating to ten forming the material positioning of the structure in orbit and from the governing singularity all outside the Sun is the square of space (ten) aligning with one particle (seven) and not one of the other structure to the inside or the outside holds any value. Because .7 + .7 = 1.4 and 10 / 7 = 1.42 the distance doubles every time there is an aligning of three orbiting object. In this there is definite proof of influences coming about between particles sharing gravity. But then again the entire Universe shares gravity and as such then all will influence everything.

Mercury	Venus	Earth	Ceres	Mars
4 − 4 = 0 14 − 4 = 10	0 + 7 = 7 20 − 4 = 16	7 × 2 = 14 32 − 4 = 28	10 × 2 = 20	16 × 2 = 32

Jupiter	Saturn	Uranus
28 × 2 = 56 56 − 4 = 52	52 × 2 = 104 104 − 4 = 100	100 × 2 = 200 200 − 4 = 196

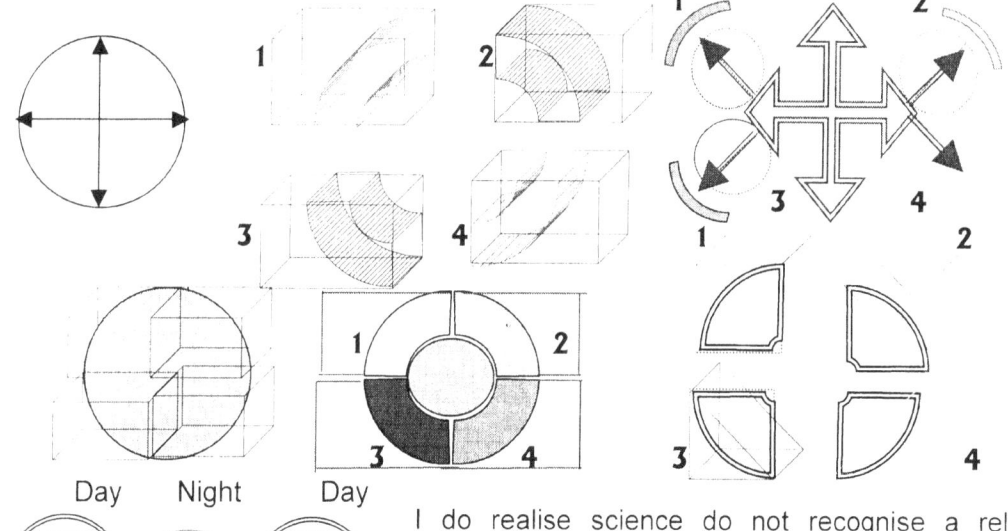

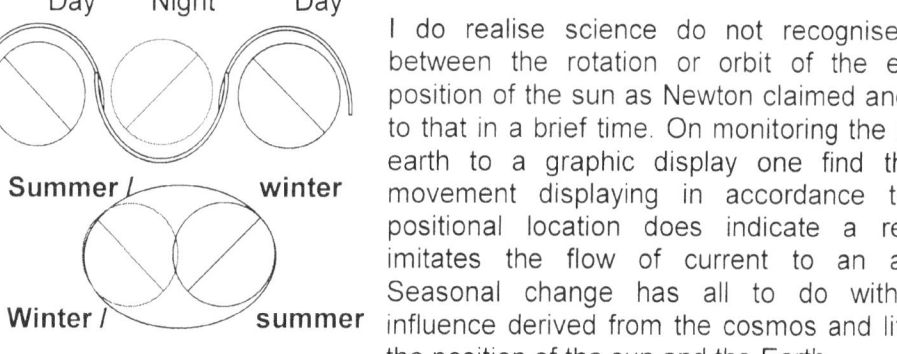

I do realise science do not recognise a relevancy between the rotation or orbit of the earth and the position of the sun as Newton claimed and I shall come to that in a brief time. On monitoring the rotation of the earth to a graphic display one find that the earth movement displaying in accordance to change in positional location does indicate a relevancy that imitates the flow of current to an almost exact. Seasonal change has all to do with the graphs influence derived from the cosmos and little to do with the position of the sun and the Earth.

It is the position singularity holds in relation to the Universe and the Milky Way forming currents and seasons moreover than the Sun shining brighter or not. The Sun in size over dominating the Earths in comparison disqualifies any positional influence that can alter the Earths heat standings. Through shear size the Sun can shine at the top and the bottom of the Earth simultaneously without effort from all normal possible angles. I show a relation between singularity in different positions maintaining seasons and north/south polarity, not only as far as concerning the Earth but also outside influencing polarization. This has to do with the second position singularity holds in accordance to matter and space and is an "*electromagnetic*" (used for the lack of a better word) sustained positional opposing derived precisely from the graph in the manner when calculating electricity.

In this it is clear why the Titius Bode ([10 + 10 + 1 + .991] / 7) and the Lagrangian 5 \\ 7 systems part their ways when applying the different processes they hold. With all the differentiating, the observer must also consider the dual massage that light uses in travelling through the vastness of universal space. The thought of nothing is just what it is, a thought of nothing and although it is in the human mind common nature to present nothing as a value in the recalling of something, nothing is a presentation of the figment in the human mind. There can be no number such as nothing and that was (possibly) Newton's biggest error. Nothing represent non-existing and that is just what nothing is, it is non-existing.

The Titius Bode influence in a manner that on the one side holds the matter-to-matter relation of 7+7/10 whilst on the other side during the same time holds the space-to-matter relation of 10/7 forming equal and opposing values. From this the orbits of cosmic structures are always oval favouring the singularity dynamics of the one structure at one point and switching the favouring to the other structure on the opposing side. Because the structures can never be equal in size (singularity will not permit that where the Roche principle will intervene) the shape is always "off centre" as well.

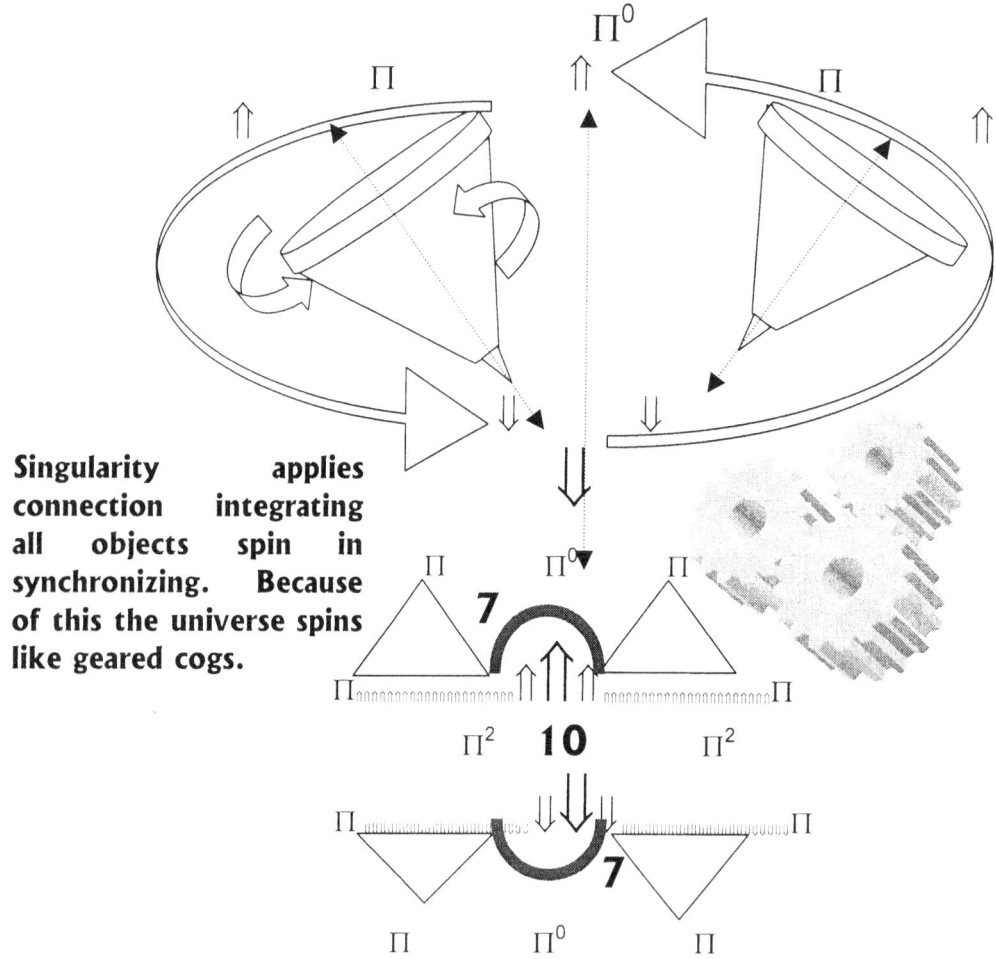

Singularity applies connection integrating all objects spin in synchronizing. Because of this the universe spins like geared cogs.

The ten dimensions I named the atomic relevancy is also showing the double value of singularity as singularity extends into as well as beyond space. The atomic relevancy is $(\Pi^2+\Pi^2)(\Pi^2 \text{ X } \Pi \text{ X } 3) = 1836$ that is the mass relation between the electron (3) and the proton. Proton = $(\Pi^2+\Pi^2)$ Neutron = $\Pi^2 \Pi$. The atomic relevancy holds the dynamics of singularity control. In the ratio and dimensions we find in the atom, all space-time derives from the atom, whatever the atom is.

It started with a dot, because that is the only form, size and dimension mathematical logic will allow our brain to accept. From the one dot had to come a second dot and a third dot. The dynamics of such a dot is smaller than we can understand because such a dot is in negative relation to what we see Π to be, and the deeper we delve in finding the smallest fragment where space started, in the spot where time is still eternal as much as we can accept eternity to be. This we find in the aligning of planets. Where the one dot from which the aligner stem becomes the reference too the distance applied between the aligner and the original dot, or governing singularity or structure in charge of holding position to all orbits following.

The reason why we should first locate the spot is because we can only work from that point forward. By working forward we have to work backwards to locate where we are heading. The cosmos started at a point and where such a point is, we will find the Universe. Every one knows where the Universe is, because we can see where the Universe is, but if we can see where the Universe is, then we should find the centre of the Universe in that spot. Einstein theoretically positioned the point of beginning at a place he indicated where singularity should be.

With the cosmos the size it is and space so large compared to our smallness we have no chance in finding the centre of the Universe. The Universe started where singularity is and singularity is the sure indicator of the Universe. With all spinning objects holding singularity we then have located singularity in as much as finding the centre of the Universe. The Universe started with a dot forming. That answer arise from taking mathematics back to a point of being the smallest possible position, far smaller than we may be able to calculate form.

My approach might seem unconventional but through the abandoning of the accepted, it enabled me in locating the precise location of a universal singularity forming a connecting basis of the Universe (this I say with some degree of confidence). The smallest figure there can be must be a dot. The dot is the only form that leaves all the options open to extend in any and in all directions should the opportunity arise. The only mathematically sensible option about extending a line from the dot will be non-bias progress in all directions equally in order to give a meaningful flow of mathematical equilibrium.

The Pythagoras mathematical principle is the proof and that I explain. The obtaining of singularity is in my rejecting of nothing by replacing it with something being the dot. With the clepsydra or "water thief" Empedocles deducted that air was composed of innumerable fine particles, braking the thought that what we now know is air, was also believed to contain nothing being altogether a space filled with nothing until proven to be wrong so many years ago. Never did science take the lesson learnt back then to the future and out onto outer space. If there is space, there cannot be "nothing" as space is something. The claim becomes obvious when observing the connection between the half circle, the straight line and the triangle, which could also promote all the qualities lurking behind the pyramid. Consider the connection between 180^0 sharing and then one may realise much of the pyramid mystique becomes less spectacular in considering the very basic in mathematics being the Law of Pythagoras on which all mathematics are based. Once the water thief was eliminated by some human intelligence the matter was left at that. Nothing shifted out to an area we think of as outer space. In outer space we now find nothing. There is nothing but an atom here and there and even the atom is covered in nothing.

I wonder why the nothing landed there? Could it be that the reverse came about and because there was no visible "water thief" the very limit of man's suspicions came into practice. Man has always been extremely good in flying from one outer edge to another and if the water thief proved something was present, then the mere absence of a water thief must therefore prove that nothing must be in outer space. But what is space as such. What can space be, because with explosions we can clearly witness space created from heat. Our culture prevents us from admitting our vision, but the release of heat produces a *"shock wave"*. That *"shock wave"* is nothing less than space created from heat released. We have to brake free from culture of the past and a rigged mind set narrowing our vision.

Einstein's Critical Density lacks the accepted matching facts we need in proving the critical mass factor. But our inability in securing such required evidence defies the most basic logic. It seems all new evidence we receive from outer space is disputing all Newton laws findings that disprove Einstein's Critical Density as the answer. The Universe will not reach a point of contracting, not withstanding whatever dark matter astronomers try to locate in the vast space.

Why would the expansion turn around and do a reverse by going back to where it came from. Consider the momentum alternation such a change will bring about.

The Sun is not a gas-filled sphere holding hydrogen in its "natural gas" form, but it is all fluid and is in a liquid form where singularity is liquid- freezing hydrogen at 6500^0 C while outer space is boiling over at $- 276^0$ C. This book explains the Roche limit in the practical sense… when applying cosmic laws instead of improvising cosmic laws uncovers that reality then becomes awesome. It becomes clear the Universe is as much expanding as it is contracting and contracting by expanding. As there is no hot or cold, no big or small, no grand opposing but relevancies in ratio to one another. If you do not believe me, then believe your eyes when looking at the picture. What ever the Sun is it is fluid falling into fluid.

Consider the time it took from 10^{-43} to 10^{-5} seconds to create a cosmos the size of a neutron. Compare that to what is happening now and see how many events took place by the creation of every lepton and every hadron and it is true that that period took longer to complete than it took the Universe to create the solar system. The flow of light through the density that space produces heat gives the speed of light the relevancy of time in space. The thicker the "soup" of heat is that space forms, the longer it will take light to cover a distance. It is very important to note that the speed of light is a relevancy between time (seconds) and space (kilometres). The speed relies completely on the value **k** holds on space –time.

The speed of light is forever a constant but the constant is part of the relevancy of space-time

If one looks at the transmission of sound, it too depends on the relocation of matter, but to a very small degree, and in this process lies the transmitting of sound. To make the error of judgment in confusing the process with the breaking of the Doppler rings are quite understandable.

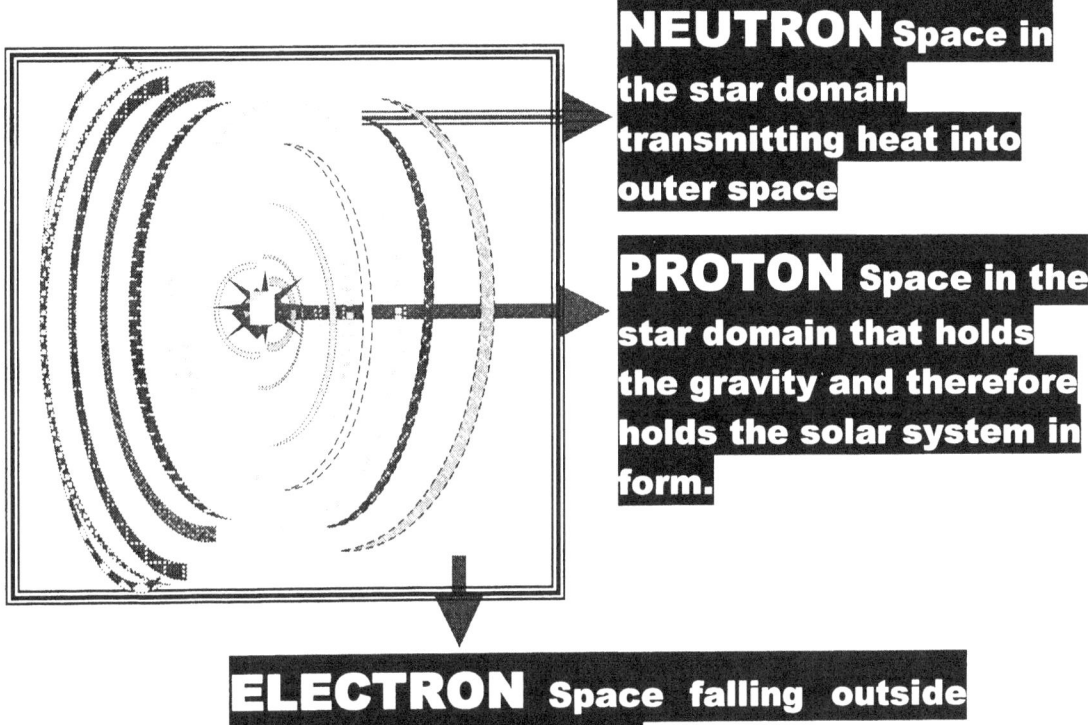

NEUTRON Space in the star domain transmitting heat into outer space

PROTON Space in the star domain that holds the gravity and therefore holds the solar system in form.

ELECTRON Space falling outside the domain of the star

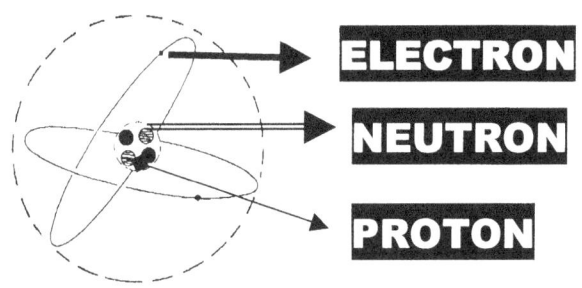

ELECTRON

NEUTRON

PROTON

The Universe connects in a way Kepler established through his relevancy theory. Those not convinced answer this: where would the Planets be if not for the Sun securing planet positions. The relation proves the ratio of one in all cases to be valid. It proves much more than merely connections at liberty of holding positions where ever the randomly opportunity placed the structure. The structure does not come closer by a pulling and tugging. Kepler's figure must still be around and by repeating the task but this time made much easier with the help of computers and telescopes of magnificence compared to those which excited the likes of one Tycho Brahe in his time. Science should become serious about science and not about self-protection and self-preservation. I found on all and every campus I went that any remark about Jesus Christ supposedly making a mistake generated immediate interest with even the most adhering Christians coming to hear the argument. Making a remark about Newton making an error gets you marched off the campus by security. Why not test Newton's $F = G\ (M.m)/r^2$ from figures Kepler left us and see how far did planets shift closer. I guess this will again make this book as successful as the others with my openly criticising Newton and Newtonians but Universities are not about knowledge but t about protectionism. Universities protect their own without any willingness to test that which it protects. It should be clear in confirming that the basis on which the entire world science union is founding all their policies and beliefs are correct and not only that, how far did the structures move closer. From that we then can see what we are waiting for and how long before the big solar clashing will begin. The absence in they're just mentioning such possibility confirm to me they know as well as I do there is no tugging and the Universe is in synchrony more than any person may ever be able to prove.

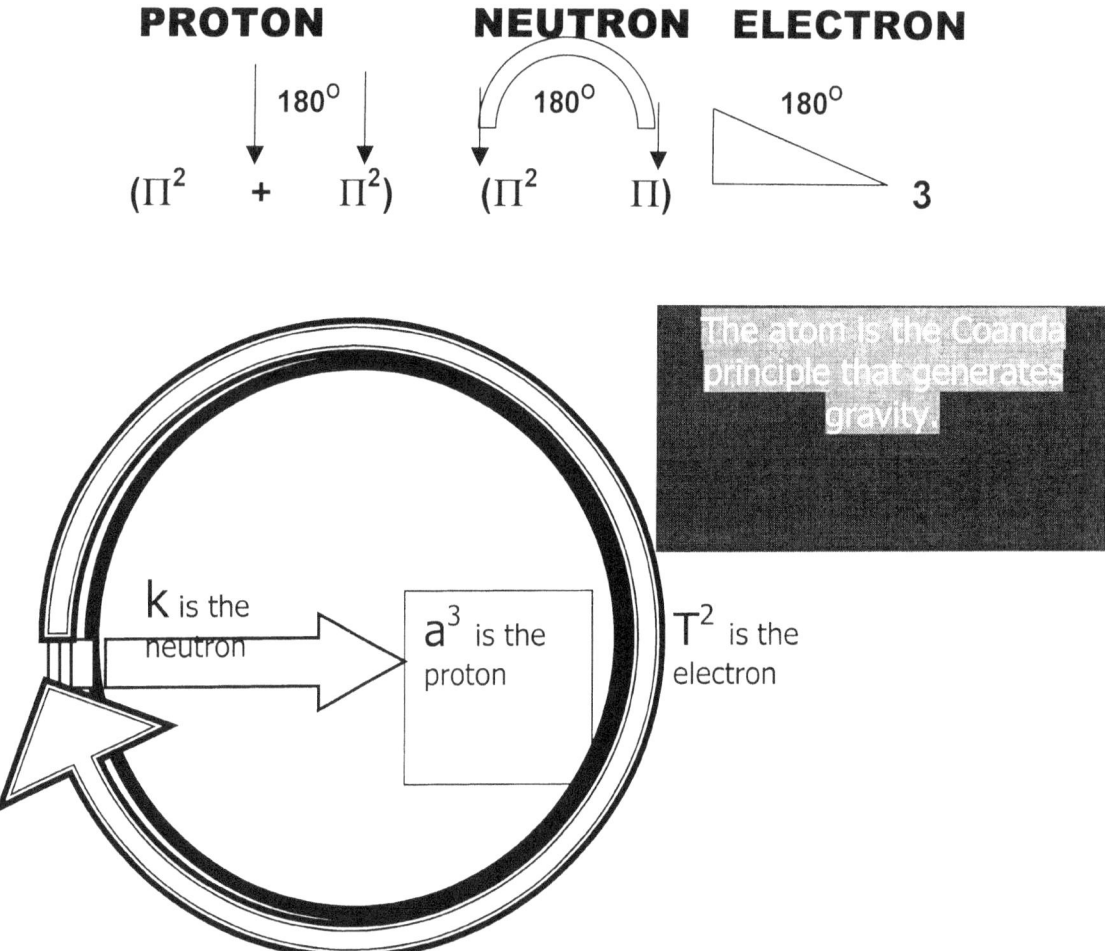

PROTON   **NEUTRON** **ELECTRON**

180° 180° 180°

$(\Pi^2 \quad + \quad \Pi^2)$ $(\Pi^2 \quad \Pi)$ 3

The atom is the Coanda principle that generates gravity.

k is the neutron a^3 is the proton T^2 is the electron

Everything in the cosmos is moving, either by own individual accord, or under the influence of some other singularity dominance. In explaining we return to the top.

When the top is in a state of motionlessness on own accord it is everything but motionless. The motion it adapts are synchronised with the Earth in harmony with the solar system and according to the greater picture of the cosmos. When an energy source not related to the cosmos called life intervenes and energises the top's motion, the singularity in that top suddenly jumps to life. By adopting a rotation energised to an unnatural state of energising because of life's intervention, the singularity of the top is not in charge but as it applies more and more energy, it will begin to find a means whereby it can escape and apply individual singularity as the top starts to separate from the singularity the Earth holds. The singularity holding the Earth would then allow the singularity of the top to rotate within a specific band where that a specific band of being active before the Earth's singularity will start to destroy the singularity in rebellion. The top on the other hand will try its outmost, when the singularity it holds gets by individual spin is too strong to remain in domination of the Earth's singularity. The motion of the top is an attempt to begin applying an individual singularity space-time defying and standing apart from the Earth's gravity. That action we see as the top starts rotating in a manner where the top does not align with the Earth's singularity, but est6ablished a driving singularity independent from the Earth's gravity. With the adding of spin, the time the top holds becomes unrelated to the time the Earth holds and the top will start a campaign too escape from the singularity domination the Earth has on the top. When the time or spin of the top exceeds the limits the Earth places on the top, the top would emerge by trying to escape from constrains placed by the Earth. The view I represent at this point is known to science for almost as long as science knows mathematics.

Not long after the law of Pythagoras was understood where Pythagoras introduced mathematics Eratosthenes of Syene made as big a discovery as Pythagoras did. But in the one instance the world took notice because the world could see and understand and the other instance the world disregarded the findings because the world did not see what the implications was. The same apply to aircraft flying and when the aircraft wishes to escape the

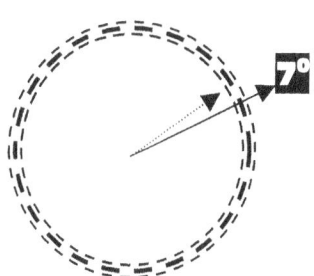

earth's singularity hold it has to comply with the laws laid down by the earth. The seven becomes as big a part of the concept as does Π as it all interacts

If we wish to find the future we should locate the past. If the cosmos is contracting, where to is it contracting? The direction of contracting must be in the opposing direction the direction of expanding. If we wish to locate the past from where the cosmos came and through that in what direction the cosmos came, we must take an effort to backtrack the direction it came from. Should the argument come about that all came from nothing, then everything either still has to be at nothing, or our understanding of nothing leaves much to desire. Nothing means not existing, not being, never found and unable to produce any multiplication of any growth.

The above questions, but mostly the fact of what is more nothing and what is less nothing draws me to the realisation that there can be no such a quantity in space as nothing because even space has to be something. Heat expands as the levels rises and clearly it is for any one to see that the releasing of uncontrolled heat creates space, which is no more evident than by releasing a nuclear explosion. The wind is shock waves, but what is the shock wave other than new space coming into prominence. In that way it is clear that releasing heat brings about the expanding of r as part of the sphere forming space. Hubble proved the Universe is expanding. Then by backtracking we have to set about reducing the sphere constituting the expanding Universe. If r in the circle is growing we have to reduce r to backtrack.

When the circle reduces, the value located to r will become implicated because r determines specific size. Not so in the case of Π, because Π in the true sense only indicates that the circle is a square without corners and therefore Π dictates form and not size. By reducing size only r comes into contest and will point to such reduction. By reducing the circle radius r by half continuously will lead to an infinite small circle but Π will remain because the circle as a form remains even being infinitely small. In the past, and even in some quarters today, science is on the search for the 100% efficiency machine. That theory runs on the surmise that a machine can drive as an output delivery without receiving input of energy. A few hundred years ago many Kings were fooled by such notion and some scientists truly spent a life in honest search of just such a device. Mostly the accomplishment came from cheats that very well new their machines were not up to the task, but in fooling a rich investor, brought about wealth to the inventor. As science progressed the no input giving all output machine became less and lesser a feature of the honest inventor. But the idea does not exclusively come from crooks finding a way to cheat the world.

The practise of receiving without giving comes from science in the form of physics. It is physics taking the world on a wild goose chase in the way physics presents the cosmic motion. Physics propagates that the cosmos is all about running without input driving energy. The cosmos is all about wasting matter to a supply of motion. This idea prevails even after the world of science saw clearly in the past that there could be no such machine anywhere. Even the cosmos must be a machine driven by an input and an output. It is the input / output driving energy that must be located and the driving ability we have to locate. Science holds the mass drawing power to prominence, but what if it is not the drawing power of mass that holds prominence, but it is the reducing or contracting of space that is the driving motor behind the cosmos. All energy we humans at present use to accomplish matter motion, holds some form of heat redistribution. Even electricity is a form of pure heat. I say that in mind of what applies when the energy of electricity becomes over abundant and the machine overheats. By overheating it means that the motion the machine creates comes about from heat control and precisely planned heat distribution.

When I realised that it is not me that is drawn towards the Earth, it is the space in which I find myself that reduces, and that produces the effort bringing me closer to the Earth. The formula $F = G (M_1.m_2)/ r^2$ suggests driving, moving in a direction and contracting. It suggests the reducing of space and not merely drawing or moving closer. When looking at any machine in practice, the machine draws power from space reducing whereby heat increases. Not releasing the heat to form space will lead to the destruction of the composition forming the machine. There is no form of matter, or element strong enough to resist matter deformation brought about by overheating. Having this in mind that matter does not resist heat, it is of importance to recognise that it is heat that is allowing space to give matter form. Looking at the manner in which energy is utilised it is space and heat forming matter allowing motion that allows work to achieve value.

At this moment science is all about a body falling where the two bodies are producing a force whereby the bodies draw one another closer. The bigger the mass, the bigger the drawing that comes about from the force unleashed by the mass of matter. The idea about this practise was phenomenal in 1602, it was impressive in 1802, but it is really ridiculous in 2002. Why would Boron form a solid having 5 protons weighing 10.811 g / mol and Argon a gas having 18 protons weighing 39.9 g / mol, but the "heavy" element with the biggest drawing power is a gas and the lightest element is a solid. That denounces the contracting force theory. The way we compile and use energy must be in a similar manner to the way the cosmos uses energy distribution. **We humans can create nothing, but nothing is all that we humans can create**. The rest of our achievements are by duplicating whatever nature provides. To establish what drives the Universe except for blaming some medieval magical force coming from nowhere going nowhere we have to find what drives us. The energy we use in all forms is producing heat in space by either converting space to heat or heat to space. Explosions are about converting heat to space. Compressing is about reducing space to heat. That is all energy composing work and is the only method of producing energy notwithstanding the immeasurable many names we use to express the same function in different forms.

Arriving at the question about locating the space and time forming the centre of the Universe one has to realise the centre of the Universe are in every singularity forming matter be it is big or small, size carries no significance. It is the impartiality of singularity that is claiming the value and not the differentiation of matter. One must realise there are no big / small or hot /cold or near / far. It is all relevancies between matter claiming space and space is heat in a turnabout manner. Every aspect in the cosmos is locked-in Universes, sealed off from other Universes and inclusive or exclusive depending on singularity holding relevancies relating to one another. The relevancies rely on inter dependence and inter linking, but there are no differences according to human sizes or standards. Accepting that principle unlocks the "so called mysteries" of the Universe and brings about clear understanding. It is all about accepting, acknowledging and interpreting the role singularity maintains on matter.

One should not try to focus on an image of such a spot or dot because there is no image. The line dividing the cosmos and that run through every particle, no matter how large or small is beyond our vision. Such a small line, so small it is not even noticeable is large enough to part the cosmos into sectors. It splits the biggest there is into particles and we are not even able to notice the precise location of such a split. In truth there is no top or bottom that we living in 3D can see. We shall have to use a general conception brought about by intelligence. Your intellect tells you about such a spot, but that is all because that spot is on the other side of the Universe (quite literally). From the centre of the dot there is a top and a bottom spot. From those points there is connection with four quarters. That produces six connecting points that are all aligning to the centre. Because it serves big and small, hot and cold equal and alike, and it is the smallest cutting the biggest into equality, size is of no issue. Size is what man makes of it. In the Universe there is no size in hot and cold, large and small. For the smallest there is, it is serving the largest there is equally.

Our instincts, our logic and our calculating process all indicate that the sphere holds a centre point from where six evenly positioned point's position matter to be. Using The formula $F=G (M_1.m_2)/ r^2$ it indicates to a force pulling objects closer, where each force is coming from each centre point the body in question has. The contraction must commit the two bodies towards a point in each case being spot on in the middle, not withstanding what direction the force is applying, the body will draw to the centre. If the Universe spins around a centre point holding singularity, and singularity confirms the centre of the Universe, then every particle holds the centre of the Universe making the number of universal centres immeasurable many, and every atom and sub atom particle presented outside the atom in smaller bits, are all not pieces of the Universe but they are a Universe surrounded by many Universes. If every atomic particle no matter how small is holding the centre of the Universe, then the gravity is coming about from that point because that is where the gravity applying in the Universe is applying contraction.

It then is the atom in the most centre part where space and time meets singularity, that Einstein found a Universe collapsing to a single dimension, and every atom at a point post of the proton where gravity initiates in according with the proton dimensional colas of $(\Pi^2 + \Pi^2)(\Pi^2 \times \Pi \times 3) = 1836$

See the fluid push out of the Sun where it lets the Sun seem to be a bowl of liquid, as the liquid is spilling back to the Sun and not escaping into the cold of outer space as released heat should do. The liquid heat squirts into the cold of outer space and then it falls into the bowl of liquid that is the Sun . The inside of the Sun is not gas but it is fluid.
In all of nature there is no **NATURAL GAS** as much as there is no **NATURAL SOLID**.
 No element is either a gas or is a fluid or is a solid. We arrange the elements in such a manner, but that is only applying to the situation the earth grants the elements.

When an element freezes it is solid notwithstanding...

When an element melts it becomes a liquid

When an element boils it is a gas again notwithstanding..

 Hydrogen is as much a liquid as iron is a gas and neon is a solid. It depends on the element relating to the space/heat in the circumstances surrounding the substance at that very precise instant in time. We have to stop telling the cosmos to show us what we wish to find and start accepting what the cosmos is telling us to find. The culture that I am referring to is all about **nothing.** At present we find that there is something we think of as nothing in outer space. Because nothing is what we wish to find and nothing is precisely what we are getting because we think of outer space as nothing. If you accept the cosmos to be nothing, then please define nothing to yourself and find the definition in the cosmos.

The liquid the Sun has is the driving force that creates the duplication in motion. Without such liquid heat the Sun would become stationary and only depend on contraction while the contraction then passes the motion onto the heat in outer space. While the Star is in liquid all motion comes from the accumulating spin effort of the combined motion all elements together accomplish. In the case where the star is still in liquid the heat is stored in the atoms and as the star develops the heat transfers to the governing singularity, which makes the star immobile as the governing singularity takes charge of the entire star.

The formula of F = G (M_1 x m_2) / r^2 only apply in a very specific range, and at a very determinable point the formula does not effect objects in the air. After such a point one will find satellites able to orbit, be it art a definite pace that matches the rotation of the earth. Still...below such a point (B) orbiting objects will come crushing down to the earth.

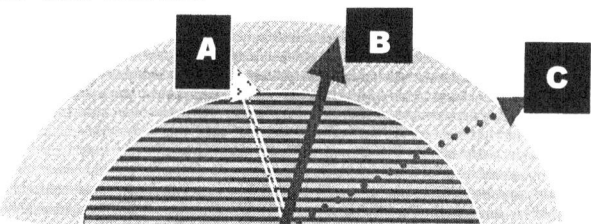

From point (B) to the earth Newton's formula apply and from point (B) upward Kepler's formula apply, but my pointing this out brings about all sorts of annoyance concerning academics. It must be clear to all persons that there are a big difference between the applying of Newton's **F = G (M_1 x m_2) / r^2 and Kepler's $a^3 = T^2 k$. When the objects reach some point they will drop to the earth and when that happens, mass do not play a part in the speeds they come to reach.**

When examining the case where two balls drop vertically, gravity, as a force does not apply and therefore gravity does not come into effect because there is no difference in speed or duration.

With out any apparent reason the formula is substituted with the following formula:

g = G(M . m) /r^2 where:

G = the gravitational constant,

M = the mass of the body,

M = the mass of the lesser body

r^2 = the radius between the two bodies.

Let us take this formula back to the accepting of the Big Bang and find sensibility amongst a lot of confusion that I can see.

There was a beginning that saw a radius between objects so small that the size will never again be repeated. The diameter of the particles were also next to nothing but that should not be a contributing factor surely...the main focus point is that particles were as cramped as it shall never again be repeated.

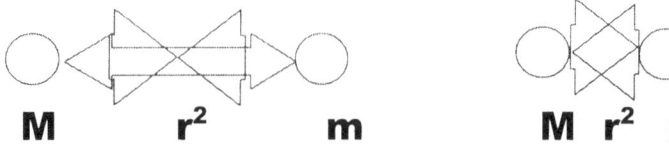

$$\textbf{M} \qquad \textbf{r}^2 \qquad \textbf{m} \qquad\qquad \textbf{M} \; \textbf{r}^2 \; \textbf{m}$$

With the radius in the square dividing the shared and combined mass of the particles the relevant mass of the particles rises by the square as the radius reduces. If the radius becomes infinite, the relevant mass that the particles will produce goes up eternal. No force in the world would keep particles apart drawing on each other with an applying force but such a force is divided by an infinitely small separating radius. This is a recipe for joining and not dividing. Still according to the Universe I am able to witness that the dividing became enormous and the joining practically irrelevant. The gravity was more than words can describe, the heat was able to melt it all in one structure, but that did not happen. It split into billions of individual atoms.

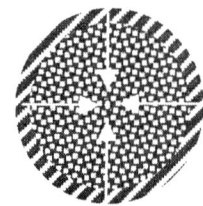

Another point I question about the Official Policy is that they as I am are in agreement that the heat melted particles onto particles and in those joining better combinations of particles came about. How it happened is another bone of contention but more about that a little later on. There was heat on the outside and there was matter on the inside. The heat was liquid because the Sun and other stars still indicate masses of liquid fluid inside. I can only imagine that that liquid inside the Sun holding temperatures as low as 6500^0 K and up to 1.8×10^6 K the heat already is in a molten form. What about the heat then when the frozen outer space was 10^{34} K and such temperatures were the general order of the day back then. If the Sun is liquid now then those temperatures raging back then must put the heat in form available in outer space at the time as thick as mud.

From the outside drawn onto the particle inside the blanket of heat came a flow of soup that became matter. That much I do understand. This carried on until...when? When did this stop? When did the Universe run out of heat? When could one consider outer space as the coldest all around? Where to did the Universe dismiss the heat that was once there but now is empty? How did the process stop of bringing from space intense heat and from that particles grew stop? When did it stop affecting the growth of a particle, the growth of space? In fact the growth of everything that grew came from this first growth. What you see or do not see grew since it was part of the Big Bang and everything in the cosmos at present was part of the cosmos during the Big Bang. I say this process of collecting heat from outer space never stopped but was an on going process we now give a nice name calling it gravity. Outer space never became empty and void but relevancies changed concepts where centres formed that should not be as it then interferes with concepts about relevancies Gravity is not and never was about particles pulling each other closer. If it was, no Big Bang was possible. Gravity is about turning space, which is released heat back to heat and concentrate the heat where gravity is the strongest and heat is the least. Space is the transverse form of heat and visa versa is also true. Should any one not believe me, try a bicycle pump by compressing the plunger while blocking the valve bit. The heat will burn your finger to blisters if the force on the plunger is strong enough, the plunger seals enough and your ability to withstand pain can last that long. Then answer your own question about where the heat came from because sure as hell is hot, it did not come from friction with air particles such as oxygen and nitrogen escaping through the valve bit. Heat is unleashed space and space is concentrated heat. Reducing space to heat is gravity and antigravity is expanding from overheating blowing into space accumulation.

 When looking at a sphere the inside has always (in a cosmic relevancy) the location with strongest heat also always has the strongest gravity in any given cosmic sphere. The centre of the sphere clusters the combination of particles forming the sphere into unity. By holding a specific centre the sphere becomes the strongest form any object can be. The sphere is without any doubt the favourite choice in forming gravity. Where gravity has the last say without other influences changing possibilities as collisions leaving debris in space or natural out burst like Super Nova explosions, gravity will enforce the sphere to be the form taken by the particle. But there is no evidence of particles of similar size joining in matrimony through gravity being the shotgun at the wedding. In cases where there is a mismatch of size outside any proportions of equality then there is a contracting of the lesser by the greater. In such cases the lesser is not qualifying as material (and that I prove later on) but the greater considers all the lesser to be heat. It is humans bringing distinction to matter in form.

The two objects should have their own value of gravity and _gravitons_ and in comparison with the _gravitons_ of the Earth; their value is insignificant. However, these two balls are in their own individual deuce to see who reaches the Earth first, and the iron ball's _gravitons_ should give it a superior advantage. This comes about because the two objects are in a position where they compare in relation to one another and share a common second factor, which is the Earth. In relation to the Earth, the gravity - motions of the two balls do not come into consideration, but this does not play a part since the Earth is a common factor. The balls, however, is put in a situation where they stand in relation to each other. When compared to one another, the _gravitons_ should give the heavier ball a sizable advantage. The sensible example one can show to prove that where some structures matching in size come into conflict about occupied space sharing. In such an event one of the structures are turned to heat as it is liquefied flowing in the space dominated by the other and larger structure. If the structure proves too large the superior structure turns the lesser compatriot into heat. Then being heat it will apply gravity and admit such heat into the ranks of its atmosphere, but not before it turned it into fragments good enough to be heat.

The Official Policy Protectors never tries to explain the relation between Newton's laws as mentioned above, and the binary star system forming the principle we know as the Roche limit. The binary stars are systems where two stars spin around each other and never collide. These stars are many times over the size of our Sun . When one applies the same Newtonian formula as given above, these massive giants must crash into each other, destroying themselves in the process. The enormous mystery is not in the apparent misbehaviour of these giants, but the fact that this is known to science since the previous century. Relate the binary once again to the comet/ Sun relation and there is a distinct similarity.

With the comet, the Newtonians regard a force to attach to the Sun in some way where this force pulls the comet towards the Sun . At the same time another force joins in that pulls the Sun closer to the comet, but such is the mass difference between the Sun and the comet, the force the comet applies never realizes. In view of this, only the force the Sun applies, comes into effect. The comet proves this force by speeding up its movement as it comes closer to the Sun . If the force did not become greater, why would the comet gain momentum?

With the arrival of the comet in the Sun 's domain, the Newtonians leave the argument to be. The Sun applying the force should remain applying the force and the force should increase all the time, accelerating the comet to the point of splash down. We must all argue that gravity is a force, which pulls an object to the centre of the larger object wherever that centre may be. The very same force that pulls the Chinese down, is pulling the Americans, and if not for the surface of the Earth's intervention, the next world war would be between the Chinese and Americans for King and country, honour and glory and to find who has the most powerful gravity force that will provide space to live in. If not for the Earth stopping matter falling right through the Earth because of conflicting forces on both sides of the Earth the Chinese and Americans will then have to establish border checkpoints in the centre of the Earth. The checkpoints will indicate where the Chinese gravity meets the American gravity and by allowing the force of gravity to find borders, we will finally have world peace. The only problem is to find the position where the Chinese gravity meets the American gravity and the two forces nullify each other. Just think if the forces of gravity, and not man, will intervene to set border standards: that must be the answer we were always finding a question for. This is a study far too complex to bother the United Nations, so we can find a more suitable group to investigate this fact to bring about world peace.

I am personally part of Africa, born and bred in Africa as an Afrikaner. I know the African solution to such a problem. In Africa, appoint a committee to investigate and then wait for everyone to forget about the problem in investigation. Therefore, such a problem is far better solved in Africa, because those in government aim to receive maximum western aid but never aim to solve problems, you make it go away by postponing the solution to the unsolvable. The African way is to ask the west for aid in order to create another useless committee to become over paid and under worked, quite capable of dealing with any non-existing issue of any magnitude that will never find an answer. Then sit around at leisure and wait each month for pay day to come for many years while the west is paying the committee to be bored until their pension dates arrive. By then no one would know the name of the committee and much less remembers the problem investigated. On the other hand, the Newtonians are doing quite fine by their method on their own using a technique they apply for three and a half centuries. To solve such problem, the Newtonians will apply a very different solution: Blame gravity's boundaries on a non-existing force, brainwash all future students in accepting it to be a force by telling them they will accept the force and forget the problem or fail the examinations and be chucked from campus, because that solved the problem so far. By the time, the student reaches a senior position he (or she) will no longer bother their mighty brainpower with the little aspects. They will advance to a point where they can move Black Holes around, travel at the speed of light, and divert time back to the past while others calculate all the mass seen and unseen in the Universe and any other ridiculous notion they may find to test their personal brilliance. If you for one second think everything about this last paragraph was silly, the silly ness started with GRAVITY ON BOTH SIDES OF THE WORLD, opposing each other, and that idea is not mine!

This is where the century, old trick of the Newtonians work best; do not think any further and no further problem will arise. Leave it at a force because with a force and thoughtlessness applying even-handedly, the problem never surfaced yet and that continued for the past four hundred years or so. So why bother with a problem that bothers no one. When a fellow like Hubble proves quite the opposite to Newton's claim of attraction, get a man who has a bigger ego than a brain and tell him to measure the Universe. It will keep every one involved occupied with something senseless while the problem vanishes through the many centuries to come. It is a force, and the way of all forces is mysterious, but never admits in believing magic. Those that do not accept forces to be of a mysterious nature should just contact astrologists and come to their senses about forces being not understandable. With everyone in agreement about forces, their nature and unpredictability, who then needs more real problems to solve?

No big-brain should bother about little issues like comets when there are so many galactica to conquer. Apparently the comet-problem just will not disappear. Something broke the force, something interrupted gravity. Let us see what happens. A force means it acts the same way as tying a rope on one object and start hauling the tied object in. The longer the rope is, the less control will be on the lesser star. As the rope shortens, the better the control will become. By implying that gravity is the force, our Newtonians tell us that we have to regard gravity in the same way the rope is hauling in a comet. It is something like fishing where the comet pulls, and the Sun pulls and eventually the angler gets his fish. One may argue that the rope is not the force because the force is actually the hauling, or shortening of the rope. I have had Newtonians trying to avert the problem they refuse to see by bringing in this argument. This manner of reasoning has the same value as introducing the African committee of investigation that will never uncover an answer. The rope is the extension of the force in a way being the sole representative of the force and the instigation acting out the force. The rope therefore is the force, extended somewhat, but still acting out the application of the force.

Weather the rope eventually broke, or hauling stopped, the effect as far as gravity applying its force, the process came to discontinuing … and we know that gravity is a force that pulls something to the centre of the body in control of the gravity. What made the force act in defiance of its nature? Why did gravity change its mind? What stopped the Sun applying its ferocious onslaught of the body holding the poor defenceless comet? Non – Newtonians will blame me for exaggerating, but I know that there is no Newtonian that can understand my argument. In that light, I ask non-Newtonians to show patience, because there may be a few Newtonians that will also read the book and to them everything said this far does not make sense.

This is where tutoring comes in best. Should a student bring up such non-academic and spiteful thinking about the mysteries of a force, then the lecturer sets a date on testing all the students' reaction about how much they accept the force. When any student shows signs of defying the force the lecturer can fail him on the spot and have a good reason to drive the silly youngster from campus.

By ignoring the problem as to why this comet brakes free from the gravity of the Sun , and continue in its freedom until gravity is at a point where it is most weak, may not bring answers, but it surely avoids nutty questions! Questions are not there to interrupt Newton's laws! Ask any Newtonian High Priest and he will either tell you that in a very roundabout way or he will simply ignore you by telling you to your face that you are incapable of understanding Newton. The best way to get out of the answer of course is to tell the sod with all the questions he has not the qualifications or the mental capacity to understand Newton. That will make the pest retract to some ditch he should be in, in the first place without bothering the greater minds with some stupid minor issue. How do I know this you may ask? I have been down that alley many times and treated with that precise treatment on occasions more than I care to remember.

Still, the comet defies the force of gravity and my questions remain unanswered.

Dear reader, if you wish to read the funnies, jokes and laughs -a- minute, treat yourself to some real good clean jokes. Read the Newtonians explanation about how comets came about; how they get to the Sun and where they came from. It is going from the ridiculous to the thoughtless and ending in the realm of the mindless. However, be warned! Only do this on occasions where you feel very depressed. The jokes will otherwise drive you in a state of laughing hysteria. Poor old Newton was considered a very dry humourless chap in his day. To think what silly ideas can come from his forces.

Hauling in and releasing something caught on a line is called playing with fish. We might say that Newtonians love fishing and confuse planets, comets and fishes when they regard the interaction of comets with the Sun . There is only one small problem with that argument and that is that fishermen and fishes form part of a second natural force named life. Life stands apart from the cosmos. Life and the cosmos only share time in space, not a joining of forces. Beside that, comets were part of the cosmos long before life had any role to play, so blaming it on someway life interacts with life does not cover the solution.

Why would the comet brake free from the Sun 's gravity? That is defying the law of gravity. Far worse than that still, is the fact that the comet's actions have the nerve to defy Newton. No one alive can defy Newton and remain alive. Does the comet not realize his actions contradict the all- important Newton and the gospel of the Newtonian - Priesthood. The best way the Priesthood of Newtonian gospel can deal with such defiance is to ignore it and no one will notice the actions of the comet. That is the scientific approach. Ignore and forget the problem. It is as simple as that.

With that let us conclude comets and really enter the world of forces at work! Let us now apply our attention to the forces of planets.

The next formula is very simple to understand. It is the fight of understanding the applying that becomes not applying it that is troublesome. If you understand the applying of the working and never spotted it not working, then forget it. You are a brainwashed Newtonian and if not, well there is still hope that you have a clear mind left. The resentment you carry with you from childhood about the formula is in, not understanding it, but accepting the outcome of the formula you never could understand. Newton said that the force between two objects depend on the mass of both objects multiplied with each other and with the gravitational constant and the derived product you divide by the radial distance square that separate the objects. I shall put this in a mathematical language for your enjoyment that will explain the life-long not understanding to better effect.

$F = G (M_1 x M_2) / r^2$. What does this say?

The greatness of the force depends on the masses of the two orbiting objects, aligning that product with the contribution of the gravitational constant. This then, you divide by the square of the distance between them at any given point.

Please, in all fairness to you, the reader, I have to warn you that quite a number of professors in physics told me that by reasoning in the manner I do, I only prove that I know nothing about Newton and understood even less about his work. Considering such allegations, I shall explain to you what I understand in as much as telling you what I know.

The Newtonian's formula states that the force between the planet and the Sun will improve as the mass of the planet increases (becomes bigger) and by multiplying that with the universal gravity constant you will get a value that will become lesser, the larger the distance are between the Sun and the revolving planet. With the reducing of the distance the mass on either side must therefore be on the increase because it holds an inverted relevancy. This means the Sun is pulling according to its mass. The planet is pulling according to its mass. The gravitational constant is influencing the pull evenly at both ends and the distance between the objects will reduce to the square value of the force's total application. I could never see what part I do not know and what I did not understand. No professor ever explained to me what it was that I did not understand either. That left me in a place where I did not understand what I did not understand and I never could see what I never could see. I shall try and make sense of my not understanding my not understanding as follows:

This is like having two balls attached by a rope on a floor that holds the same drag on both balls. When I reduce the length of the string, the bigger ball will show a greater resistance than the smaller ball, therefore the larger ball will apply a larger tug than that of the smaller ball. The rope will reduce (become shorter) at the end where the larger ball is than at the point of attachment where the smaller ball is. What is wrong with my argument? When the two balls are so miss-matched in mass as is the case with the Sun and the comet the one ball will do all the moving, leaving the larger ball stationary.

Surely the tugging at the larger end must bring the smaller object closer. By comparing the mass differences, you will find there is no comparison. The smaller object just has to come closer with the application of such a force as gravity. We know that gravity can really pull. By standing on a tall building you will find proof of this. Drop a tennis ball down from the building's roof and see for yourself how it falls. The distance between the Earth and the ball reduces by some speed. With that being obvious, the distance between the Sun and any planet have to reduce as the planet orbits the Sun each year. Even if it is small, there has to be a visible reduction after four and a half billion years of pulling and tugging! Today after wrestling this problem for the duration of twenty-five years I can say (with a clear mind) I finally know how it works. It does not work!

I was always looking for mistakes on my part. At first, I thought there are a fifth force that I am unaware of because of my slender education, a force the academics can obviously see, but I cannot through obvious lack of education. I thought that my personal ill literacy gave me a blind spot that every non-educated have and was born with. The blind spot cleared only thorough education as education removes it in the way only education in science brings knowledge. I thought the removing process similar to the way washing removes stains and spots from whites; education can remove blind spots through the process of intensive tutoring. All I wished for was some academic to help me remove my blind spot about comets and their behaviour. The comet's behaviour, I could see, was an exaggeration of orbiting patterns applied from our planets orbiting around the Sun ; in the way, we observe galactica in the sky.

Then finally I came to the point of accepting defeat. It was not I, with the blind spot; it was all the academics brainwashed into a state of having such a blind spot. Science insists on repeatedly ignoring mathematical principles, because Newton had his claim to fame with one single calculation, THAT HE, IN FACT, DISCARDED, BY THROWING IT AWAY.

He made a brief calculation as a young man that saw an apple fall from a tree. Seeing this he jotted down a formula and the chucked it away. His piers and elders picked up the trashed paper with the calculation, and got all excited by the logic implication it had. $F = r^2 / (M_1M_2)$. The mass of the two objects destroys the radius between the objects. Everyone went ballistic, proclaiming him as an instant genius, the one the world was waiting for after the crucifixion event.

I do not, for one second, deny or dispute the revelation. What I do encourage is place the event into its correct context. It was merely, and simply an apple that fell from its branch to its roots. The apple did not pretend to be a meteorite that fell from the heavens. If it were a meteorite, I am sure, with the man's genius, science would be somewhat different at this stage. However, as a young man, being very impressionable, as all young men are, and with the attention this brought about in the world of science, the matter overshadowed the fact.

I am not disputing Newton; I am disputing the relevance of Newton's scientific breakthrough. It was not two objects of cosmic proportions, colliding in a show of spectacular. It was, after all, only an apple falling from a tree. With this miracle revealed Newton found he was competent to improve on the work of Kepler and if I may dare say this, there must have been some political agenda behind this act and the accepting of it for Kepler was a German and what German can ever teach any Brit. The very same politics are still the order of the day forming international rivalry on all fronts.

Newton, and science, made one enormous blunder, from this stance. They took the radius of a wheel not to have any influence on the wheel. In doing that, they removed the very fact that keeps the universal attachment together. They put two objects in an attaching relevancy and then announced no relevancy. Doing that is breaking the most fundamental mathematical principle.

$$\frac{dJ}{dt} = 0$$

This disputes mathematics. DJ / dt can have any number from eternity to infinity, only excluding one; it cannot be 0. By placing the one in division of the other, you bring in relevance. You cannot then say there is no relevance. By doing such, you proclaim that one of the factors is non-existent.

$$\frac{dJ}{0} = dt \text{ or } \frac{0}{dt} = dJ$$

In both cases, one of the factors then does not exist. Such a claim is incoherent, because you proclaim that a circle has no radius, or a radius has no circle. When calculating a circle, you multiply either the square of the radius by Π, or the quarter of the diameter at a square by Π.

$$\frac{dJ}{dt} = 0 \text{ constitutes a circle and is also therefore } \Pi \times r^2 = CIRCLE$$

If you remove r it then is $\Pi \times r^2 / r^2 = CIRCLE$.

You cannot then say $r^2/r^2 = 0$ and therefore $\Pi \times 0 = 0$. That is nonsense. $\Pi r^2/r^2$ will always be $\Pi \times 1$, and that is the eternal circle.

When looking at any rotating object, there has to be a point of no rotation and no rotation means "no rotation", not no existence. No rotation means a factor of 1, not zero. That then is singularity. The eternal Π, the Π that may not have significance but still it is a Π of value.

When looking at any rotating object, there has to be a point of no rotation and no rotation means "no rotation", not no existence. No rotation means a factor of 1, not zero. That then is singularity. The eternal Π, the Π that may not have significance but still it is a Π of value.

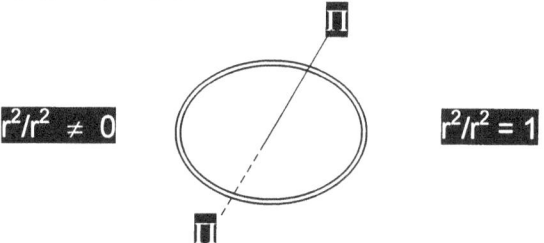

The relativity remains one, eternally one, but it cannot be zero. Therefore, dJ/dt cannot be zero.

When explaining this to any child, they can immediately see that. Explain this to any Newtonian High Priest and he may have you removed forcefully from campus. I cannot find one Newtonian, large or small to accept that. By not having a wheel rotate the rotation seize, not the wheel. When the wheel begins to rotate, you cannot state that all things remained as it was. With the wheel in non-rotation the rotation still exists forming the infinite possibility of rotation. Then afterwards the wheel starts to rotate and by the start of rotation the circumstances surrounding the wheel changes. A wheel in rotation is very different from a wheel not rotating and therefore cannot be the same thing. By establishing non-rotation, the wheel becomes the factor of one, and the rotating action becomes zero. The wheel does not disappear. But in the same manner does a wheel in rotation not remain still.

In the cosmos, everything is rotating because nothing ever stands still. Therefore the mean equilibrium, the common factor there is to share, has to be one, eternity, the eternal Π, because all rotating objects has Π in singularity, and sharing singularity, gives every object in space a relation with all other objects in space. After trying for many years to bring our Brainy Bunch the candle, I concluded that Newtonians are incapable of realizing that mathematical principle as a reality. They maintain they know mathematical principles far better than an ill literate such as I and yet....

The comet rotates the Sun , and the Sun by itself has a point of singularity where Π remains without r. The comet, holding the orbit, also has a point of singularity, but since there is space separating the two objects, they cannot share a mean point of singularity, the very point of existing. Since singularity means just that, being single, there cannot be two. The comet and the Sun have a mean point of singularity but the space they occupy divides their common singularity. That is why they orbit in an oval path, a path where the one structure holds on to more space from its point of singularity towards the space it claims. Since they do not claim equal space, BY THE DENSITY they hold, the space will not be in proportion. They do share in the common fact of singularity and singularity cannot be two, because then it will be "dualarity" or (in case there is no such a word) duplicity where both find the space they occupy, with the space they hold, will be their individual eccentricity from singularity. The two objects are holding eccentric space around their individual but common singularity forming a point of mutual singularity in accordance with the individual singularity both claim space from. That point of singularity is Π the circle without the radius because the singularity removes all forms or values of r, leaving Π to be singularity.

That is why Newton is bullshit, and his $F = G(M_1M_2)/r^2$ is utter nonsense. The moment you say Newton or any of Newton's laws, the Newtonian brain stun. For all the life in me, I could not once find one single Newtonian to see this. If you say Newton is wrong, they spiral down to frenzy, and just mention gravity and they all fall on their knees, cover their eyes in the ground, start praying and you cannot make them say anything other than Newton is correct. Dare say there is no such a thing as gravity and Newton is wrong, they have you in an armed escort patrol, straight to the department of mental disabilities and psycho diseases in preventing you committing acts of extremely dangerous life threatening behaviour to yourself and others.

What is it the Newtonians fail to see? They fail to see the relevance applying. They fail to see that the Universe not only holds the atom, not only comprises of an accumulation of the atom but that the cosmos indeed functions with gravity 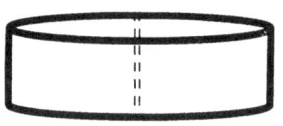 and all as one massive atom. The Universe is exactly the atom that the Universe formed which then forms the Universe. If an electron is orbiting around an atom, the inside of the atom must be a circle. If the atom was not a circle, it then had to be a cube. The electron cannot rotate around a cube; therefore, the inside of the atom is a circle.

In a circle, there is a radius that initiates the circle. The calculation of such a circle is $\Pi \times r^2$.

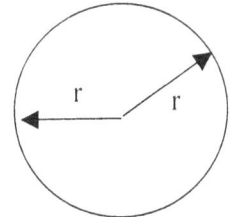

The radius r runs from the circle outwards, from a circle centre point towards Π, the value of the circle. In the centre of the circle, there is a point where the radius starts. It runs outwards from that point in all directions towards the circle Π. Technically, there then has to be a point where r is infinite and not zero, an absolute infinite. However, the circle therefore remains Π. The circle does not disappear; it remains there for all to see. It is only the radius that almost disappears into the infinite, but it does never become zero!

$$\frac{\Pi r^2}{r^2} = \Pi$$

If one removes the radius from the circle, the circle remains, only holding the value of Π. By removing the value of r, Π becomes singularity with no place to be. Singularity is the place where there is no space to be in place. However, Π remains because once r receives the slightest of space Π will find space. Then the circle will grow to Πr^2 and r would determine the space. Without space, there is no r but there is a circle with the value of Π. Singularity is in every single rotating object, be it the proton or the combining effort of all particles in the Universe. That is what light and the photon is. It is concentrated heat that the Sun (or any other generator of electricity) connects heat to singularity where the heat receives either temporary connection to singularity or a small piece of individual singularity.

At first you as the reader may think I am trying to create a mountain from an ant heap, but in scientific terms the human race is preparing for the start of the cosmic journey. By completion of this book you will realize how *Xepted* science believe they built science on a solid foundation, and, boy are everybody in for a rude awakening. Compared to the leaning tower of Pizza, science is about to start with the next section of a much bigger building adding many levels and already the view at the bottom where I am looks far worst than the leaning tower does.

If I contacted and argued with one Physics Lector or Professor about Newton, I have been in correspondence with at least a couple of hundred. What prompts me was the comet's orbit. The commit truly fascinated me from my childhood days in the way it defies all the laws of gravity. Since my very young days, I was in search of what I at first believed to be a fifth force. I have raised the argument with just as many people not schooled in the art of physics and received a very different response. The most amazing aspect was the fact that the two groups were that far polarized. The non-physics group reacted astonished, amazed, disbelieving and reserved about my view about comets at first, but with their distrust not withstanding, everyone saw my point. The non-Educated responded in the same manner that I did at first. They argued that I was missing something of vital importance because "why do the wise not see it", was their argument. They always were of the opinion that I was too little educated to understand, while the educated was of the opinion that I was too little educated to understand. Neither party had the same view about my not understanding. The non-Educated understood my argument, but dismissed it on the fact that it was so obvious I missed the rest of the knowledge behind the facts that makes my arguments too difficult to understand while the educated dismissed my argument that I could not see anything they could not see. Education brings the ability, which then made me unable.

In short, they thought I was to stupid in order knowing the rest of the story. Polarized to the non-Academic view was Official Policy Protectors where not one academic could understand my argument. The academic response was as much defending the Newtonian view as it was drawing a blank about my questions. They all seemed as if their ability understanding my view was completely locked behind some wall. The non-Educated, of which I am a member, at least understood what I was saying, but dismissed the simplicity about the argument. In the corner of the Official Policy Protectors, was no response of any kind, but to feverishly defend Newton by raising the dumbest arguments I have ever heard. The arguments, even the most highly educated brought about, seemed motorized and non-responsive. When it seemed their accepting the points I raised with my questions would demise their senses, in defence they put up a block. There is a peculiar sense of numbness in the way they could not understand what I did not understand. The academics showed no signs to indicate that they could even argue my point of view, by responding that I have an argument, and from that launched a responding argument to explain how or where I made my mistake. Their abilities in even understanding always seemed to hide behind a wall of not understanding that someone may not approve of Newton's arguments.

Newton says two pieces of rock will draw each other closer by reducing the distance keeping them apart. That we all can see by merely jumping in the air. No sooner have you lift off than you are back on the ground. That is what Newton said about three hundred and fifty years ago. Even Trying to tell the Official Policy Protectors that Galileo said mass of an object has nothing to do with the falling, seemed to pass the Official Policy Protector's sense's of comprehension by miles. I was told on so many occasions that I did not understand Newton, but there it stopped. No one could explain to me what it was I did not understand about the comet missing the Sun by miles, where it was supposedly to hit the Sun with a dazzling impact. To this point, I cannot get through to them as much as they cannot get through to me. Our understanding is so far apart, we do not share the same planet, and yet after all my arguments and investigation no one, and I repeat: not one could once clearly tell me what it is that I do not understand.

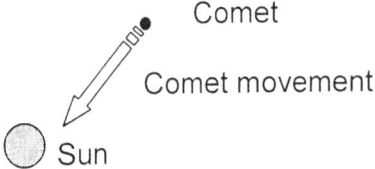

Comet

Comet movement

Sun

You have the Sun and you have a tiny piece of rock covered by water also better known as the comet. There are thousands of them flying around, but never aimless. At first Newton's formula makes pretty much sense. The Sun draws the comet towards the Sun , as Newton said it does. The comet responds by speeding towards the Sun , also as Newton predicted.

Anyone can see a collision coming ten miles away. The Sun applied gravity, the comet applied gravity, the Sun is far too massive to fly to the comet, so the comet with much less mass does the flying on behalf of both objects. Every person with even the least of knowledge about science knows how the gravity application works.

SUN   COMET

        DIRECTION   TRAVELLING

The gravity of the Sun collected the comet from no-one knows where, pulled it through billions of kilometres to the area where the Sun produces the gravity with which it pulls the comet where the comet is to find its last resting place. The mass of the Sun is obviously so large, it could produce gravity that can locate any comet hiding anywhere and collect it as a souvenir. What is there to understand?

The gravity of an object always points directly towards the centre of the object, the very, very middle point. Concluding from the fact that the comet is heading towards the centre of the Sun , just as much as the Sun is heading towards the centre of the comet, would not be out of line. The two centre points are heading for a direct collision, the collision becomes more and more unavoidable as the radius reduces by the value of the gravity that is produced the mass in accordance with the gravitation constant. The comet is heading towards the Sun , and by not even moving, the Sun is moving towards the comet by attaching the movement the Sun were suppose to have, on the comet. Newton's law proves to be exceptionally correct.

As the Sun /comet, radius reduces, the radius separating the mass of the Sun and comet effectively increases the relativity of the mass influence on each other in the form of gravity. The mass of the Sun and the comet increases by the factor of reduction of the radius separating the two objects. That will produce a growing gravity force as the comet / Sun radius becomes smaller. By the time, the radius becomes one the mass will grow on either side by a relevancy of 100, and when the radius becomes infinitely small, the relevance to the mass of both structures will raise a force with eternal power.

$$\frac{M_s \times M_c}{100} = 1 \times F \qquad (r^2 = 100)$$

$$\frac{M_s \times M_c}{50} = 2 \times F \qquad (r^2 = 50)$$

$$\frac{M_s \times M_c}{25} = 4 \times F \qquad (r^2 = 25)$$

$$\frac{M_s \times M_c}{5} = 20 \times F \qquad (r^2 = 5)$$

At a point, where the comet / Sun apply a force of immeasurable strength, the comet brakes this immeasurable force. Remember the direction of gravity always point to the centre of the object, and that is where the collision is heading. As the objects draw closer, the distance reduces, but in accordance to the relevance the objects also become that much bigger in drawing power. It depends how one considers the relevancy to grow by the approaching nearness diminishing the distance between the objects.

Then out of the blue, the comet finds the ability to eliminate the eternal powerful force of gravity, and keep at a safe distance around the Sun . At this point, Newton goes sour. Nothing Newton predicted is happening. The comet and Sun not only stabilized the force, the force begins to decrease as the radius between the comet and the Sun is on the increase AT THE POINT WHERE THE FORCE IS THE STRONGEST, THE COMET BRAKES FREE AND SLIPS AROUND THE SUN , UNSCATHED.

UP TO THIS POINT I STILL SEE WHAT THE BRAINY BUNCH AND NEWTON SEE, BUT IT IS FROM THIS POINT ONWARDS THAT THERE COMES THE POINT THAT OUR MUTUAL POINT OF CONCENT DIVERTS POINTING OUR MUTUAL POINT ABOUT THE POINT OF AGREEMENT TO THAT OF OPPOSING POINTS WHERE OUR VIEW SEPARATE BOTH HEADING IN OPPOSING DIRECTION ON AN ETERNAL DIVERTING PATH.

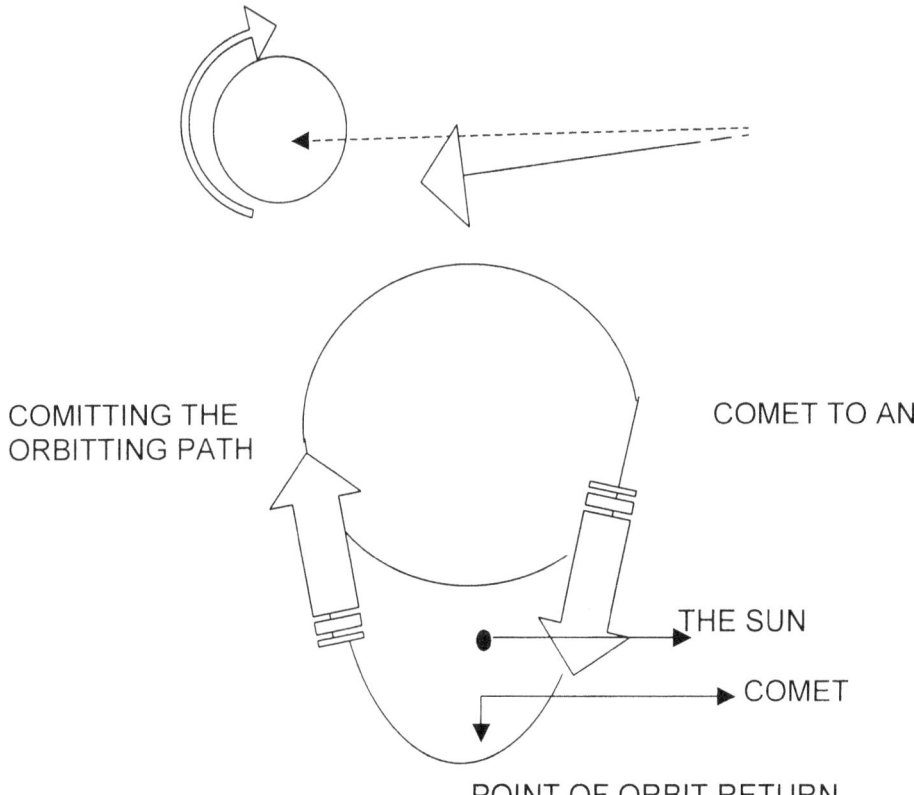

COMITTING THE
ORBITTING PATH

COMET TO AN

THE SUN

COMET

POINT OF ORBIT RETURN

Then, in complete defiance of the Newton Law on gravity, quite the opposite applies. At the point where the radius that is separating the two cosmic objects is at its strongest, it will also bring about that the gravity force is at its weakest. At the point of almost no ability the gravity force suddenly releases enough strength to break resulting in the parting of the two structures. The force now curbs the rebel comet on its way escaping the Sun 's gravity for the very last time.

At the point where the force was the greatest, the comet overcame the force, but where the force was the weakest, the force overcame the comet's rebellion.
The correctness of my argument is no longer the issue. It was twenty-five years ago, when I still held the impression that I was missing some point here. I do not state this phenomenon any longer in the hope of bringing across some flaw in my understanding. The flaw in my argument is not there because the flaw is science as a whole.

I could never understand the reason why "the ordinary", like me and others with my development level, can see what I can see, yet academics that has more brainpower in their heads, than I have life in my body, were unable to see such an obvious conclusion. You; those Official Policy Protectors are my superiors in every sense a human can have, with the brainpower to break a wall, and yet you cannot see how far the tower of Pizza is leaning over.

I make the point to help you, the reader, to judge yourself. If you are able to see the validity in my argument, you are not brain dead. Education has not yet bashed your thinking ability out of your scull. However, if a cloak of not understanding roll over your brain, and a numbness sets in on your ability to reason about this phenomenon, beware, you are a Newtonian. Newtonians should read this book very slowly because the effort you are about to launch, may be the most painful you shall ever experience throughout your academic career. You are going to suffer from reconditioning and Newtonian withdrawal, not that dissimilar to that of an addict in rehabilitation. You are going to reject me, hate me, despise me, loath me as you never felt about anybody else. If you think I am sarcastic, I am not. You will reach a point where you will abandon the reading of the book. You have my sincere sympathy and with all the soothing it may bring, know that you are not the first I saw getting such painful Newtonian withdrawal in rejection of Newtonian doping.

Once more, this phenomenon should not occur with Newton's presumptions about gravity. These bodies will and must collide and destruct, without a doubt. When the formula $F = \dfrac{M_1 M_2}{r^2} G$ applies, there should not be any force which is able to keep them apart especially when r reduces to almost infinity compared to what it is at maximum. However, they do exist and what is more, they maintain a certain distance apart.

With the "force" of "gravity" "pulling" the stars closer using the accumulative mass of the stars and multiplying that value with both objects by the mass component, this will reduce the radius r^2 progressively until r^2 reduces to zero. Seen from this view, it is little wonder that the significance of this was lost in the notion that this is yet another "mystery" of the Universe. The scientists of the day (and the past) lost the importance, which this holds for us as Earthly dwellers.

A most surprising aspect of this is that it is not that an unfamiliar or rare phenomenon. However, any answer to this would clash with Newton's presumptions, and before the scientists allow that to happen, they would much rather ignore what is obvious. However, what is the obvious?

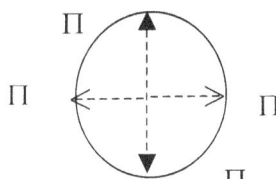

The cosmos works in relevancies that have a ratio to singularity. I am desperately pressed for printing space but I will show a small example of the implications of this statement. The ratio of space to time or then the more commonly used terminology is phrased as the curvature of space-time inclines by Π From the Earth's perspective matter holds space at 7 and matter in rotation is 4 giving a allocation where the moon should be at 4 X (7 =Π) = 28

The moon holds its singularity in relation to the Earth in singularity. Because the moon is in a Roche limit (near enough) the proton value the moon accepts is $\Pi^2/4$. The neutron position the moon holds, relating to the Earth is 10 and the moon has an own point of singularity, forming the electron edge of the Earth.

The Earth is $4\Pi^2$ = 28 days to rotate to one moon cycle of 1. The relevancy of the Earth, taken from its point of singularity is 7 (matter in relation to space will always be 7 to the space value of 10).

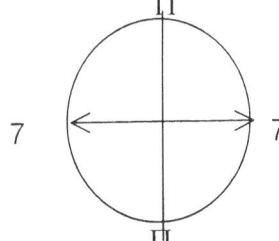

Therefore Π in singularity will accept 7 as a factor forming Π^2 giving the space holding matter on one side the value of 49 (7^2). Adding the one of singularity and half of the time factor will then be (49 + 1 = 50) on one side of the Universe

To get a relative position that the moon would hold would be the 4 quarters material ha in relation to time by ratio of the material holding singularity in maintenance and that would then amount to (4X 7= 28)

OBVIOUSLY ALL RELEVANCIES HOLD TWO SIDES TO EQUAL PROMINENCE. THE RELEVENCY THE MOON HOLDS TO THE EARTH WILL BE DIFFERENT TO THAT WHICH THE EARTH PLACES ON THE MOON.

The Earth holds singularity and the Earth's singularity holds the moon's position in singularity. To the moon's relativity, it holds value to space from the Earth's singularity of 1 x $(\Pi/2)^2$ (the relation between the Earth and the moon) x 10 (the fact that the moon is within the space of the Earth, as the moon has an individual point holding singularity (Π). Therefore the moon holds (1 x $(\Pi/2)^2$ x 10) + Π = 27,8 days as one and the Earth holds to moons single day of individual singularity at one to 28 days.

$$(4(\Pi^2)$$
Earth ← $(\Pi^2/4 \times 10) + \Pi$ Moon

Behind this principle are the sound barrier and the reason why aircrafts break the sound barrier. It is all a relation in different positions of singularity. From the time, I wrote the first few pages I was on a quest to find a more suitable person to take over the work. I, more than any one else, know my limitations and limitations they are. I knew that anyone of the more than one thousand five hundred academics I eventually contacted held more knowledge in the tip of his (or her) little finger, than I have in my entire body. I tried in desperation as I tried in vain, to convey my message to the right person that could see what I could see. It all came to nothing because the academics all sat on the same mighty Sear Tower far too high to even notice me pointing at the cracks down below. From where the Official Policy Protectors sat, they did not notice me. Those to whom I drew some attention be it personal, by mail or on the Internet saw me as a nonsense proclaiming nonsense.

From where I stood, I could see the mighty tower they sat on. I could also see what construction held that mighty tower together. I could see how much that mighty tower was leaning over like the tower of Pizza. I could see how the tower will fall one day if not soon then later, because I was at the foundation of the tower. From where I stand, at the very bottom, I see the foundation of this enormous Petranos Towers collapsing from the misconceptions it holds as a base. Down at the very bottom where I am, I could see whatever other one holding a position in this tower could not see. Those that are at the very top, ARE so high, so very secure, they would not even hear me or take notice, and yet they are the only ones that can do something about the inclining tower.

After attempting for seven years to make myself heard to get the High and Mighty to notice the insignificant me down below, so small in relation to their greatness, I decided to show the world what holds them that high. I decided to show everyone how great "they" are, especially about the greatness of their misconceptions that put them at such a dizzy height.

To every one of the Official Policy Protectors, I say this: Every opportunity you had, you thoroughly rubbed my nose in the knowledge that we are not in the same class. I accept that fact as I accept my academic qualifications being so very poor. I shall never be your next-door neighbour or the bloke living down the road from your house, because I am not in your league and I shall never be. I do not begrudge you your academic position, your mighty achievements or the height you have reached in your sphere. I shall never enjoy your company as an equal, because I can never be your equal. Your brainpower puts you light years ahead of me, and for that reason I do not even wish to have the honour of your company.

All I ask, is listen to a mere mortal, a mindless illiterate compared to you, one sod down here at the bottom where you are at the very top and that may just see things you cannot see from the height you hold. I do not wish to join your company for I shall not fit. I never had or have any ambition to fit either, because I am quite happy being in the sub-minor league. All I ask is to be heard. So many times you, honoured members of the clan of Newtonian High Priests did not even attempt reading my book that I sent to you. You did not even try to pretend I had a point therefore you merely threw my book away in disgust. To you my illiterate arrogant views about science being wrong for the past few hundred years are the epitome of a mindless person. You saw me as a totally mentally underdeveloped excuse for breathing and you could not bare my company because for me to have my view such as my view is in rejecting the view of an establishment centuries old. I can understand your disgust, but that does not change my point and that does not change the incorrectness of Newtonian science!

Before you throw the book down in total disgust, first answer the following argument and if you can answer it truthfully then throw the book down. If you cannot answer it, go on reading the book and you may just set your thinking mind in motion. Hear this from a mindless: you may have the ability to learn and afterwards reflect on that which you learn, but you cannot think, and I do not know who is the most mindless, me without education or you with education and without reason.

You are addicted to Newton. You do not understand Newton because if you did understand Newton as far as cosmology goes, then you would reject Newton. You were force-fed by your superiors on Newton at the time and during the time when you were a student and being force-fed you either had to grow addicted to Newton or die from Newton. Being where you now are it means you are not dead. That means you took the second option by becoming addicted to Newton. You saw comets come and you saw comets go. You did not question the reason why the comet went because of your addiction to Newton. You accept Hubble's expanding yet you go about looking for a critical density in the hope you do not have to abandon your Newtonian addiction. You know the Big Bang cannot be possible while Newton's mass annihilates the radius between masses because then the Big Bang had to be the Big Crunch and from where we are there is only one enormous out explosion that had no Newtonian implosion. The enormous force that were suppose to contract the small radius that was in place during the Big Bang never imploded on the mass by nullifying the radius through the tremendously large mass and small radius being present at the time. All the facts modern day science accepts does not stroke with Newton. Is it because subconsciously everyone is silently admitting about Newtonian incorrectness?

Now I come and I tell you your addiction is killing you. As all addicts do, you are not willing to kill your addiction although you do realize that your addiction to Newton is senseless, stupid and all together wrong. The second choice facing all addicts including you is to hate the messenger that comes to take the drug from you. You will hate me, as the messenger. You will be willing to kill me being the messenger of evil because you see me as the evil that wished to part you from your addiction, which is Newton. You now find the bizarre feeling to put your anger distrust and vengeance on me and on the book you are reading. I have seen many academics stop reading at the part you now have arrived at. Brave yourself and throw out you fears by reading the rest. It will eventually cure you. Newtonians declare that a force brought about by the content of the mass of a body will therefore pull objects closer where the pressure coming from the weight brought about by the mass of the bodies, will lead to heat.

If a cylinder is pumped with air, the pumping is a force. The force comes about because the force is that of the intentional action of the only force in the Universe, the force of life. The more pumping there is and the longer the pumping will last, the hotter the air will become, and therefore the hotter the walls of the cylinder will get, as it transmits the heat from a point where the heat is most abundant to a point where the heat is least abundant. Heat flows from hot to cold.

After pumping stops, the heat on the inside will reduce in value up to a point where there is equilibrium between the heat on the outside of the cylinder and the heat on the inside of the cylinder. The heat reached equilibrium with the event of time. Please for the sake of sanity, do not reply that it is the molecules in the air tank that is bumping against each other and through that collision, friction causes the heat. That is as much Newtonian rubbish as one can ever find. Should you insist on that being your answer, then ask yourself what will calm the molecules down afterwards where they get so calm, there is no more heat in un-equilibrium. Did the molecules take drugs, or did the force calm them by telling them gentle nighttime stories. I had so much bullshit thrown at me wherever Newtonians defended their Master it sickens me. I may be uneducated, but I am far from mindless!

No molecule can ever, ever touch another molecule because the electrons guarding the outside are equally negatively charged, and will therefore reject any contact or coming closer to one another or with another molecule.

When we look at the Earth, we find the coldest region on the outside of the atmosphere, where the least mass, weight and gravity is. As the circle grows smaller, the molecules become more, the mass becomes more, and this will increase the weight that brings about heat from pressure. This is not my say so this is Newtonian science.

The increase in mass, bringing about an increase in weight, puts most pressure at the centre to be the hottest. There is this force in matter that pushes and pulls, until everything is boiling hot, and that is gravity.

There must be a point, where all the matter has finally found a position to bring about the overheating that occurs in the centre. The heat is energy and cannot be manufactured, but has to come from somewhere. The heat cannot be lost, because heat will transfer to a colder region to bring about equilibrium. The heat cannot continuously come from nowhere because at a point, the molecules will settle in their individual positions, find equilibrium and maintain the heat balance in that spot. It cannot produce heat from nowhere, on a continuous basis, because heat is energy and being energy it must come from somewhere as much as it is going somewhere. Either the atoms lose their mass to heat and the mass becomes heat in order to generate heat on a continuous basis, or the heat must stop, decrease and become as cold as outer space because no further heat comes about because the heat is exhausted.

If the matter as much as mass becoming weight established heat by applying continuous pressure, the Earth should decrease in size as mass turns to heat with the consequential loss of mass and gain in heat. The Earth must then deflate and be a pretty small place by now.

You may say there is a lot of mass-producing a lot of pressure becoming a constant flow of heat, but that will mean some of the mass must have disappeared to heat because four thousand five hundred million years is a pretty long time. In four thousand five hundred million years, some size diminishing should show, or the Earth should have reached a point of equilibrium by now where the heat supply is completely exhausted because again four thousand five hundred million years is pretty much enough time I would say for matter to have cooled down by now.

Yet, the flow of heat maintains in the Earth, very much uninterrupted and in no way showing signs of decrease. If the force of gravity is manufacturing heat, from where does it get its raw material and where is the manufacturing plant? If weight and pressure leads to heat, then the heat should have transferred altogether to outer space, the coldest place we know at minus 276°C. Four thousand five hundred million years have come and gone since the Earth became the form it has now.

On the other hand, if the force continued its pushing and pulling it applied when it formed the Earth, the Earth should by now have incinerated. What force of gravity is required in getting enough pressure to push hydrogen dust into a solid iron formation and release that formation as a molecule in being as solid as the Earth is. Such an effort requires a lot of pushing and compressing. If that pushing and pressure continued as much as it must have had it had to increase with the demise of particle space that is separating the molecules? Keeping that in mind as a natural law, everything on Earth should be covered in flames by now. When you go with the argument that there is not enough matter to produce such pressure, then there was not enough matter from the start to get the place as dense as it currently is. The matter did not decrease, the force therefore had to become weaker with time and the force is so weak now, it does not collapse the Earth into a smaller space than it was say three thousand five hundred million years ago. However there are very little signs of that, in fact it seems the whole thing is getting bigger, with all the lakes losing their depth, and the mountains rising, including the volcanic activities establishing new islands.

If you cannot state why the heat has not reached a point of saturation or has decreased to equilibrium with outer space, it proves you are not thinking. In that case read on … it may arouse your thinking ability once again. This I say, not in arrogance because I, of all people should know how poor my abilities are. I sat day after night, night after day, breaking my brain to find answers, or only clues to the questions in hand. I knew at each point I arrived, that the answer is right in front of me, but through the darkness of my personal ignorance, I could not see what I knew there was to see. From that feeling of incompetence I tried once more on every occasion to contact any person with more brains than I but every time I was luckless to energise interest and had to continue with my personal incapacity.

Every time I contacted an Official Policy Protectors in desperation for help, they ignored me flat. Every message I sent telling whomever I contacted, that there is no such a thing as gravity, it is all a medieval hoax, Newton is altogether completely wrong with his gravitational laws and the Bible is one hundred percent correct about creation, they would not even reply or at least respond. Well… to them I say this: I still maintain that there is no such a thing as gravity: it is a hoax.

Obviously, the obvious is that the Newtonian Order of High Priests would lead every body to believe that Newton and Einstein's findings are flawless. All these mentioned discrepancies are known to "Xepted science", yet they keep the charade going on about other planets they are about to find, only to mislead the public and milk their tax money, in the name of research. If it were not for funding and an effort to provoke general interest in skimming tax money, then why would they deliberately spread such malice?

Scientists know about this discrepancy in the Newtonian laws, yet there is never any mention about it.

The so-called "evidence of the existence of planets" is based on just as laughable principle. Allow me to explain:

The findings which science base their proof on about the location of other planets are the gravitational pull.

At first the way in which the facts are presented does not sound that unfamiliar in an argument, and one tends to accept it without a second thought. When given the second thought, the blatancy in the matter leaves one breathless.

In the evidence, the stars and "planets" are presented to be in a tug of war. This one can see in any sketch about the solar system. How this supposedly works is that the one star first pulls the other system closer with the force of gravity, and then it is the other systems turn to apply gravity and jerk the first system to its side. What they do explain in explicate detail is that they do not understand the first thing about the matters they pretend to understand.

The Official Policy Protectors know very well that all children play this game, and therefore every one will associate this explanation with familiar events in their past, and no further questions will be asked. Let us examine this principle with obvious general knowledge about space flight and how this applies in outer space. An Astronaut is capable of lifting four tons of equipment by self-propulsion and relying on human muscle. He (or she) can perform this action effortlessly. However, what is impossible to perform in outer space is to correct his position when he is not secured to a stabilized object. Every person knows it can be life threatening when an astronaut loses his grip or connection to the spacecraft. In this, the question is; how can two heavenly bodies have a tug of war under such conditions? There is obviously some stabilizing factor on Earth one do not find in outer space, and that is not gravity because such an answer is avoiding the issue.

In the past so many SUPER –EDUCATED dismissed my rejecting the Newtonian claim about mass being a factor in falling objects. Some even went as far as refusing to read my book beyond that statement, that being on page four of a one thousand seven hundred page document, claiming I do not understand Newton, but no one can explain why oxygen is a gas, and yet oxygen is more massive than boron and carbon. If mass were a factor of the essential being the claim of Newton, then oxygen would fall to the earth faster than boron and not float in the air as air. Yet boron is a solid and oxygen is the gas.

As stubbornly as the Newtonian paternity refuse to admit about the comet not colliding with the Sun in spite of all Newton's predictions, just as stubbornly they refuse to admit that Galileo completely contradict Newton's mass blaming. With the mass devouring the square of the radius when helped by the product of the counterpart the whole formula hinges on the mass being the devourer from both ends. That means the heavier or more massive the object is the faster should it fall because the quicker will it devour the radius parting the objects. That Galileo distinctly disproves because all objects fall at the same rate.

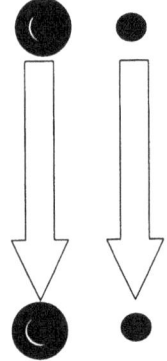 The reason why the **mass of the smaller object does not apply**, is because **it is not the object** that is **drawn** to the Earth, but **the space-time in which the mass of the object finds itself in,** that is being **drawn towards the Earth**. The most obvious proof that something applies movement to something directing the movement towards Earth is the simple pendulum. If gravity was a force, one should get the same result by applying a pull spring to the pendulum. With the spring connected to the bottom of the pendulum, both time as well as space would compromise. However, we know that the result proves the opposite, which proves that there is no force applied too the pendulum. On one of my many crusades I met one of the most influential academics on astro-physics in Africa.

I tried to explain the pendulum swing to a high ranking man of much academic importance, doing great work on behalf of NASA in South Africa but with me not knowing his superiority on the matter of the pendulum, I took this man on, on one of his specialties. This man was apparently an expert on research or lecturing or what ever on matters about the pendulum. Of course, as usual, the very first thing he asked (as all Newtonians do) before even asking a person's name, was at which university did I study and what my academic qualifications was. By replying that I have never been at any university for longer than a few hours in my life, and therefore my academic qualifications was less than zero, this highly rated person of high standings was less than impressed to spend any of his valuable NASA paid time with me. To complicate the whole aspect of my un- welcome visit was that I tried to explain to him being the expert on the pendulum that he is, about the pendulum. Boy, was the man annoyed with me.

He was very polite and very civilized about the whole issue, but his annoyance with MY EFFORT ABOUT EXPLAINING THE PENDULUM TO THE EXPERT ON EXPLAINING THE PENDULUN WAS MORE THAN HE COULD BE CIVILISED ABOUT. There is a point where a person gets a little too civilized to be true and at that point where a person gets a little too polite to be civilized about a topic. Any person can sense such a point the point when one realizes that is the point of limits. I also realized that what ever I had to say on any matter concerning all relevant matters was as good as never said as far as our Professor Doctor, was concerned. As a matter of fact that Professor from the University of Potchefstroom is one of the most exceptional men to walk this planet. I gave him a copy of The Thesis and after reading only four pages from a book containing over two thousand pages he was able to draw a conclusion and condemn my work. Of course the reason was the usual: I did not understand Newton obviously because of the lack of education on my part.

You may ask yourself: *"What was me (the un-welcome person) trying to say?"*
This is what I am saying:

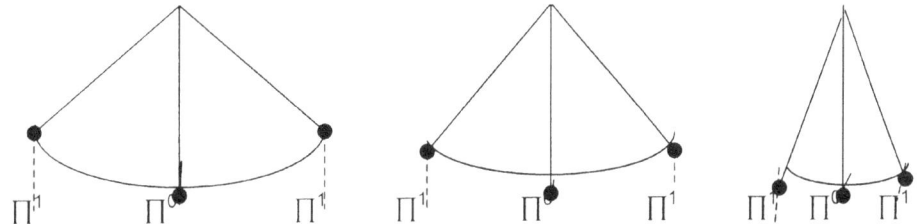

Time = constant	Time = constant	Time = constant
Space = reduced	Space = further reduced	Space = Reduced even more
Stroke $_1$	Stroke $_2$	Stroke $_3$

Time remains the same. The swing distance tarnishes. Period $_1$ = Period $_2$ = Period $_3$
Swing distance $_1$ $\neq$ Swing distance $_2$ $\neq$ Swing distance $_3$

In the pendulum principle that brought Galileo his everlasting fame, the pendulum swings at an even interval. As the **pendulum swings**, the **space tarnishes** while the **time (period) remains** the same. This is the principle on which all clocks work. What is it that Newtonians are missing for three hundred and something years about the pendulum? Newtonians are not seeing the very best example there is to indicate singularity outside singularity.

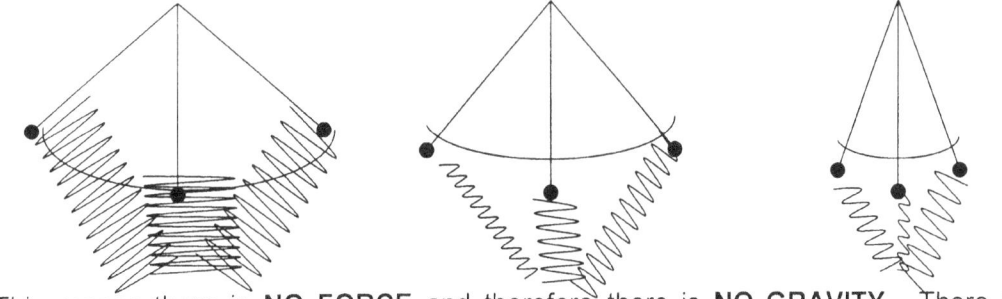

The pendulum indicates the very point of singularity the earth holds and the marks on both sides where singularity deviates in space, giving time to that singularity diverting.

If gravity, which is a force, did apply, then it would be as if a spring was fixed to the bottom of the pendulum and to an unmovable object below the pendulum. With the applying of the springtime will not remain at equilibrium but will tarnish in a vector with the declining of space. Something is holding time steady to the demise of space.

This means there is **NO FORCE** and therefore there is **NO GRAVITY**. There is only <u>space</u> (stroke) **time** (period) = **space-time.**

When a spring of 9,81 Nm is mounted to a pendulum, which is an equal force to that of gravity, both the time period and the swing distance would equally be affected, but to a lesser degree as the swing distance declines.

ALL SCIENTISTS GO INTO FRENZY BECAUSE GALILEO WAS PUT IN HOUSE ARREST FOR TEN YEARS!   HOWEVER, THESE VERY SAME SCIENTISTS

ARE STILL HAVING GALILEO'S WORK KEPT IN HOUSE ARREST AFTER ALMOST 350 odd years. HOW DO THEY EXPLAIN THAT? Galileo introduced the best devise indicating space-time and half a millennium onwards, nobody but me can see it.

What is space-time exactly? Einstein was the first to explain the existence of space-time, but Galileo was the first to indicate space-time and Kepler was the first to pin point the position of space-time.   Einstein made one big error in judgment. In his all to well-known formula $E=MC^2$, he relates to space-time as if space had a factor of one and time was the altering factor. According to the Einstein / Newtonian Order of High Priests, we live in a total dark, totally flat, and single dimension Universe.

Why would it be a total dark Universe? According to Einstein, the speed of light is the same as the speed of time. Should that be true, photons have to freeze in time, and must be unable to move through space in time! (I shall elaborate on this in due time.) This proved how far the greatest Newtonian outside Newton really were off the mark

Look around you and see all structures (**space**) are different in size. Therefore, **space** cannot have a factor of none converting to one and back to none, but relate to the size of the object, whether it is an atom or the cosmic Universe.

Time (C^2) is at an even factor as all things in the Universe relate to the same time, (although not the same duration of time). By implying that $E = MC^2$, he puts R^3 at a relative value of one throughout the Universe. Space can hardly disappear, but can compromise under abnormal star growth.

Every round object has a point establishing a very centre, a middle dividing one side from the other. That division determines the space from one side away from the other side. At one point there must be a point that does not fall on either side of the divide. Such a point will still be a circle, because from that side the circle divides into two sectors.

$$\Pi^1 \qquad \Pi^0 \qquad \Pi^1$$

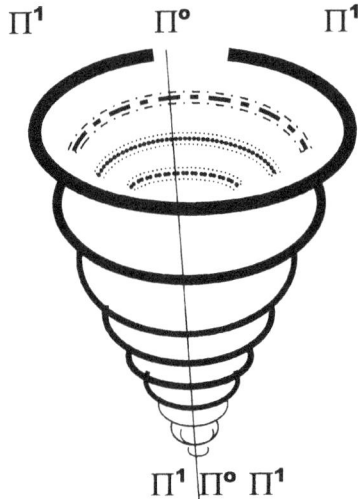

$$\Pi^1 \; |\Pi^0 \; \Pi^1$$

Every solar structure is spinning around an individual axis while the whole lot is spinning around a mutual axis the Sun provides The spin that shows on the different planets is the most crucial aspect of their orbiting the Sun . Calculating a circle involves two aspects where the one is either the radius or the diameter that

is double the radius. The other is the factor Π. Π **X D^2 / 4 = circle and Π X r^2 = circle** The point of singularity cannot be in space at large because space is not there and secondly what ever is there spin to slowly to have a connection with singularity directly. The pendulum indicate the very point where all the Universe conjuncts placing space in relation to the time-Zero singularity as indicated through the position the Earth maintains individual singularity parting from cosmic singularity The pendulum is a direct measure of space-time flowing but to this moment where I write this, I still have to find one Academic that understand my connection. Since there is no Newtonian thus far capable of seeing the comparison I draw between the moving of space in the time it takes such space to move I shall repeat it once more in the hope one might see the light.
Take the pendulum.

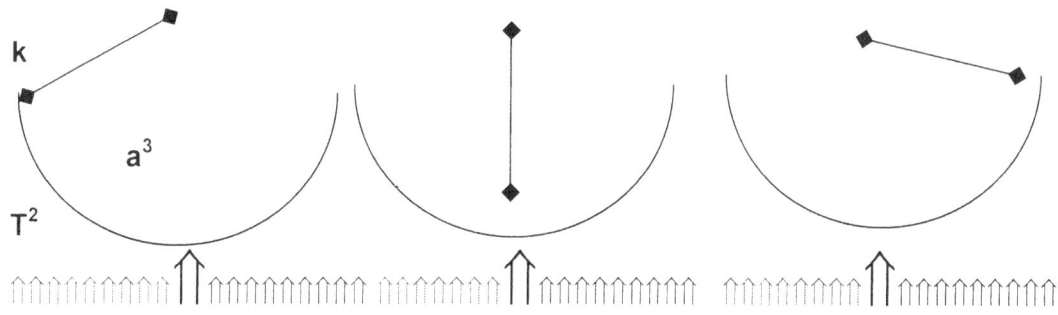

Every time the pendulum arm crosses to the other side it indicated the most important factor.

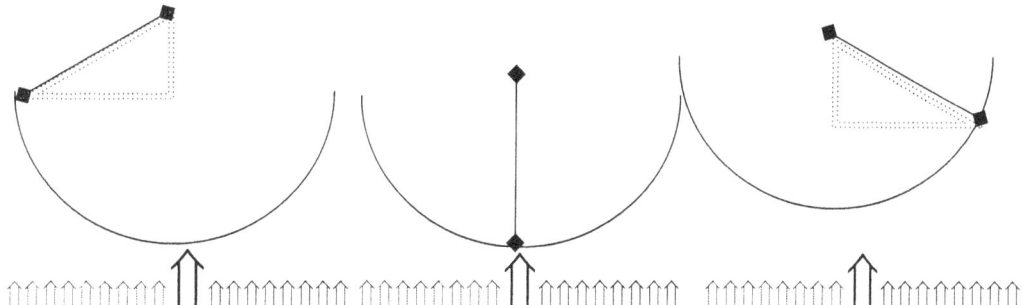

Every swing the pendulum arm does it brakes through the factor holding the universe in place.

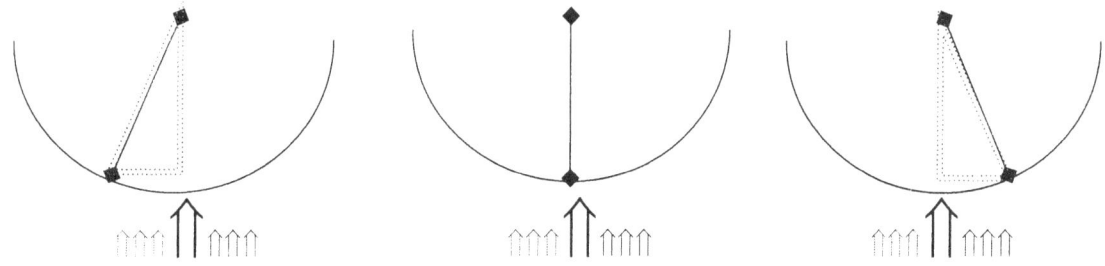

The pendulum not only crosses the singularity the earth dictates at that given time and the pendulum not only points at the factor maintaining space-time on earth.

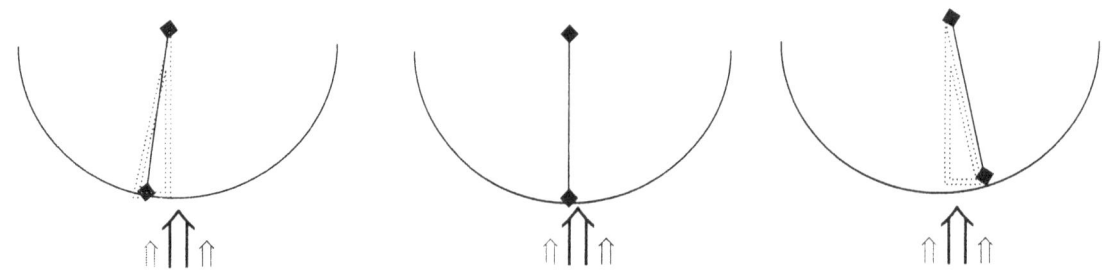

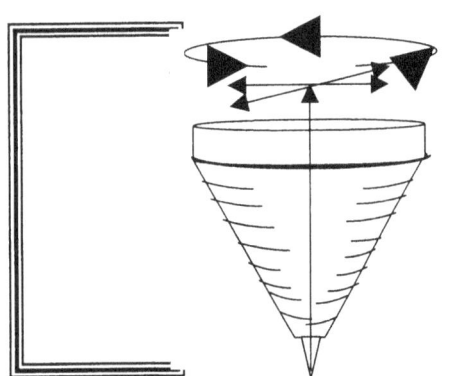

In the center runs a line called the axis line.
The line does not show any influence on managing the top when the top is motionless and bounded by the Earth gravity. However the sooner a motion sets in that is adequately strong enough to support the independence of the top the top generates enough gravity to sustain and independent attitude in relation to the Earth.

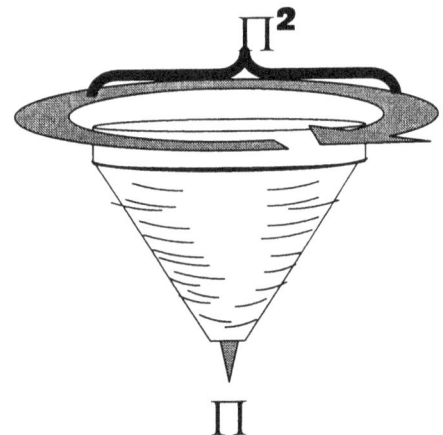

Π^2

Π

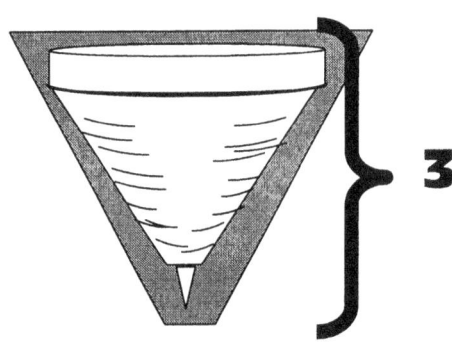

3

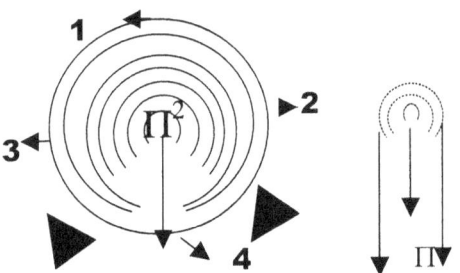

the body of the top rotating and the

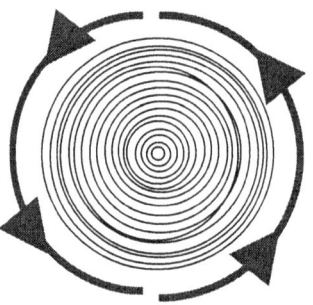

The dynamics that then support the top in motion comes from four points serving the top with time $\Pi^2 + \Pi^2$ which comes about from the circle the top forms by spinning space rotating in relation to singularity **space** $\Pi\Pi^2$.

After the top is thrown the top changes some very vital characteristics in behaviour to what it had when it was not spinning. As the top hits the ground after being thrown with it's commencing its spin initially the top starts to rotate and it is as if it wishes to exceed is spin by moving around in circle – like formations while spinning excessively around its axis.

It spins vigorously as if the top suddenly is too energetic and exited to stand still

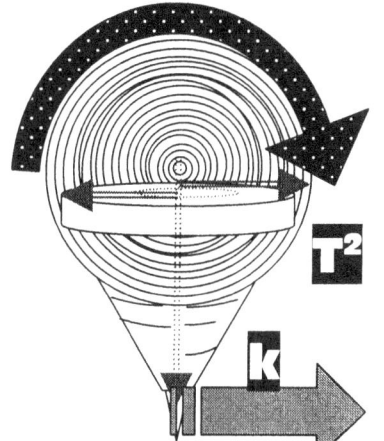

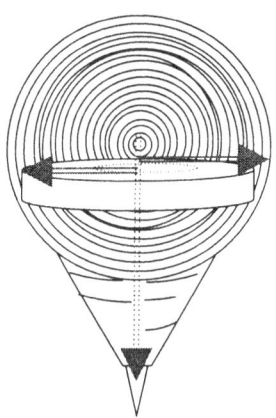

and that is precisely what happens. This surging with excitement is a charging of vitality that finds a new dynamic and is most important rule cosmic principles.

One can clearly see that it is Kepler's formula playing out as the Coanda principle. While the space is developed and defined by the motion T^2,

which verifies the independent space the top has acquired by motion of spin, it is also clear that the relevant factor of linear motion demonstrates it's presence in the moving about as the top is rotating. That puts Newton's claim of motion not being a factor in total disbelieve.

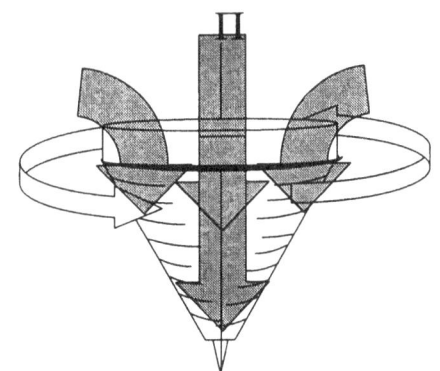

With all the excitement and no where to take the extending of the drive line runs down the developed singularity inwards towards the newly established governing singularity that keeps the newly formed Universe erect. That is why the top is spinning in the first place. The more assertive the spin is in velocity the more reaction there is from the lines running towards the centre and extending through the expanding outwards.

In real terms the space of the top expands as the spin is in contact with more time in motion in quicker presentations during the same time in period as a bigger material unit fills the space because of more material duplication in the same period that is allowing the top to spin. In this the space in which the top spins has to expand as well as the

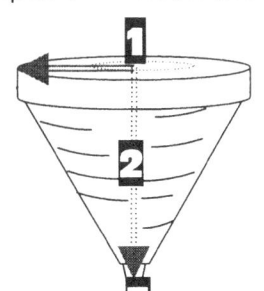

process of filling time by means of duplicating as well in order to compromise for the material relevancy growth to fit the newly acquired singularity governing the space-time and being erect by the motion.

The support that the spinning top finds in it's task to establish a governing or controlling singularity keeps the top spinning in an upright and erect position that is only supported by the motion putting space between infinity in the centre and eternity in which the top spins. By placing differences between infinity and eternity it not only charges singularity to life but also charger space-time into the Universe.

Through the behaviour and characteristics that the top display it is possible to find answers to the cosmos that the greatest mathematical minds was unable to solve. It is a case where their genius was to great to find solutions and the search of the final results proved to complicated.

Looking at what takes place as the top start spinning we find a line coming from what was a mathematical point. What is there is not there but for those with intellect to find that what is there being present. What is there is not part of the cosmos and yet what is there drives the cosmos. It generates motion by not turning and places what is, in contrast to what was and what will be the very next instant. The facts are indisputable even to the most ardent mathematical Newtonian disbeliever and what the top parts is how the Universe came about. Motion parted eternity and infinity by developing space-time as a partitioning screen.

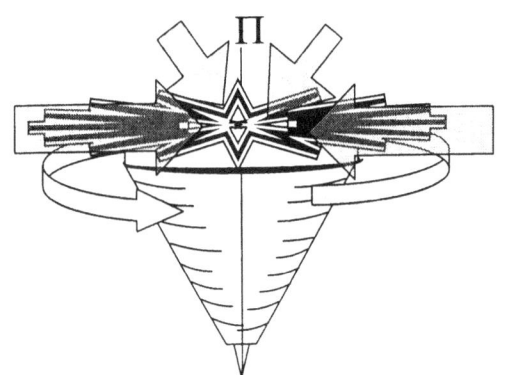

More spin increases both lines that force gravity by the increasing of T^2 extending k, k^{-1} as well as a^3. The space wants to exceed its boundary because the motion suddenly allows the space to become extended. The gravity line running to the centre wants to extend for the same reasons and so does the gravity line running towards the liquid that should be there and that should be enforcing this sudden living up to better standards.

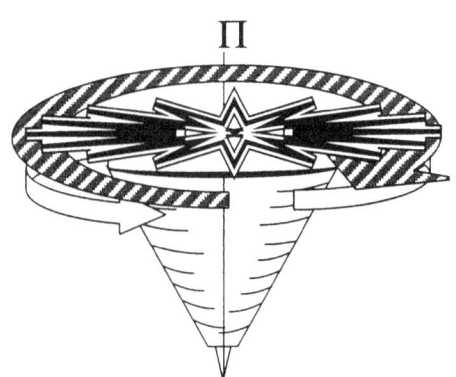

The spin under normal conditions can only come about as a result of more heat. With that aside the spin normally caused by heat will bring on a linear gravity running towards the centre of the top. This is then a product of $k = a^3 / T^2$. But to counter this (Newton's law on action and reaction), another balance comes about where $k^{-1} = T^2 / a^3$ that centres the material inline with the

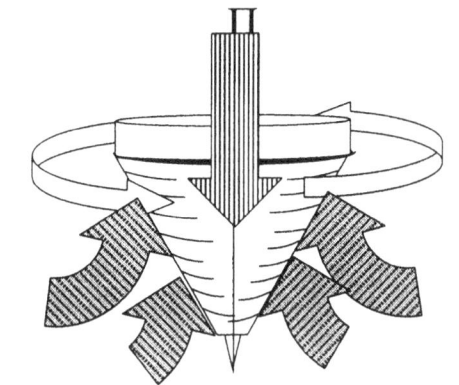

progressive spin and the extending of the motion that should be because of a liquid heat adding to the material.

The support that the spinning top finds keeps it upright and performing as if in a fighting mood. However, again I have to press the point that it is life that initiates the motion and for this motion to start as a natural flow of events requires a lot of nourishing by the independent singularity that starts to drive the object through a combined effort of rotation of all the included atoms accumulating assuasive heat to bring on such motion. When this process is in a natural occurrence within a star within a galactica it is the indication of the coming about of a newly developing in the heat centred cradle of a galactica. However it can only be gravity that is able to fight gravity by extending the Earth gravity and by extending the Earth gravity we find some part of the Roche limit also applying.

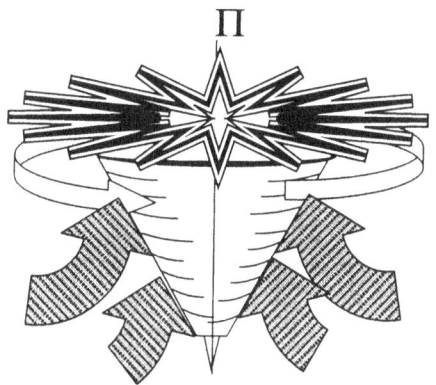

The heat that should supposedly under cosmos law drive the spinning top will come from the governing singularity, which is accumulating the heat in concentration by the contraction or cooling ability the top singularity acquired. However in this case the spin is a result of life's ability to manipulate space-time and alter cosmic events to the free will and interfering nature of life. The heat that would establish such a drive in motion in real cosmic terms would require a lot of nourishing a sustaining from a large number of maintaining atoms that produce a large flow of space-time and can concentrate much excess heat outside of the atom sphere.

With sufficient energy the top gets into a fighting mood that makes the top very reluctant to give up this newly established freedom. Be behaviour now attributed to the top is normally the manner how a star develops in the galactica cocoon and how the fledgling star gains it's birth right to leave the nest of the cradle of the galactica. The atoms form a sum total of space-time displacement that can support the generating of the required gravity in securing the heat that would unleash such a drive. Such singularity in governing come to life and release the new star from the blanket of heat that covered the star up to the time of its release.

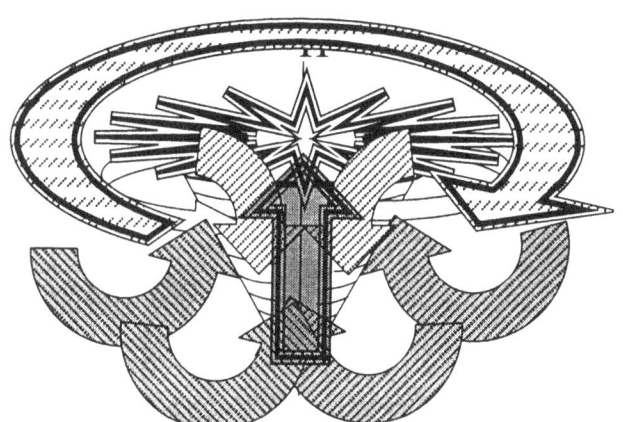

This example that we can gather from the top's behaviour shows how desperate the governing singularity can become when starved of motion and how such an excited singularity can put up a fight for life and independence. The top is in a fight for independence while the Earth is restraining the independence. The fight goes on until the Earth suppresses the last bit of motion that the top has and the top uses the last motion it has to defy the Earth's control.

When the motion exceeds the level of the Earth gravity, the top shows an eagerness to rise to a higher level of independence in the same manner that an electron reaches into higher rings of energy because the top with motion is in an electron or expanding relation with the

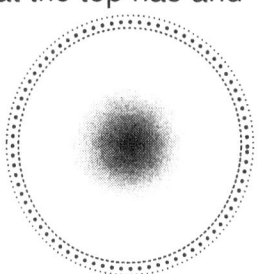

Earth which is filling the proton or contraction role and the atmosphere being in the neutron role or gravity-motion supply.

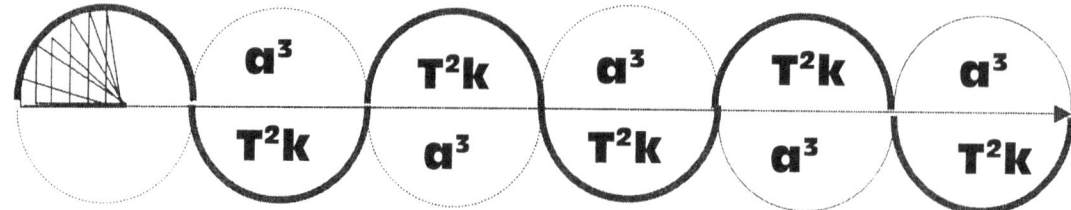

Let's quickly establish events as they translate singularity from a dot to a controlling entity that is demanding space-time through the establishing of a

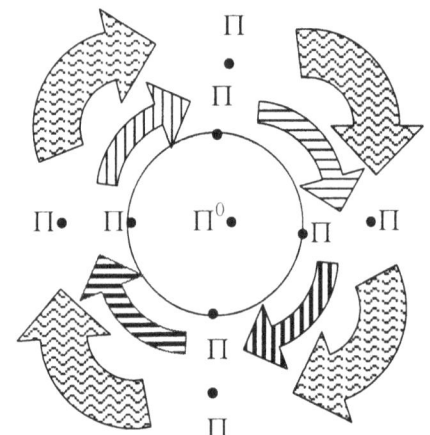

separate individual drive. The motion comes about which prove to be that which generates the gravity that drives the individuality in the top.

In the sphere centre is the spot that has to be there mathematically by measure of (Πr^2) / $(\Pi r^2) = (\Pi^0 r^0) = 1$. In order to provoke the line forming singularity into existence, motion is required, just as Kepler indicated where the space becomes equal to the motion and the motion is equal to the space $a^3 = T^2 k$

The inner four is singularity points equal to and is singularity being in direct contact with singularity charging time. It is singularity charging the presence of singularity by four points that would form the four forming time. Then one further point to the outside would form space with an indirect link to singularity. This formation goes on as long as time distorts to form space and space will forever have four to the inside connecting time and three in equivalence forming motion.

The Coanda effect is proof of gravity coming about through space forming motion. In the case where water diverts the normal directional flow the space that translates to the motion is deflecting singularity with the flowing water charging the motion. In the centre of the object having the round form, singularity is duplicated and by transferring Π to form Π^2 and the motion of the water creates a line of gravity that pushes the flowing water to follow the direction that the newly gravity applies to the water.

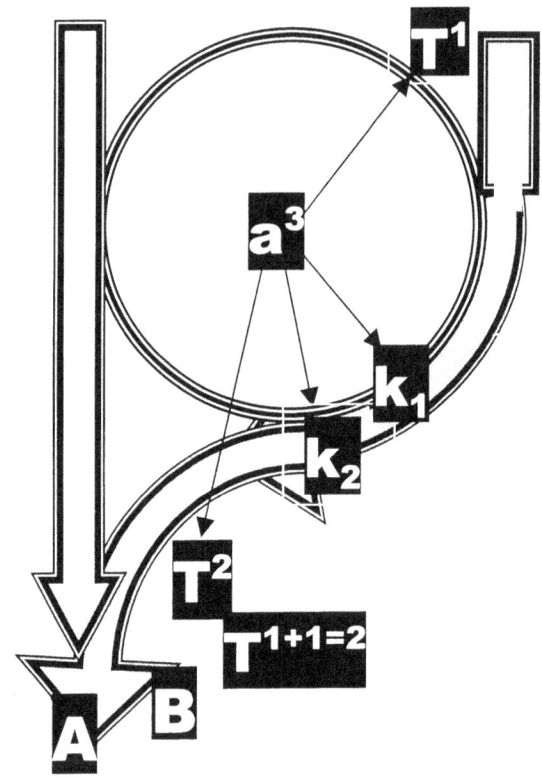

Normally water will run down to the centre of any gravity point, as A shows. By allowing the flowing water to come into contact, an object of a specific form the flow will divert (B) from the normal line and follow the contour of the object presented. For that to take place there is one condition that has to come about.

This again proves Kepler's statement of $k = a^3/T^2$ that specifically states that space (in this case the object transferring singularity to a new position within the round object) and with the motion of the water redirects the gravity flow of the water to new space in new time. Only Kepler can explain the phenomenon but only when Kepler stands alone, correctly interpreted and divorced from Newton's opinion about Kepler's statements.

The motion we detect as part of the Coanda effect runs through all spinning material. It is part of the atom as much as it is part of the sound barrier and the sound barrier is just another atom having Lyman series lines, and work also by the principle of adding heat which puts the object expanding in a higher relevancy than there was before.

However much noteworthy as it is it is prudent to consider that only when the atom unit is broken and the Roche limit is crossed does the sound barrier come into affect. It is the breaking of the bonding unit ($\Pi^2/2$) that becomes the sound barrier although the breaking is never completed ($\Pi^2/4$) as long as both objects share concentrated liquid time that the earth supply.

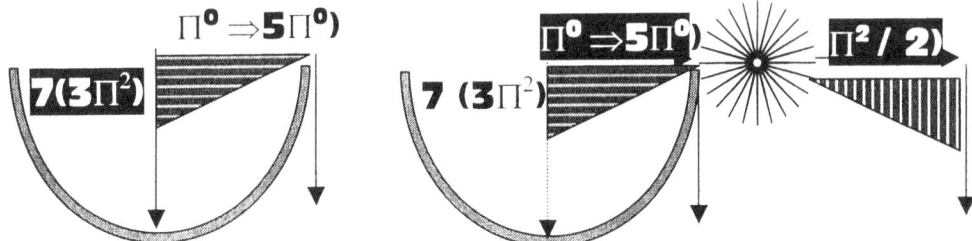

It is as if one then must claim in affect that Kepler held $a^3 = T^2$ $k = 0$. If the Sun and the Earth have a rotating relevancy of zero either the Sun has gone away or the Earth stopped existing. One cannot claim there is a wheel and then remove the spokes because according to you taste, too do not like the spokes

With everyone of the four rotating points spinning around a centre while duplicating the value of Π in relation to the centre Π^0 at a measure of $\Pi/2$ and where Π^2 is responsible for establishing as well as relocating Π by duplicating Π through the motion thereof, therefore $\Pi^2/4$ becomes a limit in relation to the development from the centre. One has to remember that a star of the present takes characteristics of the form from the ear before space was a factor.

As the absolute master of motion **Newton** should have placed emphasis on the motion aspect when he as a young man that saw an apple fall from a tree he made a brief calculation but he used the mass instead of the motion while Galileo proved that mass has nothing to do while the falling occurs. Seeing this he jotted down a formula and chucked it away. Newton however insisted on mass in spite of the clear evidence brought by Galileo to the contrary of mass playing a part. However most surprising to me is that most Newtonians are not only incapable of seeing the facts my way but they get sometimes pretty unpleasant in a very coldish pleasant way about my view. If mass had a major part, then the more massive must fall quicker because the mass will provide the drive and the drive will excel the velocity. While it is true that all things fall equally, then mass has no part to play while the dealing is occurring and that is in spite of all the Newtonian abstinence about the matter.

In this matter I am disputing Newton's honesty. He placed the relevance on mass when he was a young man and retracting his former claim would have tarnished his reputation as a genius. His glory was worth more to his mind than what the truth was. As young man he drew instant fame by claiming mass as the driving force and when as an older man he found he had to retract the first genius, his fame seeking would have left him a scar on his reputation. In that I can forgive the man for the man was human and as Cecil John Rhodes said, all men have a price by which the man can be bought. Newton's academic genius was his all-important vice. The problem is that the incorrectness stuck with science fore almost four hundred years on and no brilliant mind since than was able to make the Galileo mass connection. What happened to the many wise that walked the path after Newton had gone to better grounds. Where is the honesty in those that were supposed to search for the unblemished truth? Gravity is motion and mass is the restraining of the motion of gravity. The top shows the truth. Let us reflect once more

What is it the Newtonians fail to see? If an electron is orbiting around an atom, the inside of the atom must be a circle. If the atom was not a circle, it then had to be a cube. The electron cannot rotate around a cube; therefore, the inside of the atom is a circle. The cosmos is one big atom imitating all atoms as all atoms produce one Universe

In a circle, there is a radius that initiates the circle. The calculation of such a circle is $\Pi \times r^2$.

$$\frac{\Pi r^2}{r^2} = \Pi$$

If one removes the radius from the circle, the circle remains, only holding the value of Π. By removing the value of r, Π becomes singularity with no place to be. Singularity is the place where there is no space to be in place. However, Π remains because once r receives the slightest of space Π will find space. Then the circle will grow to Πr^2 and r would determine the space. Without space, there is no r but there is a circle with the value of Π.

Singularity is in every single rotating object, be it the proton or the combining effort of all particles in the Universe. That is what light and the photon is. It is concentrated heat that the Sun (or any other generator of electricity) concentrates to connect the concentrated heat to singularity where the heat receives either temporary connection to singularity or a small piece of individual singularity. All spinning matter has the point where the spin is still there but the radius is to small to measure by any means. That point is standing still in relation to the rest of the spin. In relation to that logic I do not accept Newtonian science holding the radius of the spinning object unrelated to the spin, whether the spin is applying or not.

Applying Newton's second law F=ma

One arrive at the formula

$GMm / r^2 = m (\omega^2 r)$

By replacing ($\omega^2 r$) with $2\Pi / T$ we obtain Kepler's third law

This law predicts that $T^2 = a^3 r$

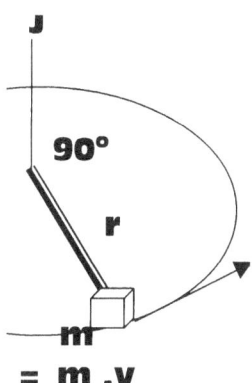

$= m .v$

The mass (m) multiplying the speed (v) forms a new value J AND THEREFORE j CONTINUOUS TO IMPLY $J = I \omega$

$J = r X p$ where $p = (v = r x \omega)$

$J = r.m.v = m.r^2 .\omega = I. \omega$ and becomes interpreted as $J = I \omega$

This establishes that $r = dJ / dt$

Since this is the absolute crux that Newtonian science pivots around I feel it is important enough to return to the whole issue once more in similar detail.

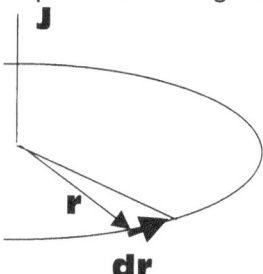

$r = dJ / dt$ In the case of planets in orbit around the Sun r forms a value of zero because $dJ / dt = 0$.

Since Newton became an institution forming the King bee of the academic cartel world wide The Brainy Bunch had Newton's vision written in the minds of the future generations almost at gunpoint...well definitely at an academic gunpoint.

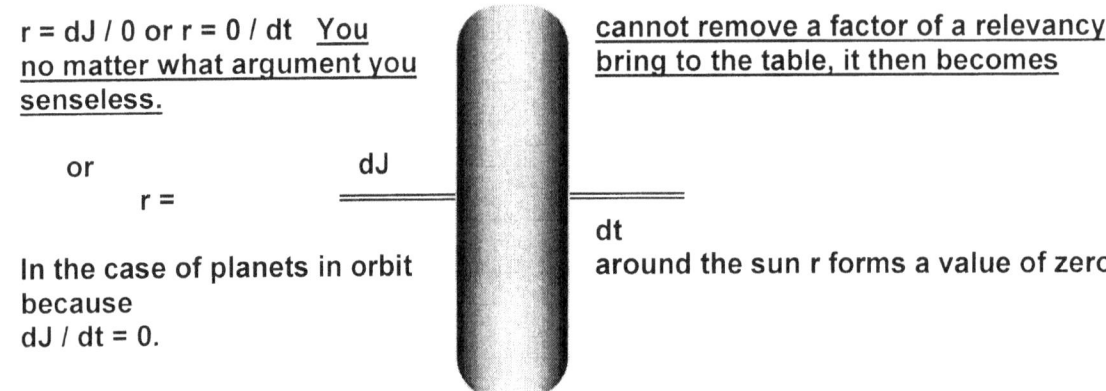

r = dJ / 0 or r = 0 / dt   <u>You</u>
<u>no matter what argument you</u>
<u>senseless.</u>

or

$$r = \frac{dJ}{dt}$$

In the case of planets in orbit because
dJ / dt = 0.

<u>cannot remove a factor of a relevancy</u>
<u>bring to the table, it then becomes</u>

around the sun r forms a value of zero

I am not the brightest in the world that I admit, but one thing no one can do, not even if you are the one and only Isaac Newton, is that you cannot place any relevancy in a relevancy and then claim it not to be in a relevancy because such a relevancy does not suit your taste.

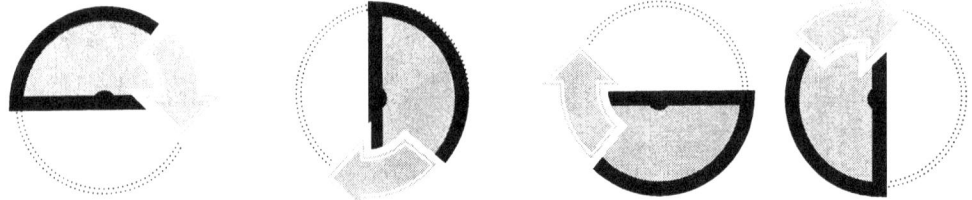

I wonder where would one put the zero part on the spinning wheel and what part must be excluded from the wheel. What Newton suggests, is a wheel has one side on top and no side at the bottom. While the wheel is spinning one may not remove the one side and then claim there is no attachment between the top and the bottom. That would mean in a graph the top is not connected to the bottom because a wheel spinning is a graph moving against time. It is the principle all driving is done and not the least electricity. Every quarter of a rotating body is opposing the opposite sector directly and completely.

Any Newtonian that wishes to justify any form of support about Newton's claim on rotation not establishing work must please explain what happens when the Coanda principle draws water by motion and how that motion cannot be gravity.

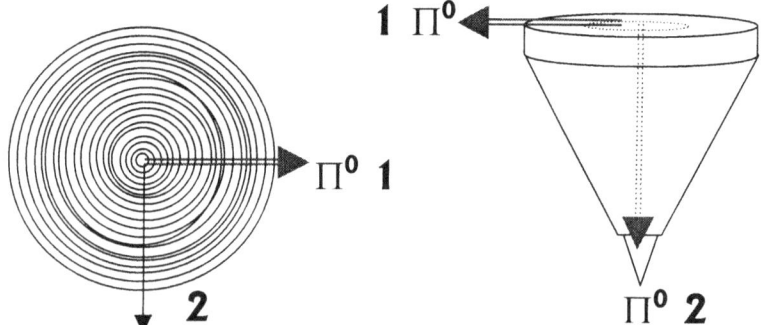

$1 \; \Pi^0$

$\Pi^0 \; 1$

2

$\Pi^0 \; 2$

If there is no production through motion how would a top find a balance and what then inspires singularity to charge an erect stance through the generation of motion? These are legitimate questions in search of answers.

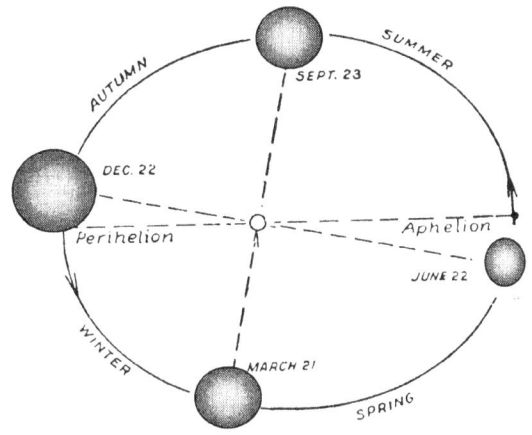

If gravity was mass inspired it would have the result that the Earth must be at some point during the year more massive than during other periods of the year. This we know is not the case and therefore the claim on mass is somewhat silly and a little middle aged. There are two equal but opposing gravity directions counterbalancing and both is the same that works independently to achieve a mutual goal. The relevancy from one side is about claiming space by progressing time and the other is by containing space through reclining time. That is why the comet never hits the Sun . It is because the Sun and the comet are in four different seasons in relation to each other while they are going through the quarter motion of time.

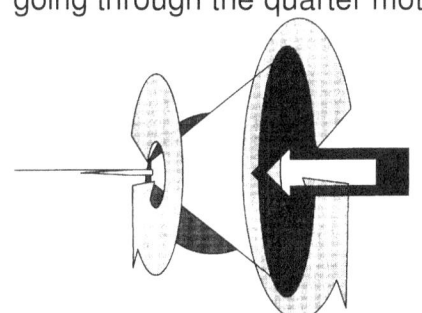

Gravity is motion and space is the blocking of the motion. Any object must have either gravity or mass but cannot have both. An object can be with gravity or the object can be with mass but it cannot be in both conditions simultaneously.

The motion of the neutron (2) covers the gravity (3) that the neutron has while the space (4) flows unhindered from the time (1) position through the location (5) fitting the neutron to the location (7) fitting the proton (8,9,10). The electron has mass because it restricts the flow or gravity and the proton has mass because in constrains the flow of gravity. Only the neutron has gravity because it flows unrestricted. That is what Galileo's work tries to prove but no one listens even to someone as

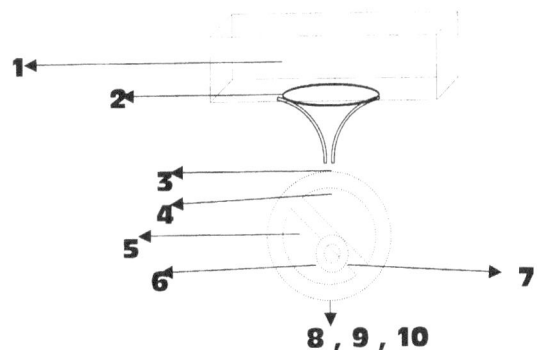

important as Galileo because every one is mesmerized by Newton's while Newton was absorbed by the lack of understanding the difference between gravity and mass. If he did understand the difference there are then he would have realized what Galileo was trying to say. While his little apple fell it had gravity, but once it landed it had no more motion and therefore the containing part took charge as the apple then had mass. Galileo said all things fall equal (meaning all things are equal in gravity) while falling or while being in motion notwithstanding the difference in size or mass. That Newton missed. That part all Newtonians that came later also missed. That is why Newton's first finding $F \, \alpha \, \dfrac{M_1 M_2}{r_2}$ being

$F \, = \, \dfrac{r^2}{M_1 M_2}$ is most true and most accurate however $F \, = \, G \, \dfrac{M_1 M_2}{r^2}$ is nonsense. All objects must be in motion $a^3 = T^2k$ where it will be $k = a^3/T^2$ in relation to one location and in another at the same time it will be $k^{-1} = T^2/a^3$.

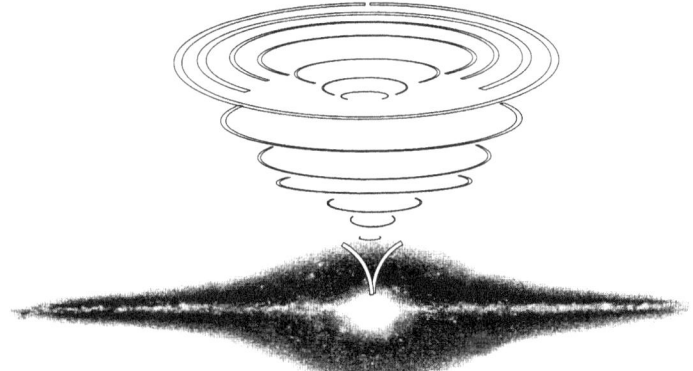

The motion within a galactica even generates sufficient gravity to re-enact a Black Hole within the centre of the galactica. This comes about just like the motion by which the top rotate, but the centre singularity being charged in the galactica is truly then a product of cosmic proportions.

The aircraft has mass when standing still but the motion that the heat of the engine produces, that expanding of heat to space converts the space to motion or gravity, which then relieves the aircraft of some of it's mass as the motion converts a part of the mass into gravity. It will always be some of the mass since the aircraft cannot be all motion. In the motion coming about from the engine that is converting heat to expand into motion the aircraft converts part (not all while it is in the Earth atmosphere) into gravity, which is independent motion from that of the Earth. While the aircraft is within the Earth, the Earth provides the motion and thereby serves the mass, which the aircraft (or all other bodies for that matter) will endure as the bodies remain a part of the Earth atmosphere.

The ship has mass but the buoyancy of the liquid in the water sustains the mass factor in order to provide the ship with another factor and that is displacement. The water holds motion in place and since the ship being on the water becomes part of the water in relation to the Earth it holds a part of the water in mass. However since the ship then holds part of the air or atmosphere in relation to the water the ship becomes part of the air in relation to the water and therefore the ship holds a relevancy of air in relation to the water. Some of the mass the ship has is regarded by the water as air and some of the mass the ship has is regarded by the Earth as water. It is locked in relevancy as the factors establish a ratio.

By $\wedge$ enlarging the ratio of air (wind we call it) onto a part of the ship, the motion takes up a part of the mass of the ship into the realms of the air and the air contributes to the motion that then find the ability to go beyond the breaking power the mass has and convert some of the mass into motion. Again the wind is merely heat expanding and the expanding provides the motion that contributes to the duplication of the ship. An army battle tank is all mass in our thinking because of the iron composition providing it with such a solid and heavy structure. When the tank is thrown from a flying craft, some of the structure goes to mass because the tank requires more than one parachute to slow the descent down making the fall less destructive in nature. If the tank is left to fall with any other body and without restraining, the tank will not fall faster than any other body because the gravity the tank has is equal to all other bodies. It is the restraining of the gravity that produces the mass that requires a larger effort to contain the decline of the tank, but that again is the interfering with nature since mass is the interfering with the normal   flow   of   nature.

The fact is that motion is the duplication of the same in ratio of the relative flow of time and when the duplication starts to claim the same position at the same location during the motion in time, the motion of the duplication of the space converts the part being restricted to the same location as mass, while the rest is being converted to duplicating gravity. By duplicating the mass converts to motion and while the duplicating is hindered the restraining goes into mass. But in all Newton's claim that motion results in nothing is nonsense.

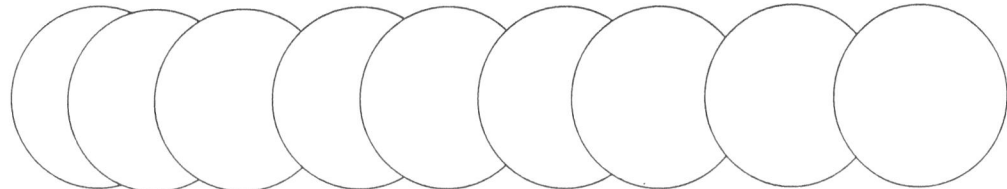

The fact that rotation does not produce work is impossible since rotation brings about motion changing the principles of the location.

The same relevancy we find in the motion applying to atoms. One do seem to get the impression that little changes in line with the rotation will bring some forward motion and some returning to the original position.

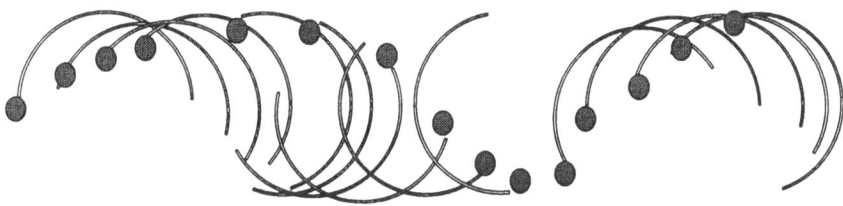

Even by using half a wheel would still bring considerable confusion but one can clearly see that Newton's presumption does not quite match reality.

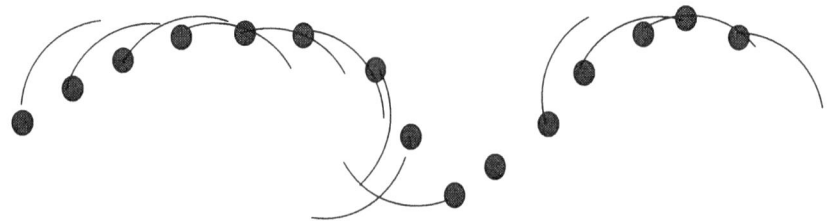

Shortening the arch changes the complexity considerably as one can then see a changing of the arch does not nearly bring the return of the dot to the previous spot.

When placing arrows pointing to a direction that is indicating the direction the line of movement, it becomes clear that there is a complete mismatching and the cosmos changes as rotation progresses. The behaviour, which I describe, is a flow of space through the line of time. Electricity is charged in this manner. The same generated force keeping the top upright is what is used to generate electricity. The flow of a charged conductor through excited space-time brings about the flow of a current.

An object in outer space has limited motion, which provides a part of mass and another part in gravity or motion. When the same object is in a Black Hole it is limitless and infinite in mass and has no motion. Outer space however is all motion as it provides motion therefore outer space is without mass. There is a mixing of mass or motion being gravity but having both is not having the same.

Even the electron serves the line of time, in the same manner. As the Earth spins through time by repositioning space in time singularity is re-applied, repositioned and re-aligned with the entire Universe in the manner I describe. The relation of the proton moving has to effect the following location

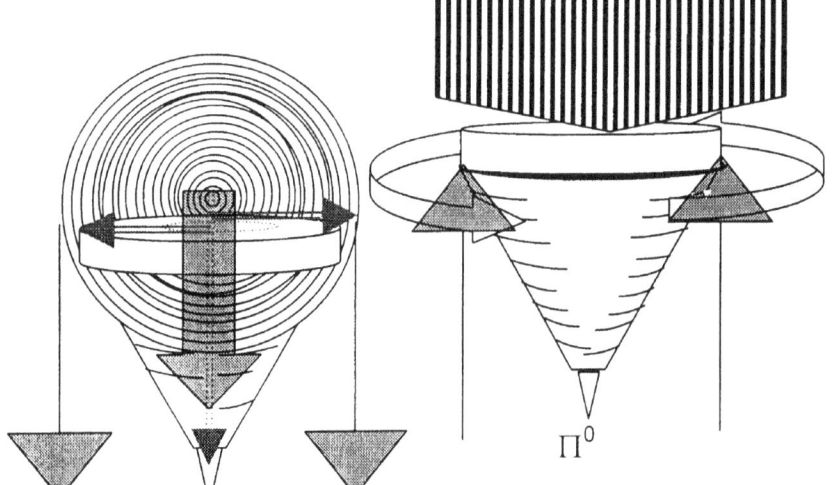

Π^0

of the electron since the electron is relevant to a position in space in time by a continuous motion through time.

Notwithstanding the motion that space forma as space is moving towards the centre of the Earth, the top finds a way to counteract the motion by producing a motion that is stronger than the motion restriction or in other words the mass, that is the restricting of the earth's gravity that fights to dissolve the impendence of the top altogether and thereby form mass which then is the lack of independent motion of the top.

By spinning there is no force pulling the top down and restraining the top to the surface of the soil. The mass is still there but that mass the top try to combat with vigour and the needlepoint holds the top spinning as the top is fighting the mass. On the needlepoint the top rides out whatever force the mass would enforce to restrain the mass.

The total restriction the mass control over the top has all but disappeared because the force or mass that the needle point of the top generate multiply the normal mass of the motionless top many times over because of the intensity that such a small area does increase the effectiveness of the top. Yet notwithstanding even more restriction by an increase in the mass restriction the motion still generates independence by motion evoking a defying erect stance.

$$T^2 = a^3 / k$$
$$k^{-1} = T^2 / a^3$$

$$T^{-2} = k / a^3$$
$$k = a^3 / T^2$$

In the spin there is relevance within the unit that forms all the principles we attach and associate with gravity. There is the expanding as well as the contracting which forms an integrated part of the rotation principle. If there were no rotation of a body, which installs the contrasting we, finds associated with rotation then gravity by principle would not have been possible. However the linear aspect also shows strong influences, which is as much part of gravity as gravity by rotation is a factor.

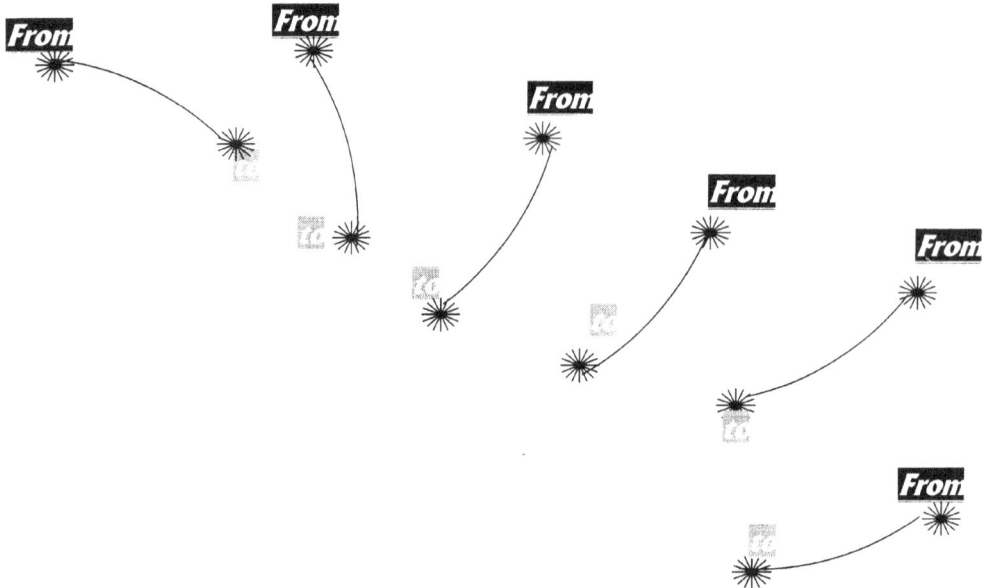

Although the electron is orbiting at the speed of light it is still in motion vertically and that also becomes a product of time as the whole structure is repositioning the relevancies, it had a moment before to what it will have the next moment. Such motion will again have an influence on the relation in the position the electron forms with the rest of the Universe while the lump of metal is now travelling as a spacecraft destined to other galactica. It is if we use the logic those intellectuals calling them Academics show and those Super-Educated that advocate how we may travel to far away galactica while we go on skipping the

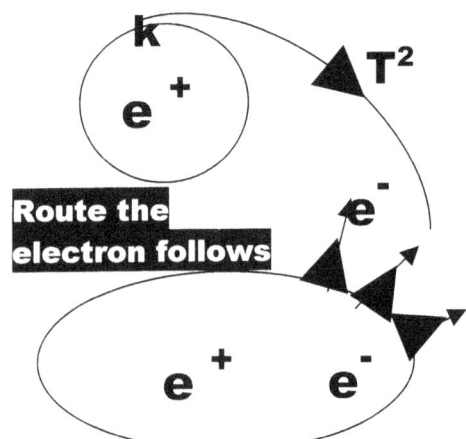

nearby galactica that is only two to twenty million light years away. Since the electron is duplicating by motion the motion links the electron to a time constant. The time constant is linked to the speed of light but time as such, is part of the speed of light. The faster we take the electron to go straight in the motion man produces, the less time there will for the electron be to circle around the atom. If we make **k** bigger in relation to increased motion, the smaller will T^2 produce a usable space.

There is **k** that forms the distance between the proton and the electron while the electron is spinning T^2 around the proton k^0. While all this action is going on, we think of the atom as being very still and satisfied with being a small part in a lump of metal we call iron. It could be any element but I use iron just as an example this time. The lump of iron is as motionless on earth as anything can be while being.

Even by coming erect through motion the generating of this stance finds its roots in the relocating of the rotating (T^2) in relation to the alignment with the line (**k**) in relation to space-time a^3 in time-space T^2k. The generating of the top and of gravity and electricity is provided in the very same manner by the Coanda principle.

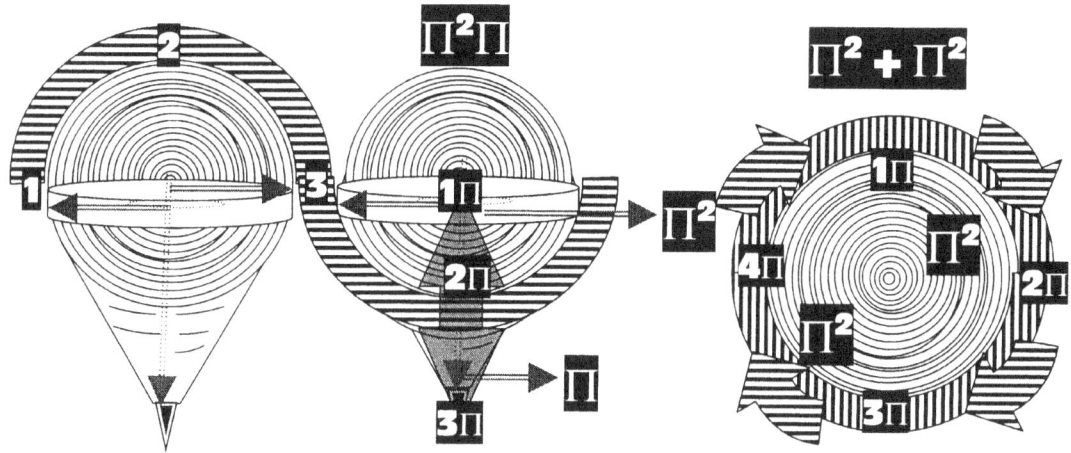

The motion that evokes singularity charges a graph from where the graph runs along the line of time. It is said that the spinning top is in balance but there the explanation ends and all parties are satisfied. Never is the question raised about what comes into balance? The balance is a control of space-time that is established as space is duplicated by time while time support space in duplicating. The space is limited by the rotary action of **4** points in relation to singularity where this generates **3** points serving infinity that creates a division between infinity holding it's centre space and eternity being **3** active positions in time and the three is an eternal motion that never ends. By setting the division between **3** in infinity and **3** in eternity the containing that comes about is creating a cyclic space in **4** points.

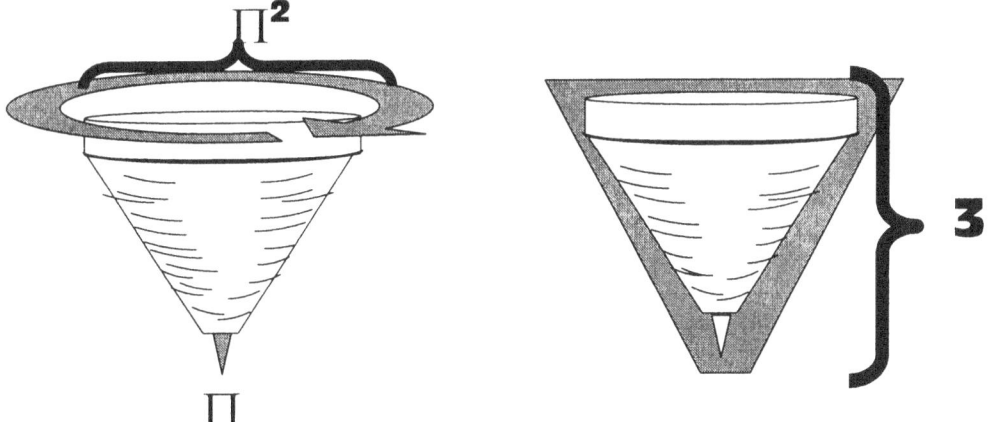

There is a something (if you wish I can use the term force although I strongly hesitate to use such an outrageous term for the most common aspect of the Universe) that is generating the power that keeps the top upright while rotating. The energy that is charged that is charged has the dynamics to stand its ground against the gravity of the Earth where the gravity of the Earth would under normal conditions depress the top into submission. However by rotating the top seems inspired and is reviving singularity by motion. The top is fighting and rebelling against the Earth's gravity when in spin. The top is performing the same way as an electric motor would. The difference there is between kit and an electric motor is the origin of the source that produces the drive. After all debating, there is one source that drives all forces small medium and large and that is the containing of heat and the distributing of heat.

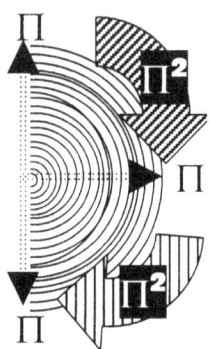

The Roche limit came in place at the time when all the phenomena came about. When the phenomena came about that action brought us a Universe to have and enjoy. It was when singularity Π^0 heated to form Π and that had to involve motion. When Π^0 expanded and formed Π it had to cross Π in doing so. In order to establish motion Π^2 it had to go from Π all the way to where Π duplicated as Π. This involved the initial motion at moment – Alfa when space formed time by forming space.

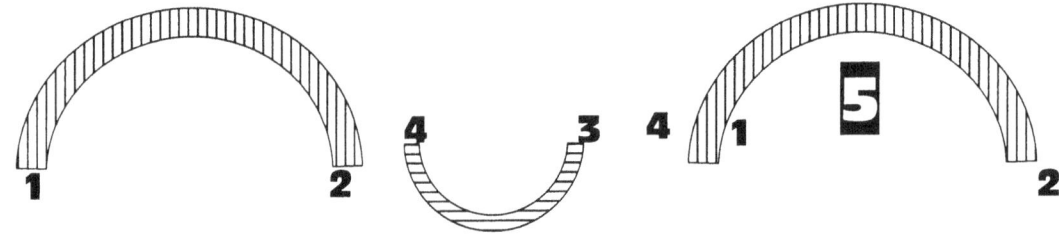

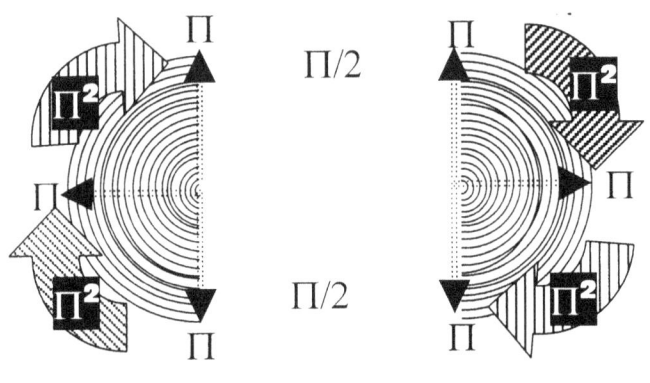

As the rotation was a change of directions involving four aspects, which was a duplication of the previous along the present going into the future, a division came in place that parted the one unit from the next unit. As the forth-spot serving singularity landed where the first developed, that made the first spot the fifth spot.

However this was accompanied by a rule, which today still apply in the cosmos.

With every four rotating points duplicating to reproduce on unit, a parting had to be devised to separate one Universe from continuing into

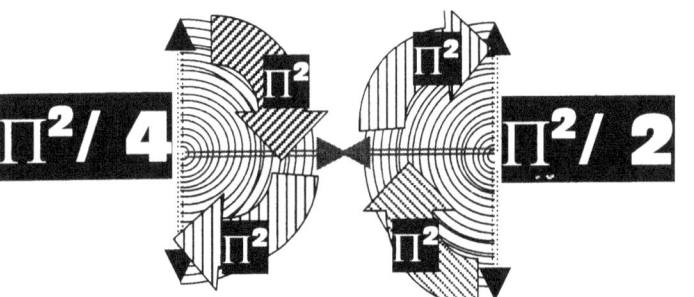

the next Universe. The four points duplicating the value of Π in relation to the centre Π^0 crossed the centre at a point measuring $\Pi/2$ and halfway where Π^2 lands the next Π by motion thereof therefore $\Pi^2/4$ became the limit that brought space in dividing point four from point five in relation to the developing centre. One has to remember that the Universe presently holds the characteristics it once enjoyed because once anything is part of the cosmos it has to remain part of the cosmos since there is no other place to go but to remain in the cosmos.

Even the motion of innumerable stars relate to a singularity in the centre that plays the part of the generated governing singularity and every faintest and slightest motion of every individual object plays a significant part in the generating of the governing singularity.

The Sun is on the outskirt of the Milky Way and the Sun is in an ova orbit around the Milky Way. The law of orbit is in principle that all orbiting structures follow an oval path.

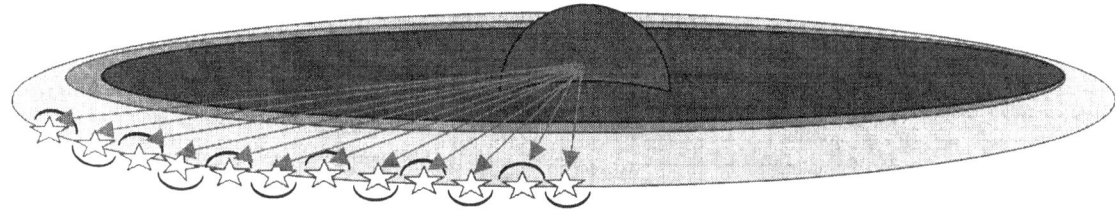

Exaggerated to a large extend the influence the Milky Way has to have on the Earth orbit comes to focus when a pattern comes in pace as the Earth follows not a circle but a wave around the Sun while the Sun sets its motion around the Milky Way. The fact that the planets orbit the Sun and the fact that the Sun orbits the Milky Way indicate an influence undeniable. The fact that the Sun is heading farther away from the influence should then lead to a variation in the planets orbiting wave. The Earth never, not once lands on the exact same spot by the completion of one more year cycle.

By not having a wheel rotate, the wheel becomes the factor of one, and the rotation becomes zero. The wheel does not disappear. In the cosmos, everything is rotating because nothing ever stands still. Therefore the mean equilibrium, the common factor there is to share, has to be one, eternity, the eternal Π, because all rotating objects has Π in singularity, and sharing singularity, gives every object in space a relation with all other objects in space. After trying for many years to bring them the candle, I concluded that Newtonians are incapable of realizing that mathematical principle as reality.

The comet rotates the Sun , and the Sun by itself has a point of singularity where Π remains without r. The comet, holding the orbit, also has a point of singularity, but since there is space separating the two objects, they cannot share a mean point of singularity, the very point of existing. Since singularity means just that, being single, there cannot be two. The comet and the Sun have a mean point of singularity but the space they occupy divides their common singularity. That is why they orbit in an oval path, a path where the one structure holds on to more space from its point of singularity towards the space it claims. Since they do not claim equal space, BY THE DENSITY they hold, the space will not be in proportion.

They do share in the common fact of singularity a point away from their individual singularity proclaiming their cosmic individual reason to exist in the cosmos. That point of common singularity holds space between individual singularity and that point of mutual singularity saves and protects the points of individual singularity. Since the start of time at moment-Alfa where both found the space they occupy, in the space they hold, maintaining a time to that space in accordance to the singularity they hold that point will be their individual eccentricity from singularity.

The two objects are holding eccentric space around their individual but common singularity. That point of singularity is Π the circle without the radius because the singularity removes all forms or values of r, discarding r to infinity and leaving Π to be singularity.

That is why gravity is a fixation of Newton's mind making Newton bullshit, and his $F = G(M_1M_2)/r^2$ is utter nonsense. The moment you say Newton or any of Newton's laws, the Newtonian brain stun. Not once did I find one Newtonian surprised at this, I could not once find one single Newtonian to see this. It always leads to an argument and the argument is about Newton being in use for centuries. One Professor even answered me by saying that I should realize Newton's formulas placed man on the moon, and if that is not proof of his correctness to me I will never obtain proof. That is beside the point. That is miles from the issue. If you say Newton is wrong, you commit the worst blasphemy possible. One may swear at God and all is understood but mention your not accepting Newton's gravity and they all fall on their knees, cover their eyes in the ground, start stuttering and moaning and you cannot make them see anything but Newton. Dare say there is no such a thing as gravity because Newton is wrong, they run outside and hide the woman and children from your rage of mental instability.

Because I have had unmentionable arguments that I in the end lost because the mental Newtonian block all Newtonians hold covering their senses, where I could not reach a single spot of healthy logic within their minds, I wish to run through the facts once more and find what is so incomprehensible about the issue.

This in fact, is the very same findings that brought Johannes Kepler his own everlasting fame when he declared that the planets stand to a value of $a^3 = T^2k$ as they orbit the Sun . Never once did he mention the presence of a force or gravity. Newton came up with this bogus idea all by himself without the help of other "giants" as he called *Galileo and Kepler*. In a later stage I indicate that I might prove the possibility that Newton did not have enough information to draw conclusions about Kepler's work. Newton saw a circle in Kepler's formula and there is a Universe of information hiding in that formula because that formula depicts the key to science namely singularity.

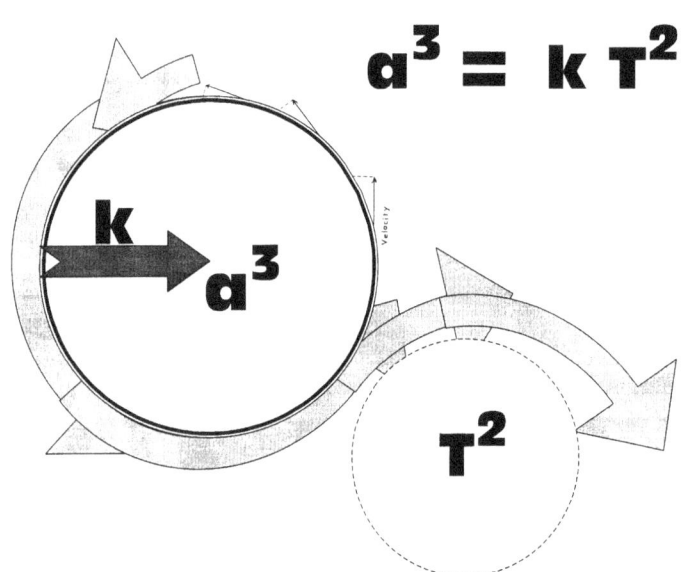

Newton made the formula one big blunder as far as the cosmos is concerned. Newton works perfectly well where there is equilibrium and unchanging in space and time as we find on Earth with the Earth forming the basis for space-time.

Taking Newton to outer space is a blunder and Newton created the blunder by re-adapting his original formula of $F = r^2/(m_1 X m_2)$ to fit Kepler's vision of $\mathbf{a^3 = T^2\,k}$. This very same bogus idea helped Einstein to ignore the space factor of R^3 and place a relative value of one to space.

This he stated (without stating it) when Einstein put the Universe in a single dimension property at the point where gravity was stretched to the limit. Einstein put the Universe to a three dimensional value of matter, space and time and then out of the blue he places space at a factor of one when gravity supposedly destroys time. This notion stands totally unrelated and divorced to reality. I do admit that Einstein is absolutely accurate when saying this, but the space he refers to, as outer space and the space disappearing in time are as far apart as the cosmos is wide.

Einstein was the one that said that space and time could never be separated because it was the very same thing, a point I agree with in all my findings. The difference between my point holding the Universe and being the Universe is within every atom because it is there where singularity is. From singularity through the atom space has the relation between Π^0 as singularity and Π forming space inside singularity holding time $\mathbf{T^2} = \Pi^2$ in relation to the triple value of $\Pi\Pi\Pi$ forming Π^3. I am afraid that Einstein made much more sense when he was still an amateur, working as a clerk in the Swiss patent offices. Then he landed himself under the spell of the Newtonian disciples and all his initial ideas that were factual, became integrated and confused with delusions of the "Xepted scientific Newtonian High Priests" called "acclaimed scientists" and their mesmerizing fantasies about gravity.

There is a way to explain space-time by finding **space-time**. Space-time is not some force well and truly out of our reach, the one we may dream about and wonder why we have to adhere to it with so much respect. Einstein the master of physics was completely lost in his physics. He went looking for a flat Universe, he saw singularity in the dark of the night hiding as obscure fairytale characters behind Black Holes, where he saw gravity lingering around stars with nothing better to do than to wait for passing light just to bend seven variations of different types of shit out of each of them. Singularity makes every atom rotate that makes every cosmic object rotate that applies the overall rotation to the Universe. **THAT IS TIME**. Each time I try to share the idea of mine with the **"Accomplished Scientists",** I do not get farther than the phrase*: "Newton and Einstein are wrong."* After completing this sentence, I get treated as a raving lunatic with extremely dangerous hallucinations indicating a murderous tendency. Why would not one academic listen to the rest I wish to say before bluntly denouncing me?

Nobody even listens or pretend to listen to the rest of my case. Every time I see the light in their eyes go blank and they sit patiently and wait for the motor mechanic to finish his senseless rambling. It is so obvious they consider me as mindless person with arrogance and having a nerve to criticize the two highest-ranking Newtonians of all time! I can assure you I am not mentally disabled! It took me twenty-one years of research and another six years in compiling and writing this book.

That is the relation matter has outside singularity. $R^3 / T^2 = 1$.

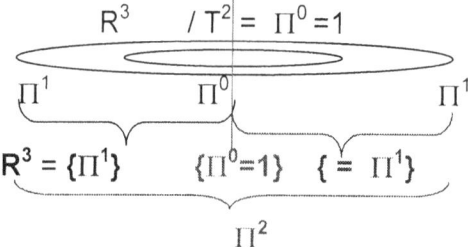

In any of the pictures on the next page one does not see space, because you see a space filled with particles. It is the atoms holding the space secured that forms the picture. Why on Earth would nobody realize Einstein was seeing the Universe from a wrong perspective? What Einstein saw was one hundred percent correct but Einstein saw what he saw in the space of the atom and not in space at large.

That is space-time holding every aspect the Universe hold to a specific relevancy The space outside singularity holds the time outside singularity because everything in the Universe is spinning. Science knows there has to be a difference because in space an object might be weightless, although it retains its mass, and no one can say the difference, except to put it down to "gravity".

Us, the tax paying public, are letting these Master Minded Academics get off the hook so easily, because every one is so scared to ask "why and how". In the pages above, I pointed to the most basic mistakes about the "gravity" which science ignores, because the answers they do not know. Even Nobel Prize winning work is blatantly misguided. I challenge any person to prove how an atom can collapse on itself, by force, by weight, by pressure or by any other means. No atoms will ever touch one another let alone compress to diminishing space, and if they do, the result is a nuclear reaction

IF IT DID NOT SPIN, IT WAS NOT ROUND, AND NOT BEING ROUND THERE WILL NOT BE SINGULARITY.

According to Einstein, the speed of light is a constant throughout the Universe. The speed of light results from two factors, being distance (kilometres) and time (seconds). This speed is accepted at 3×10^6 kilometres per second. Scientists know that it takes Sun light 10^6 years to reach the surface of the Sun , and we know the Sun is not

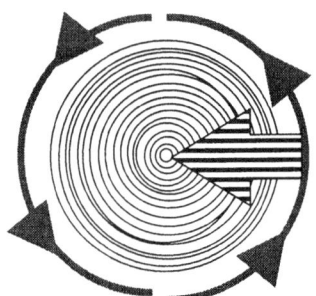

thousands of billions of kilometres in diameter. The "Xepted scientific Newton Mistaken" explanation about this fact is that the Sun light "bounces against matter" and this retards the Sunlight

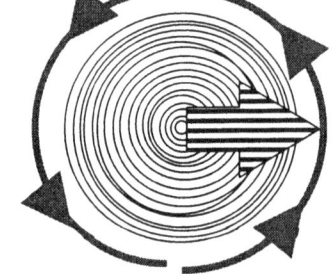

dramatically. When light hits matter, (except in the case of glass), it joins singularity immediately. Therefore, the Sun holding matter on the inside has to be all- glass, or the "Xepted scientific" explanation is not very scientific at all. It all comes down to the density of matter in space valuing the time in that space away from the point maintaining singularity. Why can nobody but me see that?

Where the arrow points we will find a spot that has no start. It is 1^1 that are the part that release from 1^0 when motion parts singularity by infinity and eternity. It comes about when motion unleashes the dot 1^1 that has no space and has no start from the spot 1^0 that has no end. Every time the top starts spinning a Universe is born in motion. The top is instigated and the motion is produced by the skills of life but in the cosmos serving nature such motion is the product have heat concentrated to sustain and maintain singularity. It is invisible, unseen and only detectable by intelligence and still it is a part of the cosmos that is no part in the Universe and from it the principle we call the Universe comes about. It is always a principal because it has no where to go but to be on call and by never being in the Universe it always is in the Universe. Sin any thing and see it is there by not being there.

Walk outside and look at the vastness of the blue sky or at night at the blackness we can see without being able to see because it is impossible to see darkness. That what you see when looking at the vastness is eternity that parted from infinity when space-time established a Universe. That which you see has no end because it is eternity in every aspect one may attach to such a connection. Standing where you are no matter where you are you are standing in 1^1 and you

are part of that which parts 1^1 from 1^0. You form part of 1^1 as you stand and being part of the centre of the Universe (because all light flow directly towards you and acknowledge you at being the centre of the Universe, you therefore also form infinity being the inner most part of eternity. That means the infinity you hold gives you with life entity that never can be disputed. You are in 1^0 that can never end as much as you carry 1^1 that never can start. You are both the spot 1^0 and the dot 1^1 and neither can ever start or end.

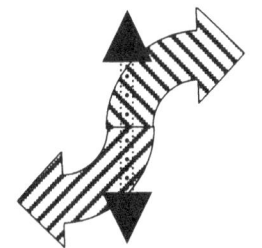

The pendulum arm covers a specific distance per time unit, every instant it swings. This is because of Singularity in position $\mathbf{a^3}$ during time $\mathbf{T^2}$ in instant $\mathbf{k}$.

The space $\mathbf{a^3}$ holds precise accordance to the time $\mathbf{T^2}$ that it takes minus the compromise singularity claims from k by reducing space to the increase in heat.

$$\Pi^1 \qquad \Pi^0 \qquad \Pi^1$$

Space-time depends on the relevancy of matter occupying space change position in accordance to all other matter relating or relevant or even only influenced by the space $\mathbf{a^3}$ in the e duration of the time the matter changes position T^2 in the instant of changing. **Space-time is everything excluding singularity diverting from singularity** and that is what Galileo recognised without realising in his observation of the pendulum. Where Π^0 is singularity and Π^1 is the diversion from singularity forming $\Pi \times \Pi = \Pi^2$ being gravity or time.

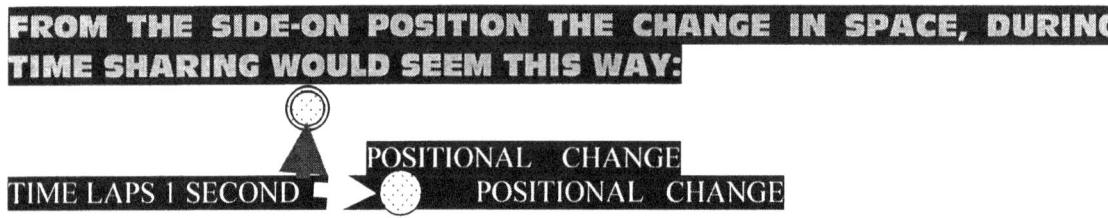

THAT IS AS SIMPLE AS SPACE-TIME IS

That is what Kepler (again I cannot say whether he wittingly or unwittingly) declared by using the formula $a^3 = T^2 k$ he announced space-time in a formula, the formula Newton raped to his advantage because in $\dfrac{M_S \times M_C}{r^2}$ $G = F$ there can be no pointing to singularity in the cosmic sense. In his initial formula $F = r^2 /$ (Mm) singularity point at every aspect because as matter falls to Earth, matter continues down a precise path that singularity provide holding that specific position that leads the way. What it does point at is the motion caused by the Earth's singularity applying on much lesser objects holding or not holding singularity. I change a to R and T to T holding **k** to Π^0, which is singularity in the instant

FROM THAT POINT SINGULARITY IS IN EVERY PROTON HOLDING SPACE AS RELATIVE AS TIME IS. $a^3 / T^2 = k$

This in fact, is the very same findings that brought Johannes Kepler his own everlasting fame when he declared that the planets stand to a value of $a^3 / T^2 = k$ as they orbit the Sun .

Illustrated it would be represented as follows:

Illustrated it would be represented as follows:

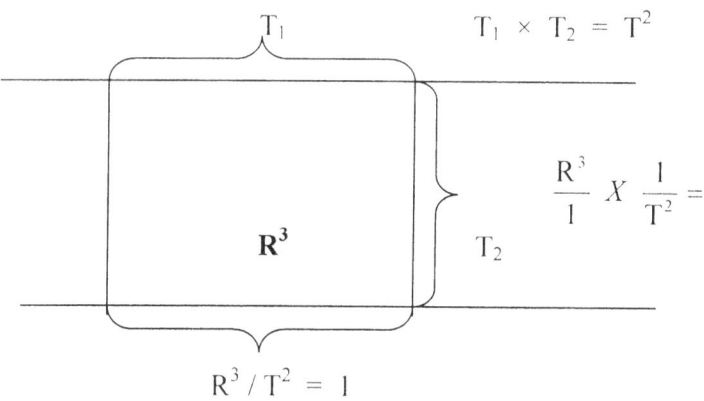

That means when the time that a structure relates to, is effected, the space will be effected pro-rata.

That means when the time that a structure relates to, is effected, the space will be effected pro-rata.

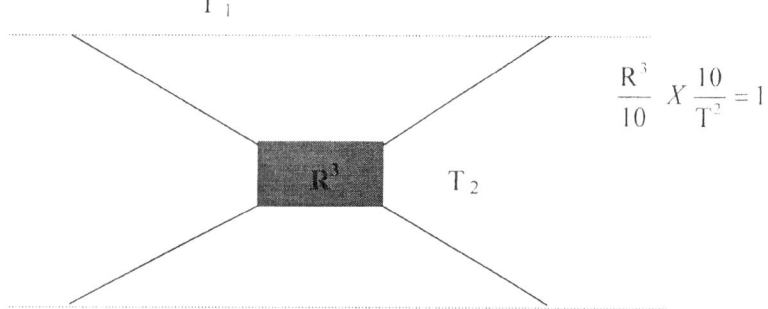

$$\frac{R^3}{10} \; X \; \frac{10}{T^2} = 1$$

If those in astro physics don't know what gravity is or how gravity comes about how can they foretell anything about gravity. If they do not know where to locate gravity or have the means to indicate where the centre of the Universe is and where gravity is centres, where then would they suggest we might be able to allocate gravity? If I have never seen a person or never heard of a person or have no knowledge of a person that know anything about the first person, then what would give me to right to write a thesis about the person in such detail that the thesis about my way of describing the person would merit say a book of five thousand pages. If the only information that I have on a subject is at best wild guesswork, what gives me the right to introduce myself as a character witness on behalf of the person?

If one can illustrate the universe and its relation with space-time, the following illustration would fit like a glove.

In the search for time in space, the most obvious place to look for the factor, time as such, must be where it is excluded from the space factor and stands alone. Therefore, one should find the place where space is zero leaving time to be eternal. Such a point would be impossible to locate and to place a value on time. However, in the cosmos at large, there is no such a place, because there is no such a thing as zero time or zero space.

Before I start with the true purpose of this letter of introducing my new method of revising cosmology, I wish to say in my defence that I chose one aspect from a wide range of possibilities to explain the way the cosmos formed. However that would constitute to a book much larger than the one you are reading and therefore I limit the development only to where matter, space and time parted and then I immediately thereafter focus on the point where the Big bang came into place.

In the beginning, there was time Zero to moment Alpha. There has never been a Big Bang, as such and there were too many Bangs too numerous to count. Everything is a variation of time duration in space. During the period of time Zero to moment Alpha the value of 1 second was equal in duration to about 1 000 billion, billion, billion, billion years (I am only stopping with the billion part in order not to bore the readers), measured in geodesic space-time values that currently applies. It could be even billions times this duration because the value of time then, was measured far beyond the speed of light, since light did not yet exist. We have no way to calculate the duration of time.

The closer time is to singularity the longer the duration would be. The method time is expanding is by heat and only heat can expand while only by reducing heat can there be a demise of space. That applied during that geodesic space-time era as much as it does today, and we must accept it as one equal to infinity shorter than eternal. It is heat in all its splendour because only heat can expand. Even boiling soup produces space that expands. When a bowl of soup is boiling, have you seen the bubbles of air rising from the soup? Has any Newtonian ever taken the time to explain that process in detail? I think not, because such explanations would be far too "everyday-like" to bother their mighty brains.

Well, that boiling soup tells the complete story about the creation. Creating is a fact of creation, however creation was not created and left on its own, the Universe is in creation being created every smallest fragmented split instant there can be. The Universe is generated as it moves and such generating of the creation is a process of creating what there is. We speak so lightly of creating and no one comes close to understanding the concept of creation. Poets and painters and writers always wishes to say how "they created their creation". That is rubbish; they created nothing. They brought nothing new to the cosmos, they only rearranged what was a small part of the cosmos into a new order, that one can detect a distinction from. Creating is producing what never was before. When looking at the boiling soup, there are bubbles rising from the soup at the top. In the soup's brew, there are only liquids and solids before the heat came. In such a manner the expanding of heat created space. No one placed air in before the event or during the event at any time. Yet from the brew of liquid and solid rises gas, or if you wish space. That space was not there previously. That SPACE WAS CREATED.

That space is energy and energy is the interaction between heat and space. As space becomes a part of the soup, a part not there before, with no room to be, it moves out. We refer to that process as boiling. That space creation is applying heat to time, and time in singularity will respond as space in singularity. The space created will vanish just as it came, back to singularity. By applying heat to time, brings forth space, and from the three components, only the heat factor is not in singularity. It removes space in singularity from time in singularity to establish room (space) for heat (time).

That is how creation started. Time in singularity overheated and the product of that was space. That is the 180 $^\circ$ of the straight line as much as it is the 180° of the half circle.

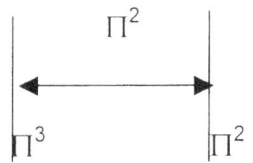

The Π^3 space from
The Π^2 is motion of liquid heat
The Π is time in space to singularity.

There is a time as a line that we find in the centre of the top, which is singularity. However there is another time, which offers material, the space in which material are able to duplicate. That too is time but it is the relevance between the holding time and the space-time where material is located.

Material uses the relevancy of time within, which developed as singularity and time without which was the expanding of singularity to commit to motion

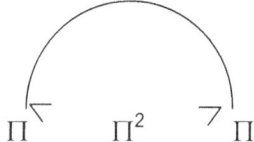

The half circle is 180° placing matter in a circle but because space only applies, to one half, 6/2 and matter holds space to value 6/2=3 only half the circle comes into effect. Half a circle is 180°. Because space has three parts in effect, it also becomes a triangle.

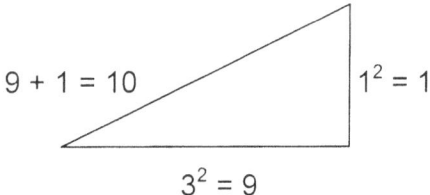

That means where space holds three and time is one, the heat within that space becomes another dimension, the fourth dimension holding space-time (3^2+1^2) = 10. That changes the matter inside space in singularity at ten and "gravity" at Π^2. This is why "gravity" Π^2 is space (10) losing one dimension (Π^2). "Gravity" is all about space (occupying matter and heat) losing one dimension back on a long journey to singularity.

As time is in singularity, and space is in singularity and both are the same thing ($\Pi^3 \rightarrow \Pi^2 \rightarrow \Pi$) the 10 of matter (heat) that affects space (10 Π) will also affect time (Π^3) and therefore time carrying heat will become 10 (Π^3) with space 10 Π. Anyone with a simple calculator can divide 10 Π^3 by 10 Π and see where Π^2 fits in. It is the doubling of matter in relation to time (7/10 + 7 /10) times the double factor of time in space (10) standing related to the line of time that refers to matter (10/7). That gives gravity its value of (Π^2).

Through the Coanda principle the motion of liquidΠ^2 confines space Π^3, to what space Π^3 confines as the atom$\Pi^3 = \Pi^2\,\Pi$) to the solid. In this containing of space by liquid in motion with the limiting or putting a border on space by liquid flowing, which confirms the space what establishes the Roche limit. In that there has to be a liquid (the neutron at Π^2 lies in the two components of space-time occupation or "gravity" manifested in the Roche limit. All objects spin and spinning is a circle Π^2 while all objects are moving in a direction $\Pi^2/2$. Again only, half of Π has any dimensional validity at any given time, therefore the dimension surrounding an object is Π. That is how gravity forms the atom as the surface of the cosmic object extends from Π^2 to Π but only half of the circle of Π (180°) can apply to time (Π^2) being in a straight line $\Pi^2 \to \Pi$, "gravity" will form at that point of $(\Pi/2)^2$ giving the Π in space the "gravity" to hold.

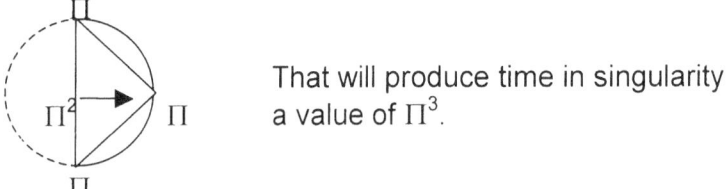

$\Pi^6 (\Pi^2 + \Pi^2) / (6 \times 10)$

That places Π in a total of Π^6 with 6 sides in space (10) affecting the proton ($\Pi^2 + \Pi^2$)

That is why space will forever comply with 7 / 10 Π^6) / 60 = 112, (the Π^6 is (Π^2 +$\Pi^2 + \Pi^2$)) and time forming the line (180°) between the half circle (Π to $\Pi = \Pi^2$) at a 180° will form the triangle of space in half (180°). The matter component of the Titius Bode law effectively applies to the value of space, therefore 7/10 comes into the calculation. That places any atom with an existence in space at a premium of 7/10 (Π^6) (6/10). The reason why plutonium at $5(\Pi^2+\Pi^2)$ $(\Pi/2)^2(3/5)=244$ is at the element limit is obvious; when dissecting the relevancy in detail. The complete element holds the very edge of what an element in space and time can endure in this era, but two or three era ago it had the function cobalt has at present

That will produce time in singularity a value of Π^3.

      Explaining the other five stages of gravity (Π^2) development is extremely complicated and for that there is no room in a book meant to introduce new ideas such as this. My motto in this book (part one) is "Keep it simple"

With time in singularity, time was eternal.

Π^O

Time is the spin rate of heat in space. This translates to heat in spin (the atom sealing time off by the spin of the electron, which then produces a motion relevancy with the proton and time in space, which brings about the time line. As we can see when the top spins the top forms the spin of heat (the top spinning) in space. That means the way the movement changes where matter and heat relate to other matter and heat in space. All the movements are relating to a circle (Π^2) going somewhere (Π) in space 3. The Π will form the radius to the circle (Π^2). Any novice can see that the longer Π becomes, the wider Π will be and therefore the longer change in the repositioning of matter will be.

Any person wishing to uphold Einstein's view about the speed of light being the limit through which matter can apply velocity, then that person should first explain how the Black Hole works. It is very distinct that whatever is inside takes that which is inside, to exceed the velocity of light. In other words the Black Hole is able to force matter into speeds that goes way beyond the speed of light. The contraction produces a spiralling of particles that takes matter into a motion dimension far beyond the speed of light. The concept involves not the moving of the particles but the slowing of time to force a duplication cycle that goes beyond the capability a photon can withstand. Inside the Black Hole must be matter, because there is no space, yet time does apply because it takes the particles spiralling inwards to the centre time to move from point to point. Matter in motion is time. However, no light can return to the surface, therefore the light is slower than the moving particles within the star. The only thing about the star is that it maintains a higher relevancy than the relevancy the speed of light can apply. By accepting the existence of a Black Hole, any of Einstein's claims about the speed of light being the fastest that matter can travel becomes fictitious.

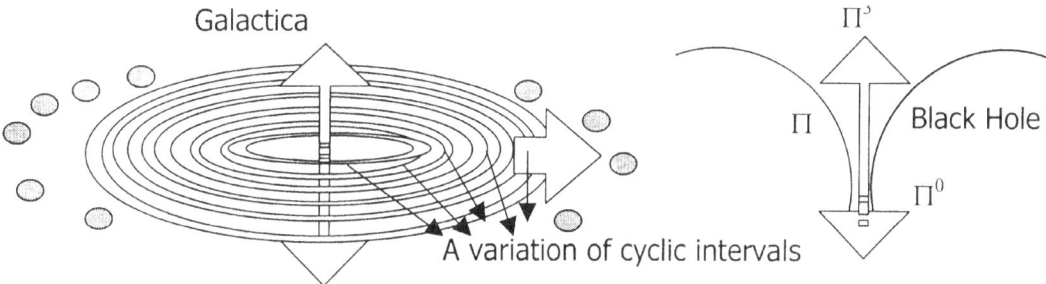

Another place where the speed of light becomes obsolete is within the centre of galactica, where the accumulative movement of matter exceeds the speed of light. That is where doctor Hawking saw a Black Hole that is not a Black Hole, but the precise opposite. Light, matter and heat, moves inward in an effort to maintain cooling as the group of proto-stars belonging to an era to the future) where they still claim their share of heat maintenance. Those particles in such close proximity, establish a time in motion well above that of the speed of light. Everything in the cosmos is all about relevancies. Particles in that phase are still very close to time eternal where motion of material took space into singularity. Time started at such a high velocity, it had to be eternal. Nothing that diverts from eternal can become more than eternal so it has to be less than eternal. It is fragmenting eternity into parts making eternity smaller.

Professor Hawking holds the opinion that there is a Black Hole centred within the centre of the star, which of course cannot be possibly true. The dynamics of a Black Hole is such that it is a star as massive as they come, that fused all the atoms into one structure. The nature and the essence of a star are to unify the singularity that was divided amongst all the atoms during the process of cosmic expanding. The star is a collection of atoms, which unite in motion that then through the unit generate motion to establish a controlling centre governing singularity. The rotation of every individual atom spinning is collected as a generated effort and the collective drive accumulates the effort to the centre of the star. The more the star develop the more is the drive of the star vested in the centre of the star and is it less concentrated in the material compiling the heat and therefore the drive. As the star becomes self secured, the maintaining of the star removes the duty of finding heat to secure the star from overheating from the atoms to the centre governing singularity. The singularity finally takes control of the motion of the star as the star evicts all space and drive time back to eternity.

Then a point arrives where the star abandon all motion. The star then achieved the main goal all stars have by producing a gravity that controls the motion of time. The spin has moved from within the star to time itself and be abolishing space all together the space became what singularity can offer. By using Kepler's formula the relevancy placed infinity in control of contraction and $\mathbf{k}^{-1} = \mathbf{T}^2 / \mathbf{a}^3$ which means the space used by the star is infinitively small as the motion producing the gravity calls on the entirety of time to move and to establish such infinitive immobility. On the other hang the space that the star then control is eternally big $\mathbf{k} = \mathbf{a}^3 / \mathbf{T}^2$ since the entirety of time establish by motion the outer limits set by the Coanda effect. That makes the controlling gravity the entirety of the time aspect because singularity being infinite commands time to motion where such command is stretching the ultimate. It is more complicated and I do explore the working of stars and the development of stars leading to Black holes much more in detail in another book I have being "*STARSTUFFIN*".

In the very opposite it is the motion of all the heat and all the stars proto or otherwise that forms the unit driving the galactica to improvise a Black Hole situation within the centre of the star. The Coanda effect that generate the singularity which control the star becomes generated as a result of all the heat and particle motion that turns about the star centre. Where the motion is valid enough to sustain a drive that would generate an equal gravity to that which the Black Hole demands, the totality of the liquid in motion in the galactica invests into a gravity that does form the drive equal to the drive of a Black Hole. But the drive forms what seems to be a black Hole. There is no real Black hole because if there was a Black Hole, then the galactica had met its destiny before any of the stars within such a galactica could journey onto a road of development. From a Black Hole nothing escape and every star is a future Black Hole on a journey of development to finally become the ultimate, the Black Hole. However there is one star more supreme than that, but there is no space to go into that explaining.

Gravity is motion. Motion is either the expanding or the contracting of material because of heat interacting with space. When material overheats, heat expands the space of the material and when heating diminishes the cooling reduces the space that material claims. In both instances it is motion applying. While moving the material overheat thus it expands. The expanding may be controlled and therefore the progress of expanding is controlled but motion has to be by way of expanding even when the expanding is under the auspices of contracting. It is the duty of the star to contract that which the galactica expanded. The galactica expanded as a compromise for the overheating but in the expanding the galactica, such expanding also develop and control the progress of young stars into adulthood. The galactica expands, expanding the stars and the stars has the role to contract the Universe back to singularity. While the galactica expands it gives the stars the opportunity to place heat stored as space of heat frozen by spin in the atom. The atom generates a governing singularity that demolished the space as it accumulates all the heat back to singularity.

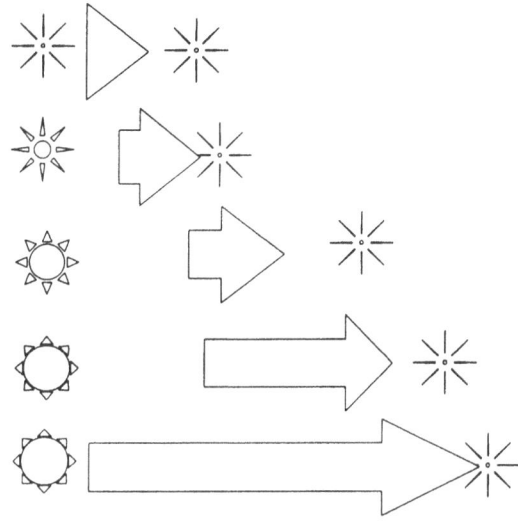

It is not only the star that reduces, or the fact that the Universe expands as the star that grows, to accommodate the space that expands by diminishing the relative space-time the star claims. It is relevancies applying more tendencies in representing the relations there are between structures in space and structures and space.

With outer space carrying the blackness in progressive multiplying, the very essence of space being space within, the atom too must be in growth claiming more space. Of all the above factors Mainstream science only acknowledge the growth of space in as much as calling it the Hubble Constant. However, that is not where the growth affects ends because it originates as much from any individual atom as it comes from Alfa singularity. Space does not expand because the space is only reducing the heat in density while producing density in space.

BACK THEN when the Universe was new

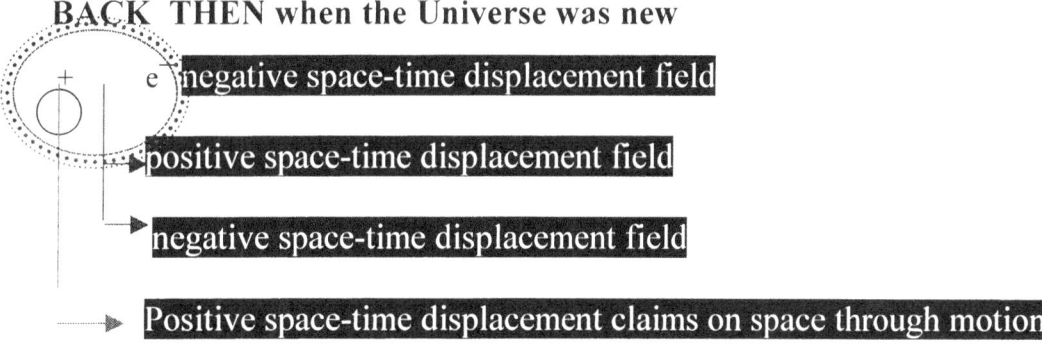

PRESENTLY we refer to the sizes we find space has in the Sun as quantum meaning they are inexplicably big

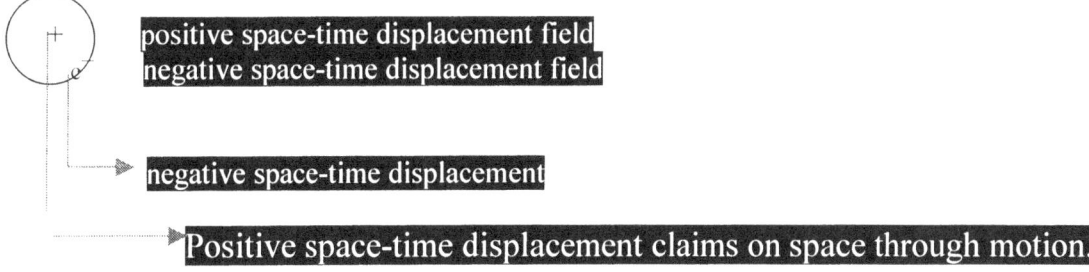

IN FUTURE TO COME they are going to get a lot bigger than the quantum size now present.

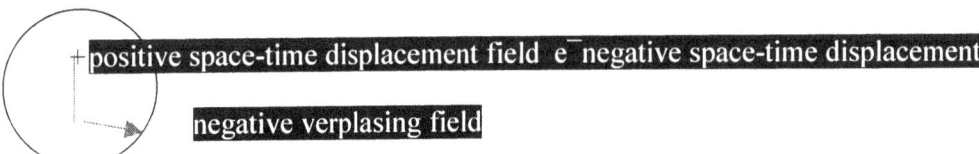

As the Universe expands the Universe is then the atom t6hat expands. The part of the Universe that Newtonians of when thinking of the Universe that expands, well that part they think expands cannot expand because that part is eternal. The part that expands is the atom and the atom is the Universe because the atom defined infinity from eternity by dividing the two points representing singularity. The star on the other hand has the task to convert the atom back to singularity by contracting all the heat into singularity. The star go in development by shedding its layers until it finally has only singularity in the centre left. Since the start is a combination of atoms forming the star the star is one cosmic atom every layer holds elements that serve the star in the particular development the star finds it in and the layers hold a certain displacement value of space-time. Therefore the atomic proton value is representative of the relevancy there is to form the need of singularity maintaining within that layer and as a layer contributing to the star as a whole.

This became the atom $(\Pi^2+\Pi^2)(\Pi^2\Pi)(\Pi^0+\Pi^0+\Pi^0) = 1836$ and the atom formed stars that still act in accordance with and to the atomic relevancy

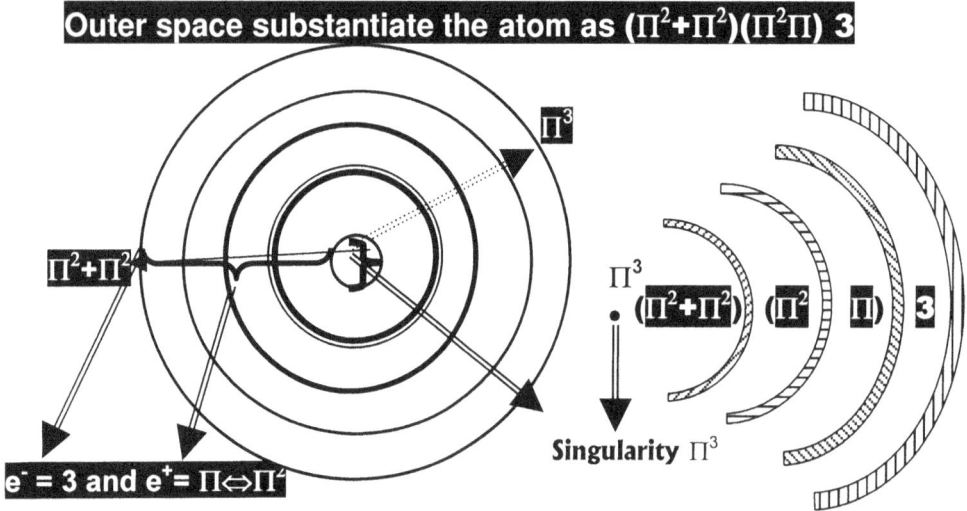

Every layer in the star represents one factor in the atom since the star is just another cosmic atom securing strings of atoms that as a unit aims for one goal and that is to secure one singularity within the star.

The manner, in which the schematic layout presents itself as follows.

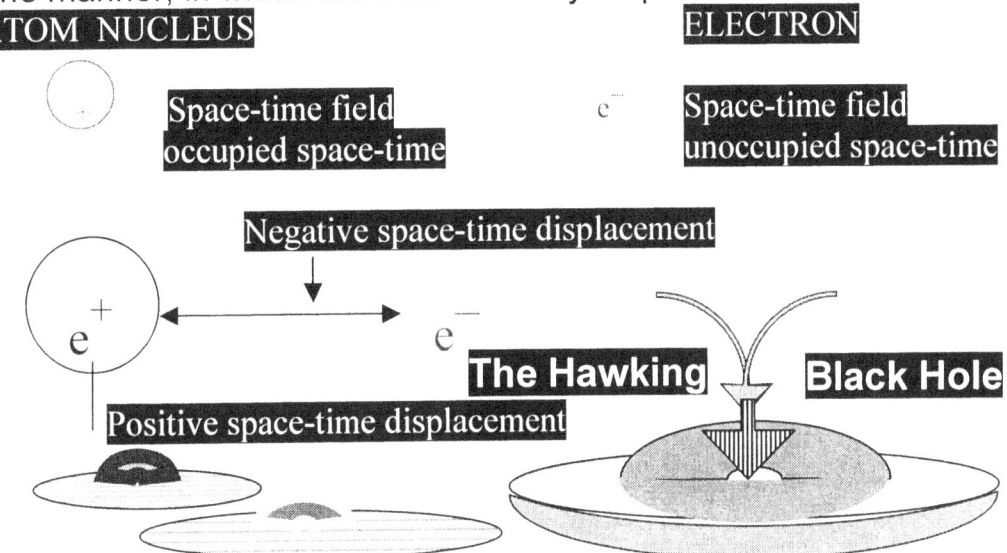

It is not only outer space that grows because $k = a^3 / T^2$ is as much the cosmic value as the value within the atom. That means that $k = a^3 / T^2$ is also in place within the atom and that shows the space within the atom grows as the Universe grows because the atom represents the Universe that is in growth. As gravity brings space-time reduction from the centre of the proton so must the growth come from the centre to the atomic proton cluster. As the atom expands in space-time, the proton can also grow dimensionally bigger through the neutron growing in stature. It expands in captured space–time by pushing the electron walls to allow the atom more space to occupy. It is pushing the electron to achieve a distance every time in the same manner that the body lets nails and hair grow. There are three factors of space-time where space-time is released. Cosmic unity and space and heat parted as singularity released the space heat holds by forming motion which produces time to set boundaries and relevancies applying.

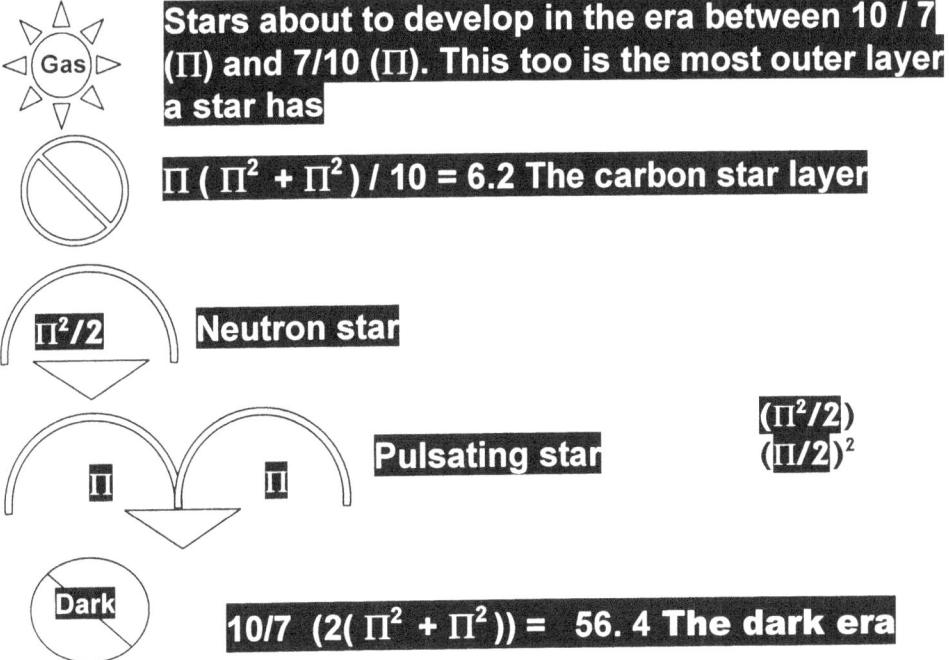

Stars about to develop in the era between 10 / 7 (Π) and 7/10 (Π). This too is the most outer layer a star has

$\Pi (\Pi^2 + \Pi^2) / 10 = 6.2$ The carbon star layer

$\Pi^2/2$ Neutron star

Pulsating star $\dfrac{(\Pi^2/2)}{(\Pi/2)^2}$

$10/7 \ (2(\Pi^2 + \Pi^2)) = 56. 4$ The dark era

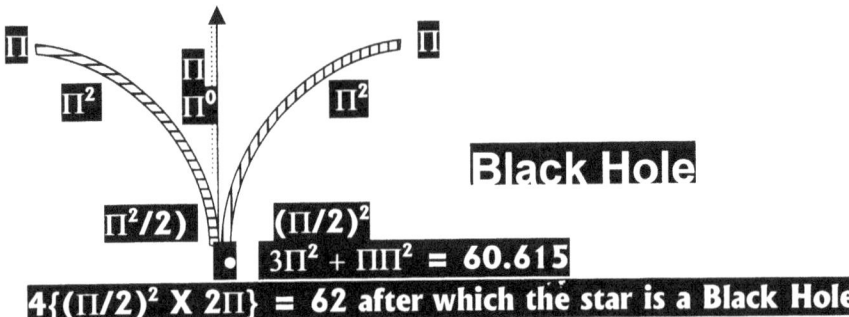

$$3\Pi^2 + \Pi\Pi^2 = 60.615$$

$$4\{(\Pi/2)^2 \times 2\Pi\} = 62 \text{ after which the star is a Black Hole}$$

Every star is on the inside many different stars because every layer holds a different (k) or relevance making the space in the star very different from every other layer in the star. This is because every layer has different motion in relation to the governing singularity and therefore has a different gravity confining space. The layer is the result of the gravity effort of all the atoms in such a layer and therefore the space in that layer will bring about the time factor that produces the proton cluster relevancy

When contracting, gravity takes place by means of lying down newly acquired heat to maintain the cooling of the structure.

However by contracting it is accumulating material that produces a build up of material in order to enlarge the existing heat surface. In that manner it spreads the heat in a wider area than what the area was the instant before and in that manner it duplicates slightly more than it did duplicate the instant prior. That means even by contracting the measure is still expanding the material by relocating the material from an uncontrolled zone to a controlled zone (the atom.) Still this accumulation by contraction is expanding by motion. When saying this please be sure of one thing: it is not the Universe that is expanding but factors within the Universe that relocate that which takes up space and provides space in the Universe. The Universe remains unaffected by all this by never increasing or decreasing. Two of the three factors swap ends and that places the third factor at a different relevancy. The Universe can expand as little as it can reduce. The Universe expands by the curvature of space –time as Einstein proved but my solution proves much simpler than the way Einstein went about.

I will in a short while indicate how Einstein is correct about the curvature of space time in his theory about "The curvature of space-time" because the curvature of space –time is the form of Π, which is the value of singularity and that forms the Universe. It is a building of what there is by the dynamics that singularity provides in relation to the accountability space-time has relating to singularity. There is the space-time complying with singularity and filling the space-time in singularity is heat and matter valuing space-time. Space-time (Π^3 to Π) cannot bend, cannot curve, forms a straight line, but what fills space-forming time is matter in motion (Π^2) and heat (3) in time in space. That part changes. The atom cannot be gas, or liquid, but is a solid, because the atom is densified in occupation of space-time. The atom is space with heat under control of directed motion. It is the heat in unoccupied space-time that produces the gas, and a liquid, is closely connected as much as part of the solid that all substances form however it is not within the enclosure of the atom.

It is THE HEAT in SPACE that produces TIME, that can and does curve, bend or whatever. That HEAT in SPACE forming TIME that forms the relevancy of space-time does bend because it can flow and flowing is changing direction or constructing by altering flow directions in space wherein matter flow but that then is part of being part of time. If, by applying the forming of gas, or liquid to the element where it is the space between the elements, of course you will get the incorrect vision of Π, where the space-time (matter holding singularity forming singularity) is doing all the bending that applies to the curvature of space (validating time) and time in singularity (a straight line) will be solid. Einstein placed the relevancy incorrectly on singularity, instead of heat.

I do admit, IT IS A LOT MORE COMPLICATED THAN WHAT I ALLOW IT TO BE AT THIS POINT, but the motto is, Keep it simple. If you wish to keep time in space constant, everything in the Universe will be oblong. That is why the Newtonians have an absurd view of the cosmos, and they present facts in the cosmos in a way nobody (least of all the Newtonians) can understand. Please allow me to explain this part first.

When material in space in time first appeared there was a displacement relevancy that was in place then that had 139 protons in relation to 138 protons relating to 136 protons. It was an atom that had many atom variations gathered and that held the construction of one atomic atom worth on the outside 139 protons, in the centre 138 protons and in the gravity zone 136 protons. There was no electron at the time because the electron came about at $10 / 7 \ (4(\Pi^2 + \Pi^2)) = 112$. This was the phase where the neutron was established as part of the Universe. The first time I can detect space / time and matter is when the proton came to a relevancy of $\$T = 7(\Pi^2 + \Pi^2) = 138$

Quoted directly from the Oxford dictionary of Astronomy the following:

The definition of space-time is as follows:

Space-time is a four dimensional position of the Universe where the position of an object is specified by three coordinates in space and one position in time. According to the theory of special relativity there is no absolute time, which can be measured independently of the observer, so events that are simultaneous as seen from one observer occur at different times when seen from a different place. Time must therefore be measured in a relative manner as are positions in three-dimensional Euclidean space, and this is achieved through the concept of space-time. The trajectory of an object in space-time is called world line. General relativity relates to curvature of space-time to the positions and motions of particles of matter.

The definition of space-time is as follows:
Space-time is a four dimensional position of the Universe where the position of an object is specified by three coordinates in space and ones position in time. According to the theory of special relativity there is no absolute time, which can be measured independently of the observer, so events that are simultaneous as

seen from one observer occur at different times when seen from a different place. Time must therefore be measured in a relative manner as are positions in three-dimensional Euclidean space, and this is achieved through the concept of space-time. The trajectory of an object in space-time is called world line. General relativity relates to curvature of space-time to the positions and motions of particles of matter.

The definition of singularity is as follows:
Singularity: a mathematical point at which certain physical quantities reach infinite values for example, according to the general relativity the curvature of space-time becomes infinite in a black hole. In the big bang theory the Universe was born from singularity in which the density and temperature of matter were infinite.

While it probably is the greatest mind to walk the Earth that produced the spectacular in the above, a much more simple mind as the one I have noticed much more simple aspects of nature that only one with a simple mind as I have could recognise because my mind does not have the capacity for the greatness of the great minds.

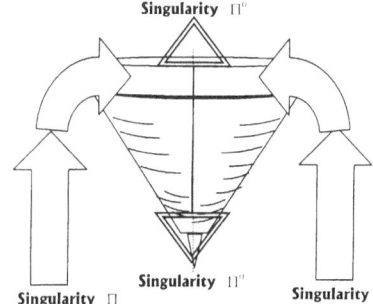

Singularity is the most vital aspect of the Universe and I can take any person's finger and show singularity. Singularity is just a mathematical point, yes, that is true, but moreover it is as clear as any person with intelligence can appreciate anything. To be a part of the Universe at some point it has to be part of the Universe at this point because from when it was up to now in our presence if it is then that which was had no where to go but to remain in our presence. If it was, it is while it is going to be because it can't be anywhere else but with us in the presence.

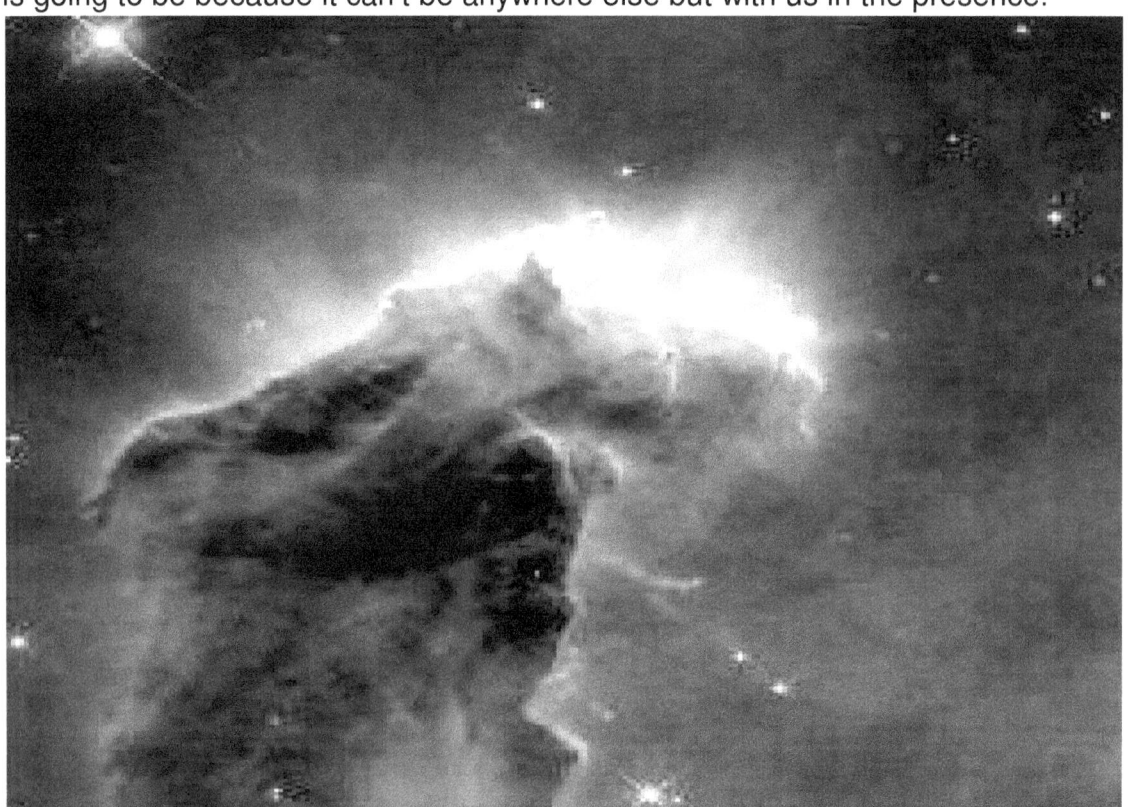

If the Universe did start from one single point and time matter and space flowed from that point, then that point must have a relative connecting base because such a point holding singularity must be eternal as space matter and time link eternal. There then therefore must be one point linking the entire Universe when regarding the fact of singularity. Then according to the theory off relativity there has to be one exact point holding time in a relevance notwithstanding the fact that time depart from that position and relate differently to all space-time away from such a point.

Every person I have discussed facts about creation recollects images in the trend depicted in a presentation as one may find shown with massive clouds of unbounded material That depict chaos and in chaos I would have no ability to use mathematics. The fact that I can use mathematics presents a Universe of order and that is just what gravity is. Where there is gravity chaos is prohibited. That would be the most unlikely way Creation came in place. The recalling of pictures representing images about creation must have form, but to mathematics it had no form. From this thought the very opposite arise where Creation came from nothing but such an idea is mathematically simply not possible.

The mathematical presence we have in the distribution and organisation of the Universe and even the calculation where able to present where chaos is abundant, tells a story of organised growth and not a blob of material with no centre from where gravity control and no even distribution of material. No wonder those Newtonians will fill a Universe with nothing and then stand back to have a view of the entire nothing they can see. Please keep in mind I am the under-educated zombie that has no brain function, which would enable me to understand Newton while they can have their view at night on what they fill the Universe with and call it nothing. How on Earth can one say one can see nothing and still point at something you see?

The thought of nothing is just what it is, a thought of nothing and although it is in the human mind common nature to present nothing as a value in the recalling of something, nothing is a presentation of the figment in the human mind. There can be no number such as nothing and that was (possibly) Newton's biggest error. Nothing represent non-existing and that is just what nothing is, it is non-existing.

In order to prove my point I wish to ask the reader to define the shortest line there can theoretically be. If he should answer anything but that the shortest line will be at a point where the beginning and is the very same spot he will be wrong. The shortest line that can ever be anywhere must have a start and finish holding the exact same spot. The line will be humanly impossible to create but we humans are capable of very little.

When the line has a beginning and an end at the very same spot and it wishes to extend the position as to further the possibility it has, which direction should it favour. Humans in the west would naturedly think of extending from left to right while in the east humans may want to go from right to left. Some persons will tend to go up or down, but all of the options are about human preference and not mathematical conclusions. Extending the line in any one direction will favour one direction without a conclusion about not extending in other directions. Such a conclusion has no sound mathematical foundation. The only option about extending will be in all directions equally in order to give a meaningful non-bias flow of mathematical equilibrium

The shortest line in the realm of possibilities must have a start and finish holding one spot and such a line will also be a dot or a circle. Not favouring one direction puts all directions at equilibrium meaning that any form what ever may be can develop from such a spot with the end and the start being the same. This reasoning prompted me to look for singularity in such a spot because if the prime spot from which all came was a spot, then the spot must hold the shortest line but more prominent it will hold the smallest form including the smallest circle.

One possibility that the shortest spot can never have is having a starting point on the zero mark. If the mark of zero holds the start it must also hold the end because the end and the beginning has the same position. If the position of zero then is the beginning, the end will also be zero leaving the line without an end as well as without a beginning.

The conclusion from this is that no line can start at zero because that will be a mathematical impossibility. A line or spot starting at zero would therefore be shorter than the shortest line possible. A line growing or extending from zero can never leave zero because of the influence of being zero disqualifies any possibility of growth. If the line then had to grow in all directions at the same pace the line must therefore be a circle. The value of the circle is Π, and that is where creation started.

That gave me the clue where to start looking for singularity. One would find singularity in the value Π and the value Π will be in all things rotating in a circle. To start my explanation about my cosmic theory I wish to firstly start with some nostalgic and the relevancy will become apparent later on. Such is the importance however that I wish to place this at the very start of the prologue.

When we were boys we played with a top we called the spinning top. I cannot imagine that there is one boy in the western world that did not hold such a devise in his hand. Tying a string securely around the tapered cone started the operation and then with a jerking or pulling throw the devise is launched in a projectile manner and the big knack to success was getting the nail end firmly on the ground and with a releasing jerk the top was rotating. The champion was always the one boy that could throw his top to spin the fastest and that would create a humming sound. The louder the sound produced the bigger the champion.

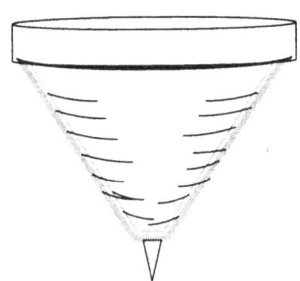

When a back braking effort produced a throw of enormity the spinning top would not only produce sound varying in pitch but also create a spin that would seem to have some instability. There are very many limitations about the spin, parameters that determine the slowest and the highest sin rate and spinning is within the parameters of such settings. The question arising is why such parameters are there in the first place?

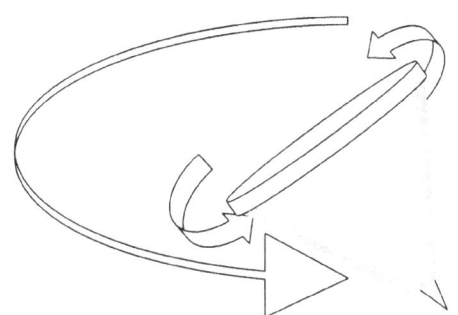 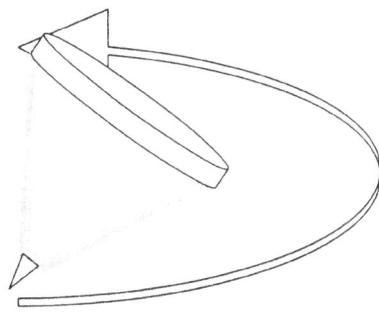

An enormous effort will have the top going oblong while spinning violently and as the pace reduced the top will stabilize by coming to an upright position. In the upright position it wall then spin for the remainder of the period where it will in the end start tilting to the side and in a last effort throw a few wild oblong turns and fall over.

Boys playing games will never realize scientific breakthrough explaining and grown ups do not play with toys. In this little toy played everywhere everyday by almost every one is the answer most brilliant of human Brainpower seeks answers about all the cosmic riddles no one seems to understand. In the spin as such one may find two vital boundaries in the motion and the boundaries are marked by a wobble coming about as if the top is fighting some other influence. Spinning too fast pulls the centre off centre and so does spinning too slow. It is the same influence coming about at both ends of the limitation in the spin. There are influences at work, but force…no; it cannot be forces setting such boundaries. From that I started per cuing what sets such limitations because that limitation must be universal as all matter is spinning in one way or the other.

I MAINTAINED DURING ALL MY CORRESPONDENCE TO SO MANY I HAVE CONTACTED, I STILL MAINTAIN AND I PROVE MY VIEW POINT THAT:

1) There is no gravity and therefore GRAVITY DOES NOT EXIST.
2) With no gravity it stands to reason that I also maintain that NEWTON AS WELL AS EINSTEIN IS ALL TOGETHER WRONG!
3) With no gravity NEWTONIAN VIEWS ON THE WAY CREATION CAME ABOUT IS ALTOGETHER INCORRECT.
4) THE BIBLE IS ALL TOGETHER CORRECT ABOUT CREATION AND FOR THOSE SCEPTICS OUT THERE I PROVE IT AND THIS TIME THE ATHIEST MUST BRING THEIR PROOF IN DOUBT. For instance that darkness filled with nothing being the night sky is light and that was the very first command "Let there be light". That is light out there! Now it is the atheist turn to prove that which is filling the night sky is nothing. Go on… prove it mathematically that $149 \times 10^6 \times 0$ = the distance there is between the Earth and the Sun .

In the past these remarks made me the clown in the courtyard and no friends came to my aid because no friends were in support of my statements. A description that would be closer to is that no friend wanted to admit any friendship because such admitting may also reflect on his or her sanity.

When looking at the cosmos from whichever angle indicates the fact that the cosmos is moving. It is forever spinning and it is going to as much as it is coming from. Everything is on the move and always encircling something of greater importance. A top can spin but the parameters of its spin are limiting the motion it can apply. By not spinning the top is still spinning as the Earth are doing the spinning on its behalf.

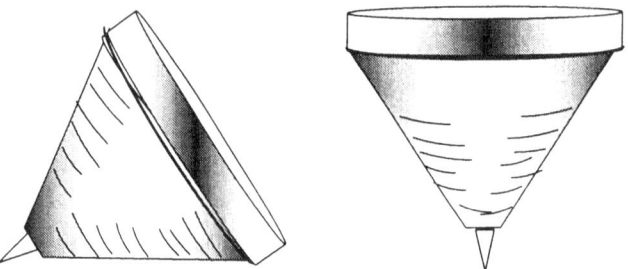

When spinning too fast the top fights something because the alignment keeping it upright starts to tarnish. The same apply when spinning too slowly but that makes sense. It is the fact that the same affect comes about when spinning too slow that triggers the questions.

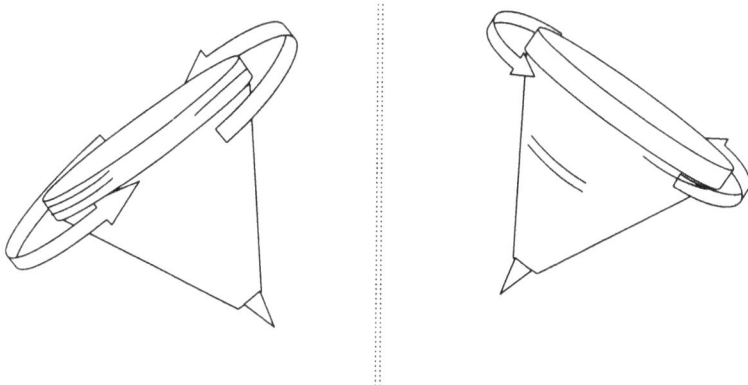

The spinning top is all the evidence any one needs to come to such a conclusion. By saying that I first have to admit (no not my mental stability), that I have no academic background and I do not enjoy any link to any university.

Without trying to not to be too presumptuous I'd say a fair guess would be that I know probably as much as any graduate about cosmology but lack certificates to prove my knowledge. I am not part of established science. In my developing of knowledge accumulation I came to some conclusions about cosmology that are unique and divert somewhat too drastic from the accepted norm. Most of the work I see the same way as the norm does but in a reverse. Allow me a short explanation

When looking at a red flower we say the flower is red. Nothing can be further from the truth. The flower is every colour in the spectrum, except the colour we attach to it. It is screaming with all might to its disposal that that specific colour it cannot accept. Yet, we maintain that that colour is the colour we associate with the object, ignoring the objects rejection of that colour. Only when looking at the cosmos from this stance, can the cosmos make sense? By recognizing a disassociation in spite of our cultural recognizing the association, can we understand the cosmos? We maintain the Sun is burning, while the fact of the matter is thjat the Sun is freezing. From our perspective on the outside we see the Sun burning as we see the red flower. What we see is not what is the truth. Only by applying the correct view to the cosmos can the four principles I introduce, make any sense and find any proof... and I do prove them. Only by telling the complete story as I do in the complete six parts of "*Matter's Time in Space – The Thesis*", can the explanation surface to a point of understanding. One cannot draw any conclusion from the outside; one has to be inside the star to see what is going on. To get such proof I had to do extensive research on cosmology. The proof lies in unrecognised and miSun derstood laws and principles science know. These laws fall outside the parameters of applied physics.

I defined gravity; I defined energy, but before that I had to prove the existence of time and time's control over the Universe, time's role in the Universe and what time is. This was up till now not yet been achieved. I had to prove what space is, that time and space is sides of the same coin, with matter forming the separation. The main conclusion that brought about such conclusions was my different view of science. It's not the explanations science at first that made me question the validity of Newton, but the things Newton cannot explain but is factors in the cosmos nevertheless. As a school going youngster I was fascinated by astronomy and in particular the cosmology aspect. In a long and strenuous process of self-education I was completely stunned by the behaviour pattern that the comet had in its relation as it orbits the Sun . Please forgive my boyish way of presenting the following but it is important that I bring it across as I saw it as a boy and as a matter of fact still see it today as a middle -aged adult.

We may start by determining the influence of gravity on planets as we find them in the solar system. **First, let us concern ourselves with a comet**. It is common knowledge how the comet relates to the Sun 's gravity. **Firstly, picture the comet at its farthest Point, away from the Sun .** The **gravity** of the **Sun pulls** the **comet straight towards the Sun** , this we all

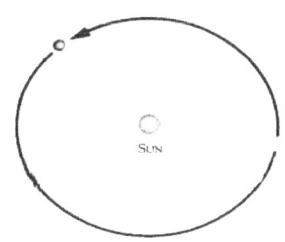

know. Gravity always pulls an **object directly towards** the **centre of a cosmic body**: that too is common knowledge. Therefore, the comet is drawn directly towards the centre of the Sun and throughout its journey the comet is picking up momentum directly related to the gravity that is cantered in the middle of the Sun , (**gravity is always cantered in the middle of a cosmic body**). As the comet is increasing its speed, the comet comes closer to the Sun and therefore the Sun 's gravity pull is simultaneously increasing as the distance between the two cosmic bodies is reducing. Each instance the comet is drawn towards the Sun , the gravity that the Sun applies to the comet becomes larger progressively. When the comet is at its <u>**closest point to the Sun** , **something odd happens which cannot be explained by Newton's gravity at all! Remember gravity should now be at its strongest point because of the proximity of the two objects.**</u>

The comet remains at an even distance encircling the Sun .

No longer does the gravity of the Sun pull the comet towards the centre of the Sun .

At this very point the gravity that the Sun applies on the comet does not pull the comet towards the centre of the Sun any longer, in fact, it seems as if the effect of the gravity has been neutralized.

1.   The comet stays at an even space from the Sumas it goes around to complete a half circle's orbit around the Sun .

2.   No longer does the gravity of the Sun pull the comet towards the centre of the Sun .

3.   At this very point the gravity that the Sun applies on the comet does not pull the comet towards the centre of the Sun any longer, in fact, it seems as if the effect of the gravity has been neutralized.

4.   The comet stays at an even space from the Sun as it goes around to complete a half circle's orbit around the Sun . It only completes a part of its rotation around the Sun .

5.   After this, an even more peculiar event takes place. **The Sun , at the point where gravity should be at its most dominant, suddenly loses its complete grip on the comet.**

6.   The comet brakes free from the Sun 's pull of gravity and speeds off towards its destiny into the vastness of the cosmic space, undeterred by the gravity of the Sun .

Then after a pre-determinate and pre-calculated time the Sun starts applying its gravity on the comet once more. At a point where the comet is at its farthest point, the gravity of the Sun becomes strong enough to bring about a complete turn around to the comet's direction of travel. **However, the gravity between the Sun and the comet is at this point, at its weakest point of influence.**

However, **this is not all**. When we regard the planets as they stand related to the Sun , the effect is the same, but not as obvious. All the planets follow an oval orbit around the Sun and therefore the same factors concerning gravity apply to the letter as it does in the case of the comet. Let us investigate the one planet we relate the best to, which of course is the Earth.

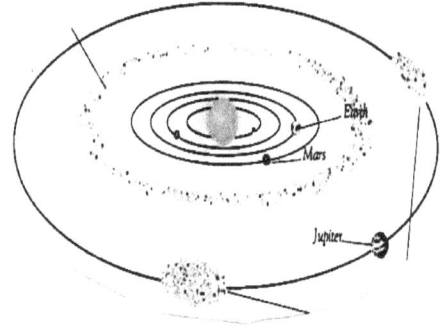

So, when the Sun 's gravity is at its strongest, the comet manages to brake loose and neutralize the Sun 's gravity pull in order to avoid its fatal collision with the Sun and when the Sun 's gravity is at its weakest, the comet cannot escape the pull of gravity. There is definitely something very wrong, either with the comets or in this case the Earth's circling behaviour or the laws made up by Newton.

Well, this is the part Newtonians are so able to understand and because I am a mechanic I am not able to Understand. The Earth makes a circle and Newton in all his mathematical splendour never provided for this circle. He even went further by introducing Π to what is the eternal circle! The Newtonians sympathise with my poor understanding because of my low intellect and education! Who should be sympathising with whom should be a better question?

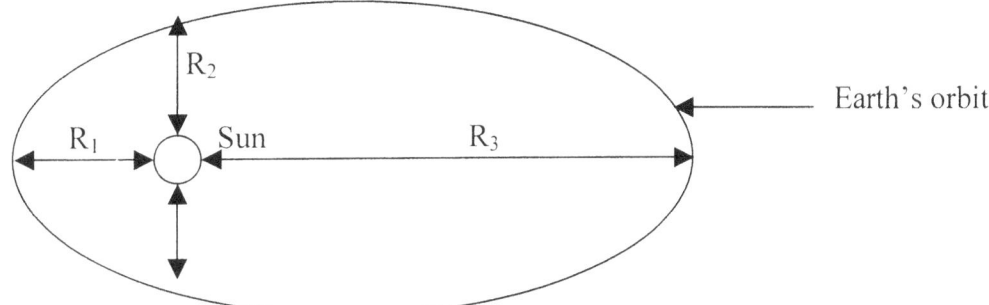

This illustration does exaggerate the radius of the Earth's orbit around the Sun , but since it has taken place 4 500 000 000 times, it has no real effect on the validity of the next statement.

At one point (R_1) the distance between the Sun and the earth **is less than** at another point we call R_3. Let us put a value of R_1 = one and R_3 = three. This means that each year, for the past 4 500 000 000 years the effect of the common gravity between the earth and the Sun has a greater effect than at another point six months later. **At one point the earth should be drawn or pulled closer to the Sun** and **after another six months** interval **the earth should stand less effected by the Sun 's gravity**, therefore it should move away from the Sun . Each cycle of twelve months would have one point where the gravity pulls the earth closer and exactly the opposite must apply six months later when

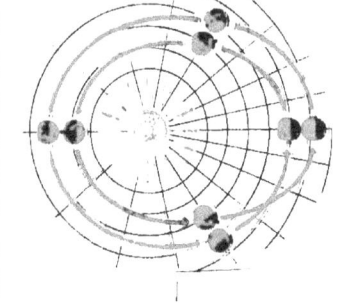

the gravity is at its least. So, for the past 4 500 000 000 years the earth has been re-establishing its seasonal swing towards the Sun and away from the Sun , which means by now the earth has to collide with the Sun in midsummer or escape from the Sun in midwinter, as it may then drift away into the unknown.
For the more mathematical minded person the argument is as follows. May I remind you, THAT NEWTON'S OWN LAWS ARE IMPLIED, and again the planets disobey these laws completely!

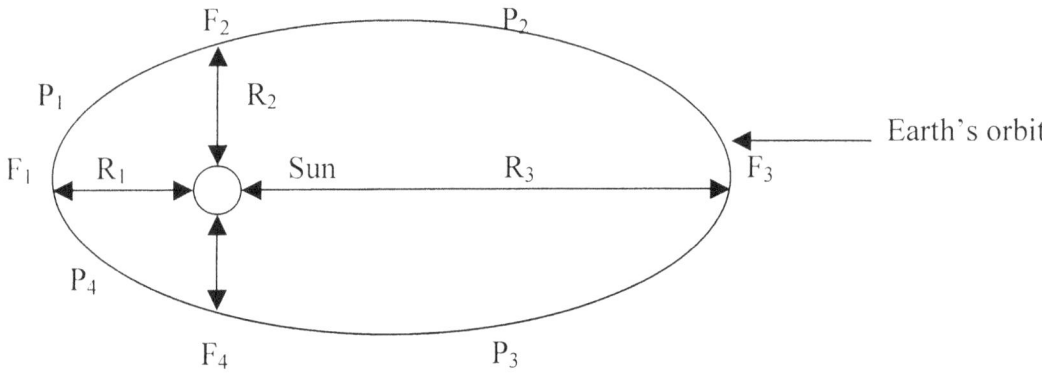

$$F = \frac{M_1 M}{r^2} G$$

We know that $F_1 \neq F_2 \neq F_3 \neq F_4 \neq F_1$
because that is what seasons are all about.

Even if $F_1 \neq F_2 \neq F_3 \neq F_4$, we know that $P_1 = P_2 = P_3 = P_4$.

Because r is at different values F could not be to the same value.
Therefore, the value of F has to be unrelated to force its value on to P.
Nevertheless, Kepler has proven that $P^2 = a^3$, although $a_1 \neq a_2 \neq a_3 \neq a_4$.
If $a_1 = a_2 = a_3 = a_4$, we would not have had season and climate changes on Earth. That means that to proclaim $F = \frac{M_1 M}{r^2} G$ is nonsense. The truth of the

matter is that Newton actually proclaimed that in an ellipse, which has an uneven circle (Kepler's findings) the value of $F_1 = F_2 = F_3 = F_4 = F_1$, but because an ellipse has no constant radius, it actually means that $r_1 \neq r_2 \neq r_3 \neq r_4$ and thereby anybody can see that Newton's calculations are wrong. $F_1 \neq F_2 \neq F_3 \neq F_4$.

In this book, I dare to prove that **there is a difference** between findings of *Galileo and Kepler* on the one side and the work of **Newton and Einstein on the other hand**. Only one of these two group's findings can be right, because there is an unmatchable difference in the concept of these two groups' opinions. Newton considers that a force exists between two bodies in space: the mass of the two bodies' product is being brought into context with the gravity constant (G). This value is then divided by the distance r calculated as a square (r^2) value. I have to admit that I have not once seemed to bring across the importance I see in the arguments above when translating it to academics of stature. Every time I introduce the behaviour of the comet I get the impression that academics either will not or cannot see any truth in my arguments. In every incident I became the accused of not having the brainpower to understand and from my perspective there is little to understand.

I stood accused by many academics I crossed paths with in the past that I am not familiar with Newton and because of my poor academic background that I am not capable of understanding Newton. That is not the case and I have to be very adamant about that. I would accept such accusastions if Newton's science explained all of science. That is hardly the case because I can state four very prominent cosmic principles that no one can explain by applying Newton's claims. The Roche-Lobe, the Titius Bode principal, and the Lagrangian five-point position and the the Coanda gravity contraction is what Newton's gravity formula cannot explain at all. Neither can the Big Bang theory be the starting point if cosmology insist on using the application of Newton. I am aware of the Critical density theory and black matter, but those arguments has not found proof in the slightest and in truth serves as an escape corridor because science is at the end of its tether with the phenomena contradicting Newton. I admit that I am lost at finding a starting point introducing the book because the issue remains comprehensive when dealing with issues of cosmic proportions I shall explain the four unrecognised phenomena I use in proving my statements. With my introduction of the phenomena, which I named the four cosmic pillars you will find it obvious why science do not accept them even if it is documented throughout the Universe and is quite commonly found. It totally annihilates Newton's formula of $F = G (M_1M_2)/r^2$.

From these cosmic phenomena I produce a path of cosmic development, preceding the Big Bang. The problem that comes from this, is that I take the reader from a point and lead the reader through the explanation of the existing principles, pointing out how they are flawed and introducing my explanations and proof and substantiate my argument. This is a path one has to follow. There is no point where one can drop in, or out and in again and maintain the golden thread of understanding. To conclude, only this: As I bring proof of existing evidence in cosmology about phenomena and of which science acknowledges the existence but science is failing to understand or explain the correctness thereof.

How can the Universe expand? Well, it cannot, and that is yet another illusion the Newtonians create through misunderstanding. What is in it is in it and it cannot grow, as much as it cannot shrink. It cannot expand and it cannot demise. It is only a consistence of changing relevancies, where the relevancy flows away from one part of eternity or singularity (space) to another part of eternity or singularity (time).

Every aspect of the Universe holds relevancy by applying time to space and the time to space first claims space from singularity then controls space from singularity and influences space outside the direct contact with singularity. In every event the factor remains the same, as it is only the relevancy re-applying a dimensional influence on space-time.

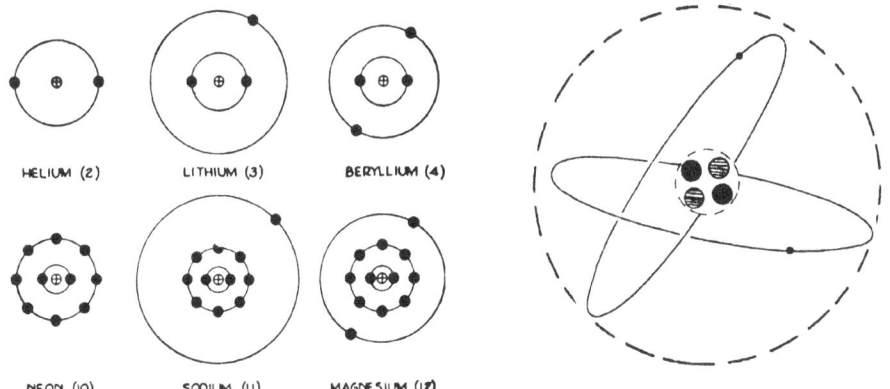

The   relation that an atom has with heat stems from the number of protons in the nucleus of the proton cocoon.

The key to the relevancy is heat and space. When matter heats, it expands therefore it takes more space. When matter is cooled it shrinks, therefore takes less space. That is the relevancy because matter in any form is heat. Heat produces the increase of space and reducing space produces an increase of heat. That is the relevancy. That is the secret of the Universe. That is the secret of gravity. That is the secret of momentum and every other aspect within the Universe.

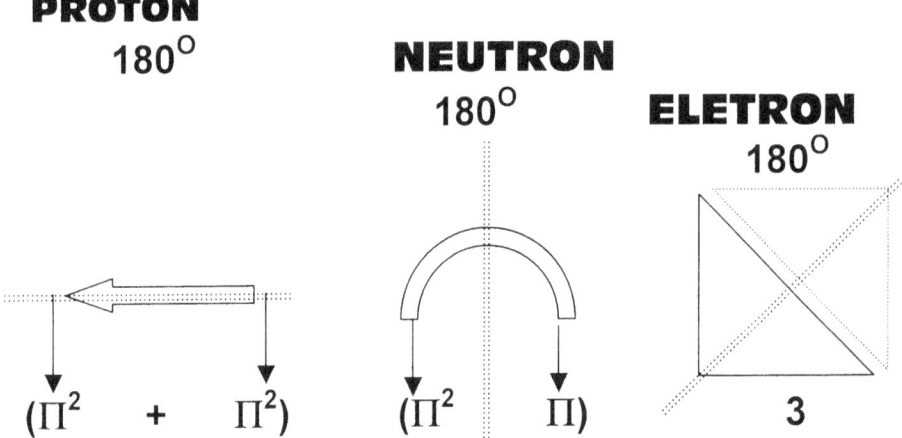

PROTON
180°
$(\Pi^2 + \Pi^2)$

NEUTRON
180°
$(\Pi^2 \quad \Pi)$

ELETRON
180°
3

Time stood still in eternity, then after a command of the Creator, time started to move by overheating and eventually formed the relevancy of the proton $(\Pi^2 + \Pi^2)$ the neutron $(\Pi^2\Pi)$ and the electron (3). As a star returns time by depleting space to the dimensional increase of heat, space destruction is in progress and the star will abandon systematically some of the dimensions the atom holds. That is the relevancy. That will be whatever position there is in the Universe. In the depleting process of dimensional re- adapting, the star shall abandon aspects of space-time. The electron (3) may become obsolete, the neutron $(\Pi^2\Pi)$ may become obsolete in neutron stars and even $(\Pi^2+\Pi^2)$ the proton will become dysfunctional as space reduction completely disappears from the star's space-time occupation. However, those stars will be dark, and beyond our vision.

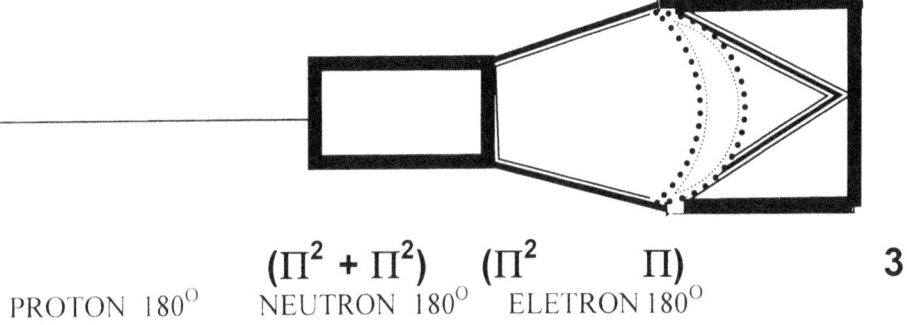

$(\Pi^2 + \Pi^2) \quad (\Pi^2 \qquad \Pi) \qquad\qquad 3$

PROTON 180°   NEUTRON 180°   ELETRON 180°

The relevancy holds value pointing the relation between the various dimensions as they are in the atom. The relevancy of $(\Pi^2 + \Pi^2)$ $(\Pi^2\Pi)$ (3) = 1836 will remain but the mass of the electron and the mass of the proton will change in every space that time applies. Cosmology thus far was incomprehensible because it was incorrect. When applying natural laws, it becomes so simple that a person as ordinary as I can understand and explain it.

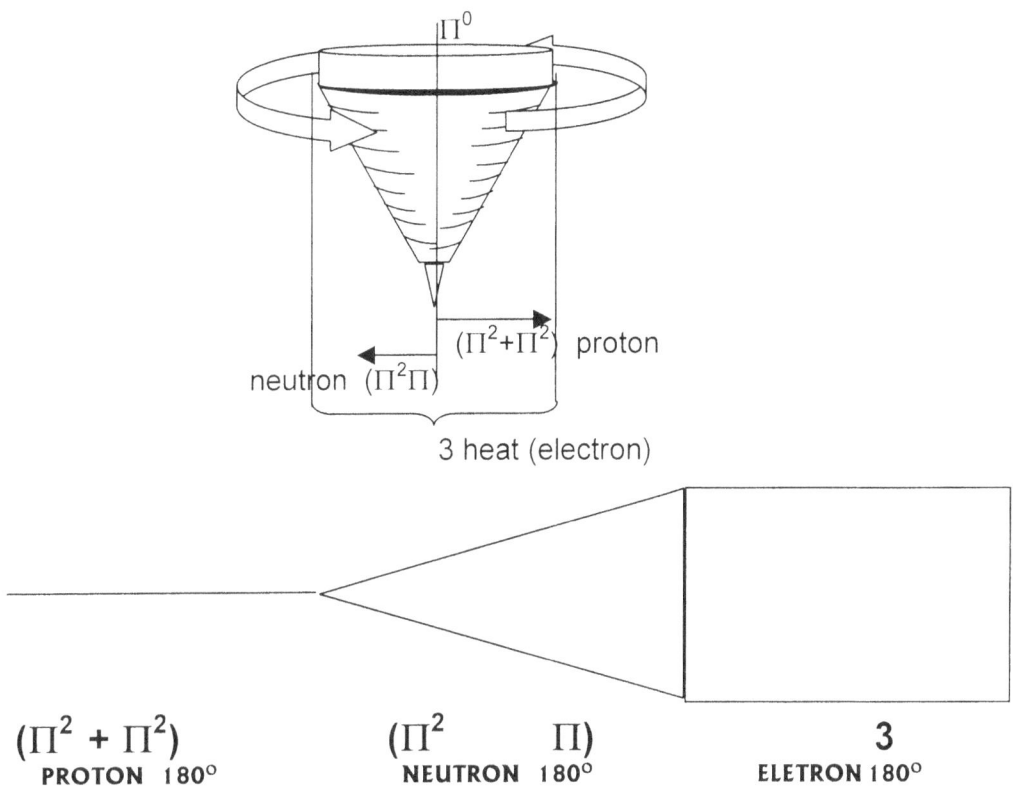

$(\Pi^2 + \Pi^2)$
PROTON 180°

$(\Pi^2 \quad \Pi)$
NEUTRON 180°

3
ELETRON 180°

The above indicates where singularity originates and how that establishes the factor in singularity Π. The Universe started from the factor in singularity Π. The entire Universe holds a spinning relevancy to all other factors in the Universe. If that were not the case, the Universe would not be there. The first person to consider the factor in singularity Π was Galileo. In the swing of the pendulum he saw singularity remain as one, that formed time, destroying space to maintain time. To prove my statement I shall very briefly indicate some barriers of motion science refer to as the Doppler effect, but Doppler used a slow moving train that at best could indicate two or three very minor moving limit.

If I can understand it, every other non-brainwashed human on Earth should understand it. The relevancy of $(\Pi^2+\Pi^2)$ $(\Pi^2\Pi)$ and 3, is a dimensional reduction of the flow of heat from space back to time. The flow of heat becomes necessary to prevent solid matter from overheating. By removing heat from the gas of space, through the neutron, to the solid of matter, space reduces as the intensity of heat flow requirements increases.

When we look at the night sky we see images of stars. I am of the opinion that our vision of stars and our interest we show in stars is just what sets us apart from other species. We are able see what we never can touch though we can appreciate what we never can have. We interpret what we see without ever making contact to confirm and that gives us external knowledge and insight. Our vision about that, which we see tell us that there is more than the animal's concept of a plain survival on Earth where it is that you can eat or you can be eaten. Fathers show their children the constellations and although we no longer attach religion to our stargazing it never subdued the bliss we find in our astonishment about stars.

The star that gives us our greatest wonder is a star we cannot see. Every one 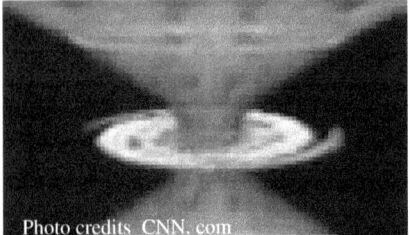 stands amazed at the fact that there can be a thing such as a Black Hole. There is so much to ask and such a lot to wonder as to why and how and where and which…yet, we cannot see any that we interpret. We see but we cannot see and that makes us wonder what it is we cannot see and what it is we wish to see. The fact that our view is obscured by the fact that our view is obscured dramatizes our sensation of wonder many fold. That is human and that is why we are what we are and why we are in terms where we are.

It is part of the human concept to believe your eyes. Seeing the sand dunes on Mars is equivalent to seeing the sand dunes on Earth by means of the television media and could just as well be of the Sahara. The Sahara is a place we can go and visit should any of us wish to do so, but the dunes on Mars are another problem. Visiting and confirming what we then see is not that simple to accomplish. The Martian dunes are not only space away, which means I can cross space in time and visit. The dunes of Mars is not even space away but is time away. There is no way I would ever cross time to see for myself what there is to see. That is what is wrong with science, amongst others. Science is of the opinion we see space. We do not see space. We see time, but it is not time we see, it is the distortion of time that we see. The "further away" we look the more time we see. However it is not time we see. It is the distortion of time that we see. The further something is away, the more it is in the distance, the longer it will take the light coming from the object to reach us. That means the longer it takes light to reach us the more time is distorted to put distance between what we see and what there is to see. It is not space that we see but the distortion or the compromising of time. It is the time delay between here where we are and there where the object is that we see and we do not see the object or the space the object has or even the space between us and the object. We see the time delay there is between the object and us. We see what was there in time gone by, however, we do not see what is there and we see space for what space represents to the Universe. We see space as time delay, time slowed down. That is what space is, space is time delay. That concept urged me to go and look for the beginning of space and the beginning of time and the origins of the concept space-time. Please allow me to explain the beginning of space by measure of time delay. At the time of the Big Bang everything was small…not so…it was as big as it is today. If the Universe was the size of a neutron, then we had no size at all. One cannot compare apples with oranges and see bananas. The space we see is the distortion time has to separate points of comparison.

In order to understand what I am trying to say I have to use a picture that is most probably not a true event. What we think we see is space. It cannot be space that we see. In the forefront we see a line that is a result from a comet travelling. Then there are pixels indicating lesser star structures and some clear dots indicating stronger light spots, which would personify larger stars present in that direction.

The rest we see is the black of night. If the Big Bang theory is correct and to my thinking there is no doubt about that, then not to long ago there was a lot less space between the objects than is the case at the present. The space was less. That cannot be the case because if the space was less it would then take the light much quicker to arrive at the spot we are at present. The light coming from what should be the comet is relatively quick in reaching my location while there may be some of the faint dots that have light travelling a considerable time to get to me. I presume the comet is closer. Looking at the image of a roving planet it shows a structure filling space at intervals. The space it fills is a constant because the space does not change in becoming bigger or smaller. However, the space it is moving trough appears to grant the roving planet another position every time it is photographed. It is in the terms of time that the answer is. It takes a different period to position and obtain the light coming from the different position where the object is located.

If the prime object were the space as it is in the case of the space serving to fill the roving planet, then no changes would come about to the space. It would take as long to fill the space between the object and where I as viewer am in position with travelling time. It does not because the motion that the light has to endure is shorter or longer by time duration. It is the space that is constant, yet the time to travel varies. It takes time to cross the space whereas the space holding the object remains filled at an even volume every time. In the case of the planet space filling with material by gravity the same space filled without changes. In the case of the dark space, that space is putting time at a different duration to reach the location I am in. The "further" the object is in distance from where I am, the more time it would take the light coming from the object to reach me. The object will appear smaller as the distance increases but I know the space the object holds is filling the same volume as it does when being close to me. In the case of the space filled, that space appears to change but that space is filling a volume at a constant. It is when the space in which the object moves increases the space the object holds then diminishes. The space that the light has to pass through to bring me the picture of the object increases. That cannot be because the space is filled all the time by the same margin. It is the time the light takes to bring me the picture that increase and it is that light that shows me a diminishing space. It is not the space between the object and my location that increases, but the time that increases and by allowing the time to increase I allow the space of the object to appear to become lesser. The space the object holds has to remain the same and the space between the object and me cannot change by motion. It is filled by volume that motion cannot change. Only time can be affected by motion and since it is motion that is changing it can only be time the motion can change. The slower the motion the longer wills the time be that it takes the motion to negotiate the space. That black of the night that I see is not space that I see but is time that I see and the space I think I see is the retarding or slowing of time that I see. Outer space is not space but it is time that space retards and therefore space is not space but a retarding or a distorting of time. That means that which see thinking it is space that we see is all the time, time that we see and being time it has no outside because time is eternal. Space, being infinite interrupts time to give time in eternity duration value.

The relevancies we are about to address are about form. It takes us into a Universe when a line had the same value as a half circle and as a triangle does. It takes us beyond space to a Universe when time formed space. It puts the Universe beyond distance. It is what came about when space interrupted time to deliver us the black of night, which we incorrectly think of, as space. The Universe did not start small it started outrageously big. It is not expanding it is reducing. When the Universe started there was no outside to that which started because if there is an outside then what is on the outside of that which started. The Universe has always been an inside that went smaller. The limits grew smaller not bigger. The initial start had no limits. That which we think of as so small and tiny, so small it has no sides is so big it cannot have sides because it is too big to have an outside and all we see and all we cannot see fill the inside.

Where we are now in the Universe we are so much smaller than what was when the Universe was the size of a neutron or whatever it was. If the Universe as one block without limits had no outer limits and was the size of a neutron then it grew smaller because what was our size when the Universe contained all it had in a neutron. When the Universe was a neutron we were not even a thought. It is easy to lose perspective but perspective is all we dare not lose. That which took al the space a neutron could offer back then has no limits now and has no boundaries. It is too big to be cooped up by limitations and boundaries. We with limits and boundaries now have measurable quantity to calculate, but what was the Universe then has no calculations art present. Where there is no boundary to shift what shifts then and yet they say the Universe is shifting its boundaries because the Universe is expanding therefore it shifts! Where no growth is possible since it captured the growth at the beginning where too can it grow. The end of such a shift by what cannot shift to where no shift is possible will eventuality be what they named The Big Crunch even before locating the Big Crunch. It is like naming a baby even long before knowing how the procreating is taking place that will lead to impregnating of some member of the specie (which member it will be is still then still unclear at the time the name giving was undertaken) where it later on will lead to conceiving the baby … that is the manner in which science dogma is enunciated but that is how clever those mathematicians are that knows everything there is to know on science. They can name a baby before even knowing what procreation is and that they do by calculating what they don't know anything about… like procreating the baby! It seems more likely that that which has no prominence finds prominence, which means the lot is shrinking. The Universe is surely shrinking to give us space to be.

When we altered the size of the moon in relation to the size Mars has what we did was change our relevance to that of Mars. We first brought Mars on a time line as close as it would be if it were hanging around in the space the moon has at present. Then we moved the time line back because it takes time to travel to the structure. Pushing Mars back does not increase the space, because eventually Mars fills the same space. It increases the time duration between Mars and us. It is not space we cover. If it were space then the time would be equal for light and for all to complete the journey in the same time. By changing the time the relevance change as to how long it would take to get there.

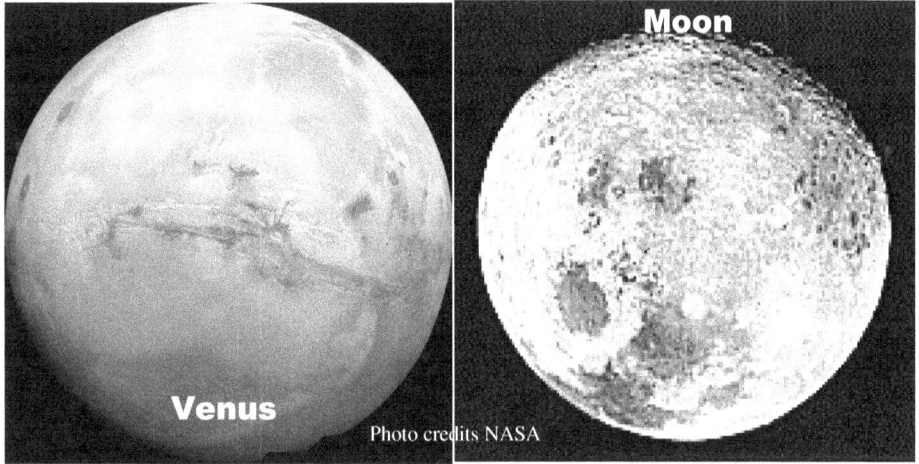

Photo credits NASA

When we look at the images of the two solar objects it is so easy to put them out of perspective and in the same size, although we know they are not the same size.

In cosmic reality the reality is quit substantially different. When we put our hand out we are able to touch...say the door we are immediately in contact with the door. It is the door we touch because it is the door we see we touch. Moving back one meter we find we are no longer able to stand upright and touch the door because we are one meter away from the door. We are one meter away because we can see we are one meter away from the door. We grew accustomed to this thought because Galileo's pendulum shows we are in time in space in the Earth timer in space. The time we will take to touch the door corresponds directly on Earth with the distance there is. Things change drastically when we leave the Earth or when we view object not confined to the Earth as we are. The truth is we are accustomed to think we are one meter away from the door we are unable to touch because we think we see the door is one meter away.

However that it is not the door we see. We cannot see the door because the door is not there for us to see. We see light banging on the door and as the light is rejected by the same door that the light comes flowing to us. We see the rejected light bringing an image of the door we cannot see. It is light we use and that we are used to of using to confirm what we see but such confirmation is what makes the most intellectual stumble. In quite the same manner we see the darkness of the night and observe such darkness as darkness. By darkness we interpret the meaning as that which we cannot see or that which we are unable to see. Reality tells me that the darkness is light that is too bright for us to see. Take an image of Mars with a close up view. Then reduce it and go on reducing it until it is so small it becomes invisible. The space filling darkness is not darkness filling space because the ratio of darkness increases as the ratio of light in comparison to the darkness reduces. The object does not go dark by moving back. It rather becomes more of the same when it blends with the darkness, which proves the darkness is not darkness but it is light.

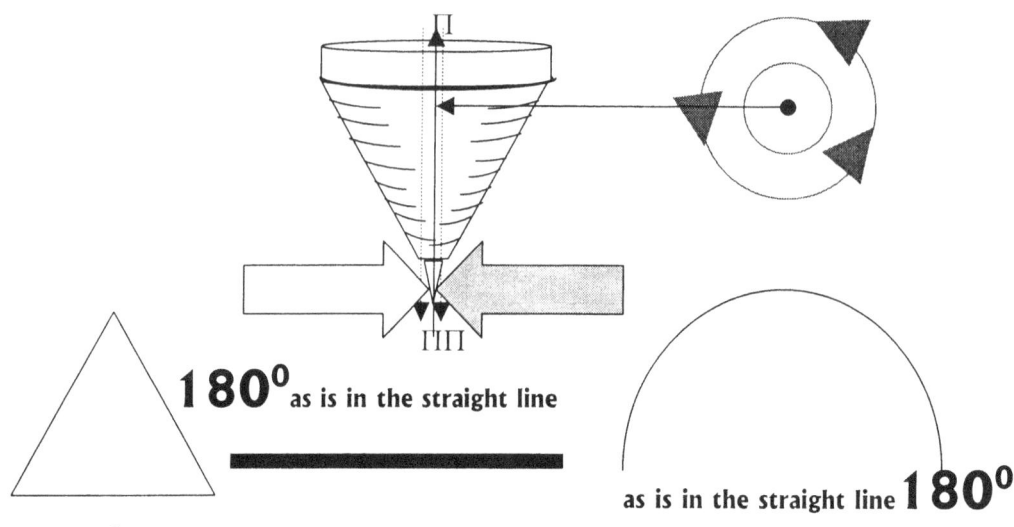

180^0 as is in the straight line

180^0 as is in the triangle

as is in the straight line 180^0

Einstein famously said there are matter, time and space and if I am not mistaken he said very little more. Is the space that Einstein would have reduced to a flat surface forming a flat Universe really space just because it is not time. Then on the other hand we must ask the question what is time then? What is our Super designer of space whirls really reducing when he reduces space to have the lot fit into not one but two Black Holes? What is matter and what is space? Looking at what the top tell us we have matter time and space a little confused. We find matter in time but also we find time in matter and that no one before realised! By reducing the space an object has the darkness becomes either more or less but the darkness promotes the object or reduces the object. The fact that large objects are close and small objects are at a distance we on Earth relate to more space and less space. The only factor that can produce more space and less space is time because time is irremovably connected to time. By reducing the share of the combination of space-time time must reduce or increase to allow space to do the opposite.

We have our focus square on the distance we find that part us from the object in our view. It is another culture thing from the time we hunted I suppose but we confuse

distance with the time the distance really represent. By the end of the twentieth century every one has become so accustomed with photographs and lenses bring into focus and bringing objects close it became part of our breathing. We reduce the space in order to see the object better. Is that really what it is all about or is the issue significantly more complex.

The pictures that you are looking at are probably the best example of space-time you will ever find. The term we use is where space stands (division) by time. That means in the context space is enlarged or reduce by the time factor that increases

or that decreases time. The time divides the space (space / time or space $\div$ time) according to the time it takes to reach the space from the location where the viewer stands. The time factor divides the space factor into smaller sectors of time components, which is part of the overall picture in eternity. It proves that everything claiming space is related to the time giving the space a relevant position according to singularity $k^0 = a^3 / T^2k$. It puts time T^2 where space is a^3 in relation k to where the space is $T^2 = a^3/ k$ and most of all the development of space-time k depends on the growth of space a^3 which time T^2 will sustain $k= a^3/ T^2$. The expression clearly reads that k is the time T^2 it will take us to reach space a^3. That is the only way there mathematically can be to express space-time in a meaning fill mathematically expression.

The realising that only time can affect space by the measure of appearance is a huge step in the right direction. Space is a constant therefore time has to influence the appearance of space to become apparently more or apparently reduce to become less. Being big is a sure sign to the brain of an object being close. That would then appear as if there is little space between the observer and the object in observation. That is culture talking because space may appear larger or smaller but it can only be a medium of space that may allow space to appear. Space as such has the same measure and has the same prominence when measured. Time is the factor that allows space to reduce and even to reduce to the obscure.

Moving the object back into obscurity does not reduce the space the object has but puts the space the object has into a much larger definition in space in relation to the space I witness. The light streaming from the object will also fade into obscurity and disappear as the definition of the object declines in relation to the space it holds by comparison to all the other space in view. The light the object had did not decline or reduce but it diminished in relation to the gross of space holding light. In relation to the space out there the space diminished the light in relation to the darkness the light then offers. That way the light could only reduce by comparison if the light was less in relation to what the light is in the darkness we see. That means the darkness is flooding the light the object has and therefore the darkness we see is light. However, our relation to the light makes us in relation to little to be able to appreciate the light because as the object retracted from the position we had, we also diminished in space by the same measure. The space we hold therefore is too little to enable us to appreciate the darkness flooding us with light.

In the presence of this there is the Universe at $k^0 = a^3 / T^2 k$ which holds space -time in an equal proportion but is equal to space-time. However and this must be clear; it is not space-time! The mathematical statement puts the Universe equal to one and one is singularity by the dimension of zero. It is not zero but one and the one takes on the dimensional position of zero. The Universe is singularity that grows into space-time. It would be well advised to go on a search to find the Universe that is going flat before again declaring there is a Universe that is going flat in gravity and not knowing what universe is going flat with intensive gravity. The universe apparently is on the one side space-time but it is on the other side it is a flat singularity.

That is the easy part to figure. By moving the object back in relation to what we view is not diminishing the space the object has because the object will hold the same space it had before. The object is as big at present as it was at the Big Bang event because what was there was there with no adding. What is present in the Universe is in the Universe and no adding or removing of what is in the Universe is possible. If the Universe grew the object had to grow in parallel with the Universe because the Universe got somewhat bigger than the size a neutron has but so does the object have much more space that what the neutron has. The size the Universe had contained what was inside the Universe at the time the Universe went bang. In that there is little to no change possible.

As the cosmos present its evidence, we can see from such evidence how destructive overheating is. Forget pressure, because Newtonians over simplify everything with pressure and exploding. That might happen to a drum they fill with gunpowder but that is not applicable in the cosmos. In the cosmos, unlike in containers, there is no retaining wall that sets limits to pressure inside the container versus pressure levels outside the container. The cosmos has no pressure or pushing or pulling. It has a flow of space-time by concentrating time and duplicating space as it is driving space-time towards the centre. In any picture about any star there is no containing wall that keeps whatever is inside, inside. There is no limit to what the wall if the structure can contemplate before bursting. In the centre of a star is a point holding singularity and since such a point has no space and is immovable, space has to compromise by flowing towards such a location. We regard what we see at night as space and how wrong can we be?

Have you as you sit reading this part at this minute sat back and gave a thought about the light enabling you to read? Such a thought brings to mind the most simplistic answer one can imagine. The light hits the page bounces from the page and contact the lens of my eye where the lens conveys the photons becoming electricity to a part of the brain that translate the electricity to an understandable message and that makes one read. It is as simple as that! Ever gave a deeper thought about light streaming across the night sky, coming from ends of the Universe we do not even realise it is there? How does the photons manage to convey one complete picture coming from as far apart and as wide an area as it does? With a few photons connecting the eye or lens no one ever noticed the wonder of light. The photons reflect a view that seems as if coming from all the billions upon billions of stars. But most is coming from darkness covering an area no man can measure. Yet how many photons can actually connect to the lens of the camera or to the eye? Still a few photons coming from a single direction directly ahead eventually tell the entire storey. It is very simple to take the process of seeing by means of photon conducting very lightly and I have never heard one of the Brainy Bunch really in sincerity uncover the process to its utter and full potential. It is impossible that light from such an array of assorted sources can simply come together at the eye lens and show a picture of objects spanning across a Universe as wide as our mind can receive where the objects they reflect is beyond human measurement and the quantity is inconceivable many.

If scientists think of outer space as geodesic zero, with nothing in outer space but space then how do they explain the fact that we can see the nothing. How can we see nothing being in between light? According to official science that blackness out there represents geodesic space. Geodesic zero means the light travels in a straight line from where it originates unhindered all across space to where the light connects the eye. By crossing the vastness of space, the intensity of light reduces. The light coming from the Sun is quantifiably less than where the light is hitting the surface of mercury. The light loses intensity while travelling through the Blackness of outer space. If light was losing intensity the intensity of light must be that what is robbing the light that which is taking immeasurable small measured light away from the mainstream of light flow must be light. If that which was collecting the light were not by own measure light the light would have stood apart from that which was collecting the light without being light. The light cannot mix with the darkness and remain apart because it is not the nothing the darkness represents without standing in an identifiable visible support of what it remains to be. It can only mix, if it was the same that was mixing.

Isn't it rather reasonable to think that if the light were different from what it went through the light would leave a luminous trail as it went along since the light cannot mix with the conducting medium but still leave some part of the light behind as the price it pays for passage?

Think how wide your view span is. Think of the enormous room, space, volume size you pack into such a small space as your eye sockets are. Try to calculate at night what the volume is at that point that you can see simultaneously. Pack that lot into your eye sockets. See how many cubic light years you put into an area where only a few electrons can go. How do you manage that! Look at on dot in the night sky and think about that one simple dot. Think how many electrons must be streaming from that spot. Every photon that reaches you must have left one atom that went fusing. As it travelled on route it went in circles for millions of years before it was able to leave the star. The photon did not increase in size, and it carried the information it obtained from the one point that went into fusion. At the outside of the star it did not grow bigger and at that point it represented the information of such a small area we have no means to calculate it even if it was

possible to put that area it represented into your eye. As it came along the information it represented grew in stature, as it became a larger part of a smaller growing space. It started to tell about information it could never know anything about except on the condition that the photons fuse together as they travel on. At best the photon originated from an area smaller that the eye socket and it represents (say for arguments sake) one group of stars. How did that information scramble unless all the photons fuse as they come along? It can only be that light is singularity overblown and represented by photons over spanning it whole context. Light then is singularity k^0 in space a^3 in motion T^2 over the distance it came k. Light we see is therefore $k^0 = a^3 + T^2 k$ or the space it represents a^3 at the distance it travelled $T^2 k$ bringing in once again Kepler's formula of $a^3 = T^2 k$. **That is the story of one photon telling us about one light source that may be one star or one group of stars.**

The picture gets even more extravagant when one think about that that one photon might represent an entire galactica with a combination of billions of stars grouped into one area.

That is not the only picture we are getting. Our picture contains much more than a few billion stars in one dot. We might see several such galactica in one area.

 **Nothing is all about not being and not "not seeing".**

We visually see two spots holding light that is in the overall mathematical picture quite close to each other. Why do we not see one spot in double vision? What is keeping the two dots apart? How do we put the entire picture in precise co-ordinates into a total three-dimensional unit that fits all into my eye socket? Surely in comparison there is an example of everything going to nothing. Try to put that measurement in reduction into a sensible and audible mathematical expressed configuration in order of simple understanding. Convert that mass into a comprehensible reduction. The cosmos is not about mathematics because though you Newtonians fool yourself in that you're in superb ability with your mathematical skills. The most obvious puts your mathematical skill to shame. When it cosmos down to true issues you and your mathematics can only degrade the cosmos.

The fact that one can see the night sky is a proven fact! The fact that it is there in every man that has the ability to see is the proven fact. The fact that your calculations fall short is the proven fact. The fact that in true issues of cosmic proportions your mathematics does not even cover the idea there is presented in the wider cosmos is a fact. The fact that by your mathematics you do not even have the capacity to think out the question, which is hardly a reason to understand the question that is hardly a reason to come to same answer proves how far your superb ness in mathematics leave your wonderful atheistic claims in shame.

Truth is that the two spots that are seemingly so close together is might be further apart than the entire Milky Way is wide. This means there is a lot of never explored between the obvious which makes the unobvious part very unobvious. I hope that makes sense in the way I wish for it to make sense.

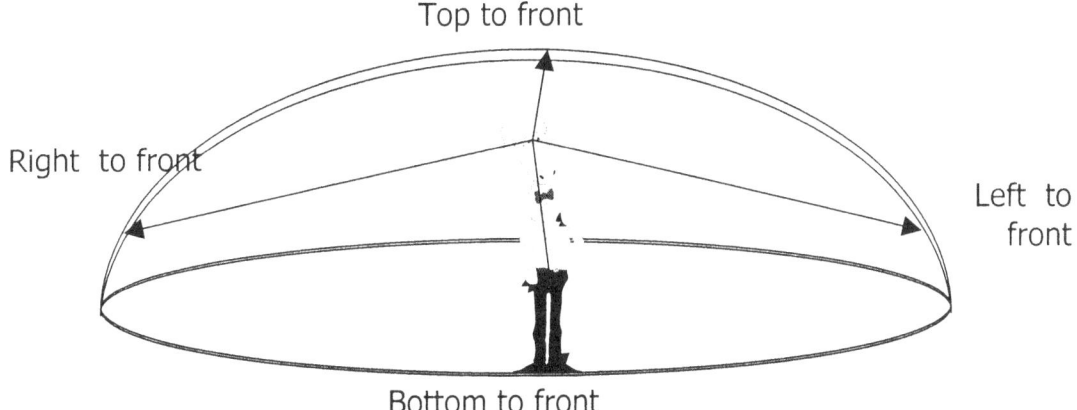

Go out into the desert and look at the night sky where the sky is unblemished of light pollution. The night sky seems three dimensions. All that is lost with man's artificial light degrading the unnatural cosmic light. See how wide an area runs from front to centre, top to centre, left to right to centre and calculate that what is in the eye range to an understatedly mathematical measured volumetric size. See how much " space" there are in the view. Then go home and start calculating so that it might enable science to find answers.

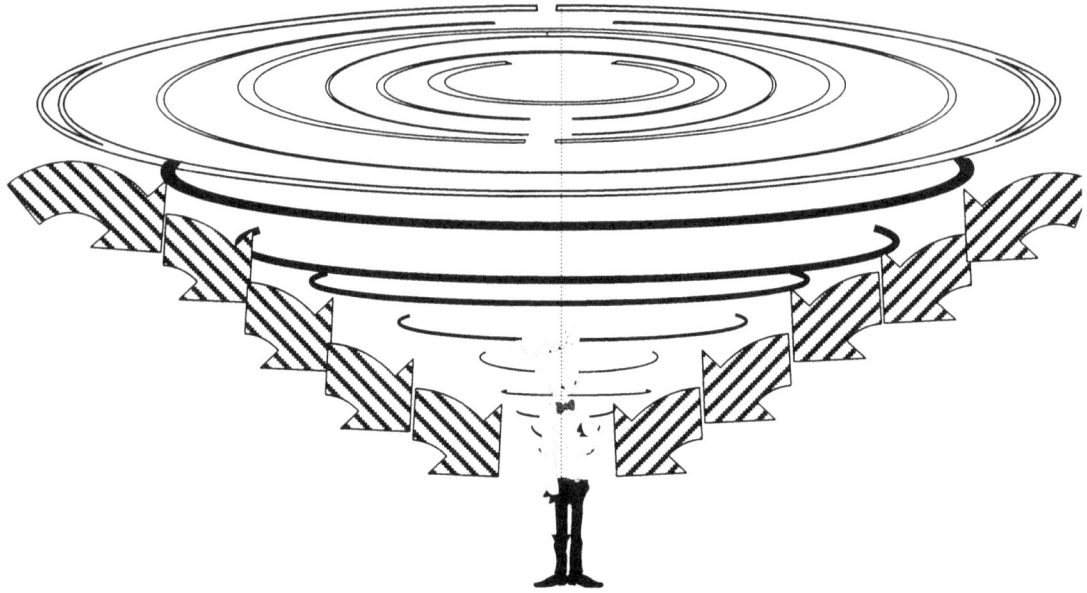

Then try to fit that calculation into the size of the eye. Reduce that lot to the size called the eye nerve and squeeze that which is visible in the desert night sky into a nerve fibre that would carry the information unto the brain. Let those mathematical Master Minds show precisely how that space goes into one eye.

If light came as individual streams of photon flurries our vision the concept would translates into proof that that as such shown in the fragmented picture above. It would be a picture unconnected bringing across some photons in the manner where every object stands apart not being related in any way and that will be what we see, if it is anything that we see. That we know is not the case but that means geodesic zero is as much rubbish as anything Scientists regard with simplicity and with careless thought. Geodesic zero means nothing and how can I see nothing as darkness because "nothing" is not darkness, nothing is "nothing" and the darkness I see is darkness showing the darkness as something.

 Such an idea by itself is outrages because the stream of photons reduce in space to such a minute quantity that taken the area the photons travel and the space in vastness it covers, the chances of one photon coming across many hundreds of light years through billions upon trillions of cubic kilometres of space and selecting my eye to convey the electricity is less than infinite. Yet such conveying takes place every second of every minute. The position of the location of the second singularity, which is the precise duplication of the first singularity but in a diminished capacity, is obvious to miss when one is not applying a detective mentality, as one should in scrutinizing the cosmos.

We may view two dots that would seem to be close to each other but seen through the lens when using a strong telescope the two dots are many degrees apart. The two dots might hold information across an area in space that covers something such as our Milky Way many times and that is done by each of those we are observing. As far as the size on the claim of space they contain go, we can calculate but the numbers such calculations arrive at is meaningless to our small minds. What is the difference between 10 light years across and 5×10^5 light years across? Now we get to the true sticky issue I am aiming to get to after concluding this approach run. Now we get to the question I wished to propose in the very beginning namely: What about the blackness in between. If a dot of one or two millimetres might represent\t an area covering a 100 light years and the two dots both being a millimetre across are some five millimetres apart, the five millimetres they are apart represents an immeasurable time span. That black stuff Newtonians conclude to be nothing. Let's dissect the "black" part first.

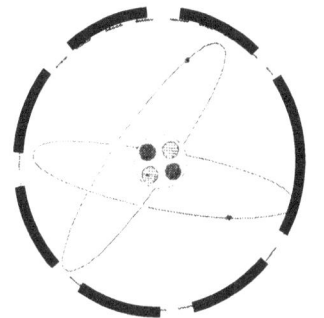

The atom in whatever form is a combination of what forms the Universe and in that to the extent that it is the Universe. The combinations may go as high as the most advanced galactica or as low as the most insignificant sub atomic particle but the end result is the atom is the Universe notwithstanding description.

All galactica forms one atom. All stars form one atom. All layers within stars form one atom. The electron proton neutron cluster form one atom and all cocoons holding an assembly of the above forms one atom. All subatomic particle groupings form an atom.

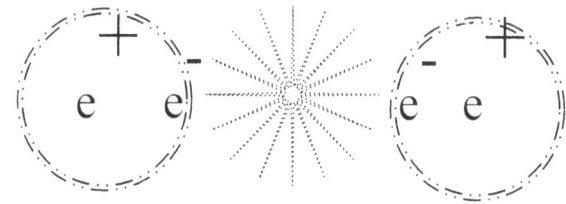

In the centre of any atom there is a line generated by all the spinning particles within that unit. The unit represents all the particles and the particles in the unit are a combination of infinitive numbers of particles joining together to produce the unit. The unit goes down as small as one

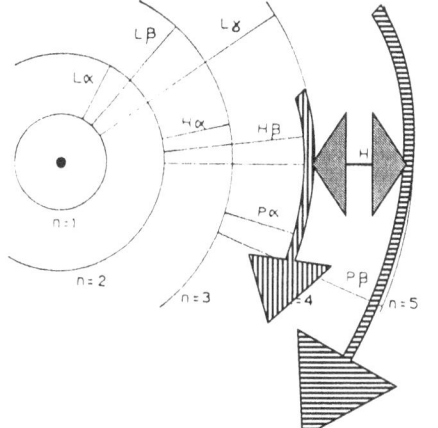

can allow reason to take the particles and the Unit goes as large as the Universe at large. In the end it is all the same to the cosmos by cosmic standards. It is individual motion that parts one sector from another sector and in the end the parting is connected by infinity.

From what ever our abilities are that part of our abilities vested in the Universe will never reach or even almost nearly reach the end of the line. The atom is made up of energy but that is as general a term as time or space is. Never can any one accredit energy with an infinitive meaning but only with a vague description.   By seeing what happens when an atom increases space or decreases space we can see what is it's final substance. By growing, the atom endorses heat and by reducing, the atom rejects heat. The atom is heat in more or less quantities. Newtonians use the term energy as a very convenient escape passage. The moment their explaining runs dry the term they grab onto is energy. What is heat...heat is energy. Every time I hear that phrase I feel as if can blow my top, and moreover using it so liberally seems to please every Newtonian. It is the same as saying that every one is willing to bluff the other one as long as the other one is prepared to be bluffed and then will bluff right back on the same terms. What is energy in the most infinite sense? Energy is heat pure and simple. Energy is heat where heat is time delayed and time delayed forms time in progress either by being space or time and mostly both.

To answer that we must be clear on what is cosmic within human realities and what is beyond cosmic in realities. At this point the atheist mind gets equal to the abilities of the animal and being an animal in mind by being an atheist I suppose the following will go far above their abilities of reasoning. The thoughts that I generate travels far beyond the speed of light and therefore my thoughts are not part of my cosmic material that I take charge of. I can move my body's parts with my mind's thoughts, which proves that my mind's thoughts control my body parts. I am not going to go into details about this but in another book with the title "Xepted Astronomical Mistakes" ISBN. 0-984410-1-4 I explain life in physics in extensive detail. Life control space –time by converting thought to electricity through the generation thereof. The electricity is not life. The electricity is only the result of what life can generate. The in electricity is command instructions that regulate and control space-time whether by moving things physical or by moving the physical body supplied to host and support life or by manipulating other life or objects life created / established / uses it all is terms and the term makes no difference. All motion we have come to accept as common practise is exclusive to our environment, which is a host to life. All other places are most hostile to life and only on Earth does life find a way to flourish. Every other place holding space is hostile to life to the point being there or putting life there will be futile to the life as part of the body holding life. There is a need for an acute and a deliberate turnabout in Newtonian standings points about life and cosmos motion. There is a need for a differentiation about what is cosmic motion and what is life inspired. However, to understand any and all motion is the most complex issue in nature. In order to understand the true concept behind motion we have to break the cosmic motion down to where motion initial start we have to return to moment –Alfa.

There was a continuous motion where eternity met infinity before moment –Alfa. Eternity was spinning within infinity because we still have infinity and we still have eternity. They are tangible entities found everywhere throughout the Universe and are in control of all aspects of the Universe. Between eternity and infinity heat moved in and heat parted that which cannot part. This partition is presented as a time delay that later (at present time) became a time delay and all of the entire Universe is heat lagging behind the time which is in front and was pulling on time that was further behind. It is space-time parting infinity from eternity.

In the motion there are three factors filled by the same substance that is the same substance although the parting is also standing in for the substance. It is the same thing that is in time and time is in front of that which is flowed by that which is behind. The three is inseparably one.

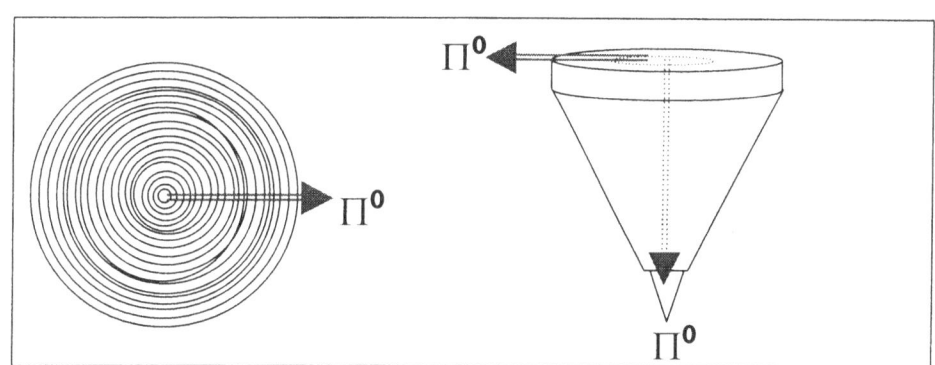

There were four in the centre that wasn't there because the four shared a centre spot, which was the centre spot.

The four spots that represent eternity was spinning around one centre that represents infinity and because the centre was on the same position as where the four in eternity also was spinning time became a repeat of three circling around on where the four was sharing one location at any one time. The same one in the centre was also three spinning on the same spot.

The three was at the same time the three holding eternity while the three was spinning around a fourth centre spot.

Time came about when the four parted by having three positions standing apart from the fourth centre

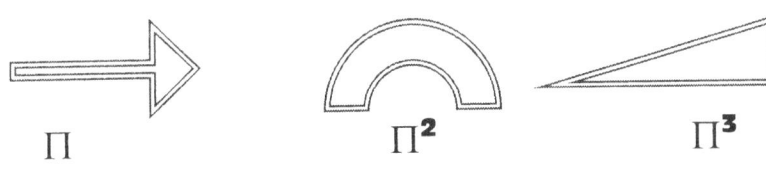

position. The reason why it stood apart is because the one point was allocated ahead of the centre, the other was in par with the centre and the other went behind the centre. A Universe was born because 1^0 moved from 1^1 to form 7 X Π^0 and that established Π. By motion of Π moved from Π^0 to Π establishing Π^2 that resulted in Π^3. The

$$\Pi \qquad \Pi^2 \qquad \Pi^3$$

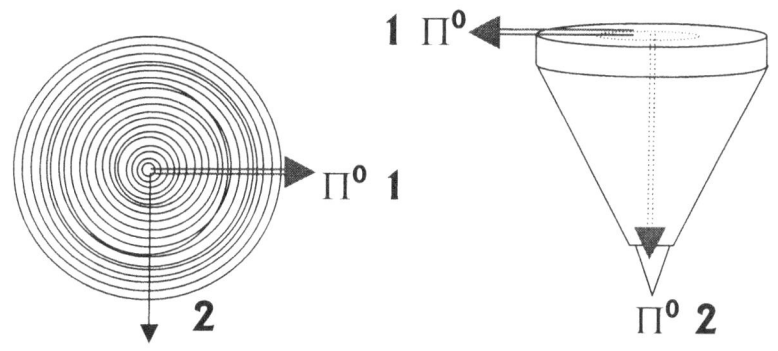

proof is still in every spinning top.

The universe came in place because infinity as 1 parted from eternity as three and because 1^0 that also were incorporating 1^1 as part of 1^0 then by motion became relevant to 1^1. A top can spin because the line forming singularity parted from the three positions forming time and with motion space-time entered between the two factors. As motion becomes part of the top the outside, which defines eternity and is representing the eternity factor then through motion associated with the individuality of the top forming one edge of the top that parts infinity within the top from eternity outside the top. It is in the spinning top present for all to witness.

 **The center singularity point expands with heat accumulating and release space as motion to establish three points serving time in the past the present and the future. It is a relevancy that it brought about.**

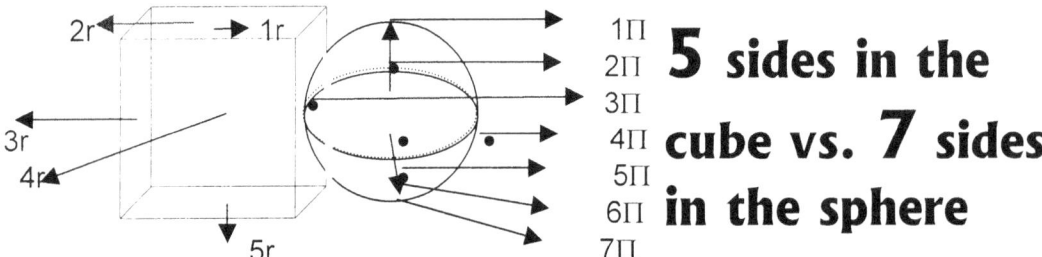

5 sides in the cube vs. 7 sides in the sphere

This is proven by the flow of space-time just as Kepler's calculations reflect. The space a^3 is a combination of motion in time that produces a dimensional quality of $7/10 \, \Pi^6/ 6 = 112$. It is a^3 that is a collection of motion (kT^2), which is an assembly of time ($k = \Pi^0$) and ($T^2 = 10^2$) which is then in the dimensional expression $\{a^1 = (\Pi^0 X 10^2)\} + \{a^1 = (\Pi^2 X 10)\} + \{a^1 = (\Pi^0 X 10^2)\} = 298$. That is the space-time ratio T^2 / a^3 that Kepler introduced as the value of k. This is more a symbolic expression than a calculated mathematical statement but it does prove that space is time by three positions and the combination of space-time by three positions in three allocated setting is $7/10 \, \Pi^6/ 6 = 112$. It is the seven of space in relation to the ten representing the square of time (5+5) in relation to singularity holding three positions in space as well as three positions in time relating to the Universe we are in having six coordinates to fill.

It is the time factor in the relevancy in form relating to the space relevancy in form that produce the Titius Bode law of cosmic proportions. What this indicates is there is a movement Π^2 of Π^0 to Π^0, which is a movement of 1^1 to 1^0 that circles about 1^0.

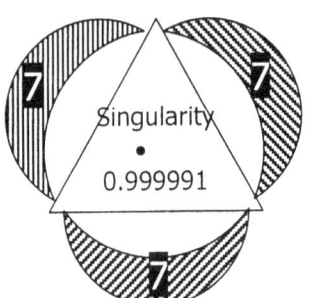

We may not see such a flow as a direction because all the direction we associate with space is part of the three sevens that becomes the three flowing with time. We see the flow of time coming from eternity towards infinity because the time is lagging in the side eternity holds and is catching the side infinity holds. What we see at night, the black stuff that we see that is holding all the bright dots in position is time in eternity. Eternity is reuniting with infinity and infinity is within every spinning particle within my body

In the relation of space against material there are always three positions of seven

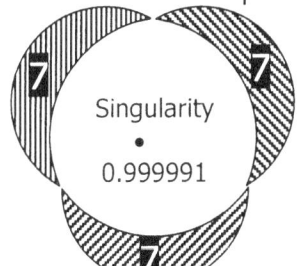

circling points forming a triangle of time to come time in present and time gone by rotating about a very specific centre. This is the flow of time and that indicator points to the direction of the flow of time. The positions in time as well as the reference the positions make during the flow of time and the order the allocations of the various positions bring set time as a cosmic controlling centre. There has to be a centre. That centre has to be because of the motion surrounding the centre. There has to be time retarded that formed heat in between such centres in relevancy. The nature of the spin promotes, as much contraction as expansion and the loss of expansion is the gain to contraction.

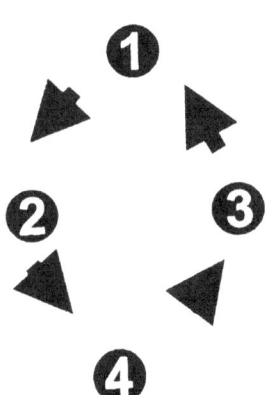

The spinning going on in the past is the spinning going on in the present, which is the same, spinning involving the same points in the future. Please note that it is the space-time within those particular circumstances that is generated and what is generated is different. That which is doing the generating is identical, precisely the same duplicating an exact clone copy.

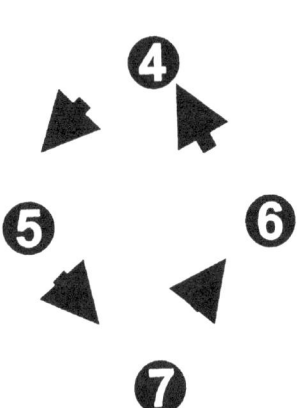

It is absolutely vital to understand that the three points coming on, as time is the very same three points disappearing into singularity. It is where time in eternity that parted from time in infinity is catching infinity to become unified once again. The unification is part of every spinning object there are and is the centre of the Universe.

Singularity by Time

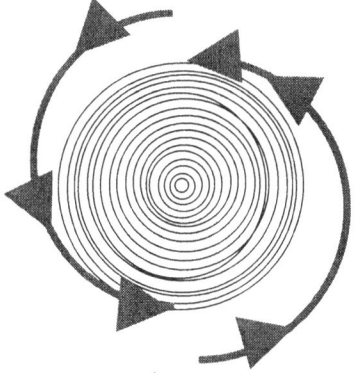

Looking at the top it seems that the body structure of the top is solid and the air surrounding the top is liquid. The top as a structure composes of solid particles that light cannot penetrate and that material cannot pass through. In that sense it seems to fit all the conditions we set for solidness. The top spins and it spins through the air that allows the top to spin seeing that the top has much more density than the air has.

What can move is liquid and stands related to what cannot move being singularity. Since everything is singularity everything is immovable but also since everything is
Every thing outside the top is liquid with the top forming a solid or so it seems to us. Well yes in a way and not that much either. The top is a pump that pumps

heat from the outside inwards just like a turbine engine. Every atom that is rotating inside the structure of the top is keeping the centre erect. The centre is totally motionless because all the atoms in the top are moving and the moving of the top circle is extending the singularity of the top to the edge where the top meets eternity. The extending of singularity is holding the air as a liquid and being the liquid the flow of the liquid keeps the top erect and spinning. The spin produces a cold in relation to the hot that the liquid is.

What is moving is liquid and what is not moving is solid. Everything has a reference in relation to another point. That which is capable of relocating is

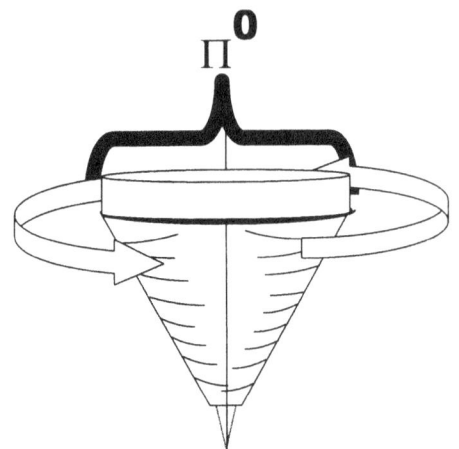

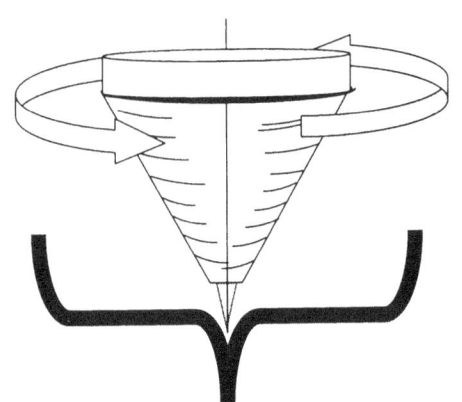

forming a liquid in relation to that which is securing the position of rotation. **Everything in the cosmos can move and yet not one particle in the cosmos can move. The cosmos stands divided between the eternal moving of eternity and the immovability of infinity**

Everything around the top is liquid with the centre being a solid. However the solidness and liquid has cosmic standards and just as it is in the case of hot and cold, big and small, fast and slow, our standards and cosmic standards do not share any measurements. So too does cosmic notions about liquid and solids have a totally different meaning in cosmic terms.

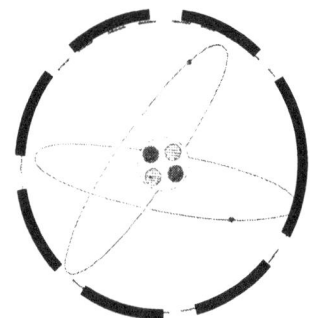

There is a pumping interaction of space-time flowing towards singularity through every point that confirms singularity. Every thing in the top that forms the material is also liquid. By providing motion the matter in the top serves as the liquid factor that extends the space that singularity provide. The structure is composed of atoms. In the atom there are a governing generated singularity around which all material rotate. In the case of the atom all the rotating material forms the heat while the generated centre, which is incapable of rotating, forms the solid factor. Every aspect that is without motion stands in a relation of 1^0 and that which is relatively moving or changing location or find a new position holds 1^1. Everything that is standing still is 1^0 and everything that is moving is 1^1.

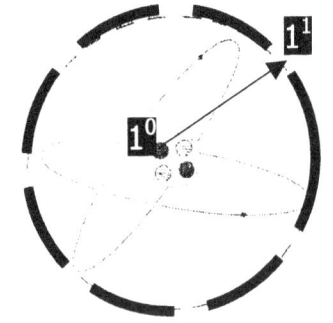

Gravity or motion is a constant relation that solids have with heat where heat forms the liquid and solids form space. There is the rotation but part of the rotation is the lateral progressing by rotation to confirm the generated centre. The generation is in the rotation but the flow towards is the lateral and just as electricity produce a flow of time in relation to space collapsing, space-time by measure of gravity is using the same system to do the very same.

There is no substance difference between 1^0 and 1^1 and it is a relation where one moves as the liquid partner and the other is the solid factor. Both are not as much equal as they are precisely the same. Infinity cannot move and eternity cannot stop moving. By parting infinity had to move and eternity had to introduce as part of the cycle a point where it stops moving in relation to the other side that cannot move but does start moving. The factor that shows motion forms the liquid while at that moment the factor that does not show motion forms the solid.

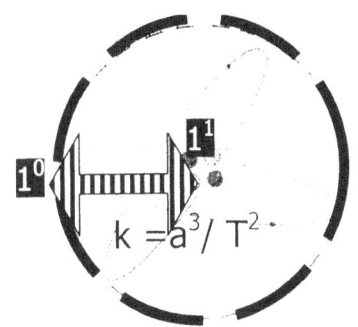

The measure of 1^0 is transformed to 1^0 and which ever are 1^1 is passing the extending of space on to 1^0.

Time spin because everything spins in order to secure the centre singularity. But also time moves and in that there is the linear that always are part of cosmic motion. The centre is referred to by heat but heat also secure the centre by reconfirming the centre in the lateral. But in both cases singularity is reinstating singularity by confirming as it is referring one another. In the manner that 1^0 confirms a position in singularity 1^0 is supporting 1^0 by generating 1^0. By generating 10 it is repositioning and reallocating a position by confirming 1^1.

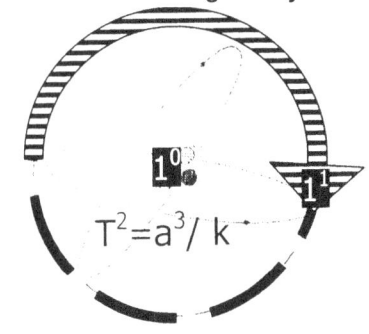

Coming down to understanding the concept in the infinite we must turn to light, which is the smallest material particle visible to the eye. Since the photon is small and fast and we are incapable of really managing an investigation into the photon, we must turn to the photon's more spectacular but far less frequent counterpart being what is referred to as "ball lightning" Ball lightning is heat or time liquefied as it is generating a point within the centre and such singularity is generating motion to concentrate time to heat or electricity or flames or whatever name one wishes to attach to the very same thing.

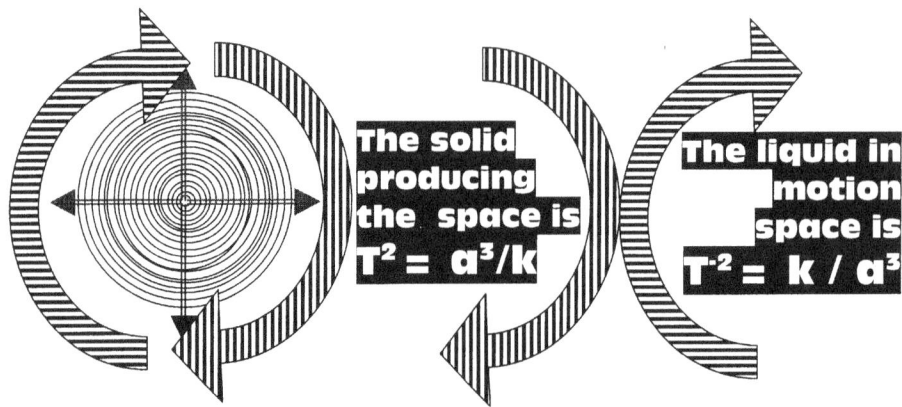

The spin centres singularity (k^0) and the space (a^3) establish a singularity centre motion by spin (T^2) as well as by relevance (k). It is the Coanda effect where the liquefying of heat into a visible electric flame produces the limits of a space created by the motion of the heat being charged by a centre charging the centre again. It is again the manifestation of Kepler's space-time $a^3 = T^2 k$, which is $k^0 = a^3 / T^2 k$. It is time catching up with infinity and the result is eternity returning to infinity as the heat is once again recovered by the uniting of eternity and infinity. That is the photon on a smaller scale as well.]

That is all concept of space-time on a smaller scale than what light is. In the Universe size is not an issue but a change in relevancy. Since the ball lightning is larger the spin factor T^2 takes most of the motion and that increases the space a^3 factor to suit the situation. But that also decrease the relevancy or linear factor k. In the case of light the spin factor T^2 reduces the space factor a^3 and that allows the relevance to match C. Light is what remained of the Big bang where space was an electron C^3 leaving gravity at C^2 and motion by relevance at C.

We stand in time holding space. Life takes charge of the material lent to life to support life's manipulation of space-time by another form of movement separated and apart from cosmic motion. By bringing about motion there is a discrepancy established. The discrepancy is by the changing of the alignment between the one position time holds in eternity relating to a point infinity holds and the next minute realigning the space point. It is realigning 1^0 with 1^1 by three locations in time. One must not look at the motion in the circle but at the motion of the circle as the relevancy reduces by duplicating that which rotates in ratio with that which contracts.

Locating and finding the presence of singularity

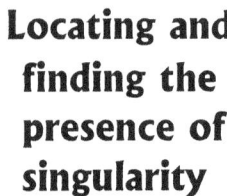 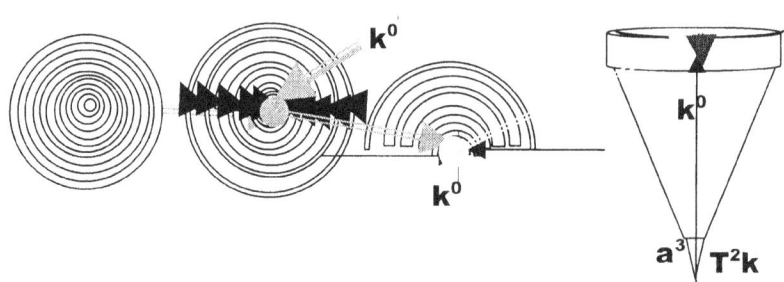

It is the rotation T^2 that duplicates space a^3 / k by reducing the relevance a^{3-1} where the relevancy is k^{-1}.

Since in truth the contraction is T^{-2} but it would lead to great confusion if we state that time moves backwards because it cannot, we have to use the reducing of time as space reclining its value **a^3/k**, which is precisely the same thing. In the end it is 1^0 relating to 1^1. The essence of moment –Alfa was that eternity parted by placing light between eternity and infinity. The essence of the atom is to remove light from eternity and place it in the atom to be united with infinity once more.

What is allocated to the position where 1^0 are in relation to 1^1 is space-time. The motion established between the two allocations of singularity is time delayed as it is returning time in delay to time in the moment. The moment is where infinity is encircled by heat in the atom. However eternity is also 1^0 just as much as infinity is 1^0 and eternity is also 1^1 just as much as infinity is 1^1. It is the same thing which heat parted when eternity moves away from infinity. Since eternity is three in time, which is the four in space and is the three in singularity, eternity is space, which flows by charging singularity. What is in eternity is not just like what is in space or just like what is in singularity. It is the same. It is a clone of the very same thing. The difference is not 1^0 or 1^1 but the motion of time putting 1^0 at a point to relate to itself at a point 1^1. The factor 1^0 fits into 1^1 in total harmony since the factor 1^0 is 1^1.

Because the factor 1^0 is the factor 1^1 and the factor 1^0 is precisely what the photon encircles. The photon encircles the electron centre because the electron is rushing the light through my eye nerve to my brain. The movement is taking 1^0 in the way of 1^1 to my brain, which holds 1^1 as a reflex of 1^0. Since what is in my brain are 1^0 is that which forms the expansion of time is also 1^0 but is 1^1 in relation because of a time constraint, I can see all 1^1 because I have 1^0 in my brain, which is 1 in time. I am as much part of eternity, which I see as I am infinity uniting eternity.

What I see is what I am is what is in the yonder of time and only by measure of time delay is there a differentiation between that which I am and that which I see that is flowing towards me. I am the end of time and therefore I am the centre of the Universe. The atom hosts the centre of the Universe and because the Universe started with the atom, the universe concludes through the atom. The atom is the gateway where eternity again once more reunites with infinity and in that the Universe arrives at a conclusion.

Because I am at the end of eternity where that which has no end starts, I represent that part of eternity that holds no start in relation to the part that has no end. I find a start in that what I can see as being the part in eternity that has a beginning. Being at the start of that which has no end I am unable to see the other part of eternity, which is the part that has no end. That is because there is no such a part in eternity as forming the side or the part in eternity that has an end. Eternity has a start through me but has no end. In the very same manner can I see the part of infinity that has no end because I represent that side of infinity while the side of infinity that has no start is also unseen by me. That too is because there is no side where infinity starts. I lock in the divide between infinity that has no start but uses me as an end and eternity that has no end but uses me as a start. When I and all of space –time are eventually remove then again eternity with no end will once more unite with infinity with no start and all space-time will disintegrate into that which has no start and neither has an end.

I can see what ever is out there in all of time because what is there in all of time is exactly what I hold in space less time. I am 1^0 and therefore all of 1^0 is also part of me because 1^0 has no sides and has no space, therefore all of 1^0 fits all into 1^1 and the whole Universe fits into my optic nerve with no squeezing required what so ever. I can fit into me what I am and since I am what time conclude I conclude eternity by uniting eternity with infinity. I am a black hole as much as a Black Hole is a black Hole. The difference between my being a Black Hole and the Black Hole being a black hole is that I still sport atoms and the Black Hole got rid of all time delay of any standing.

Culture will have us believe that when one sees a colour shining from an object the colour is associated with the object. Logic tells a different storey. A yellow dot is all the colours in the spectrum but yellow because it is disassociating with the yellow. That goes for red blue and all other colours we may visualise. I think the norm accepts this as scientific fact with very little argument or substantiating proof about that required. We have in our minds the formed concept of opposites applying. The opposing side of red is blue. The opposing side of white is black. If white is an array of all colours scrambled, then the forming of the colour black must be where all colours that are present in the scrambling of the white is also absent in the scramble mixture of black. If there was total absence of substance as far as colour goes, the heavens would be a few specks of white dots that is blending because outer space as "nothing: is parting them and therefore "nothing is parting" them. Going one step more in sane is that we then all agree the colour of the nothing is black since we can find no colour mixture would blend to form black.

What then about colours that are technically not colours as is the case with black and white? White is simple. By spinning all the colours in the spectrum the colour white shines through. Black is quite another matter. A friend of mine whom is one of the best painters I have ever come across told me that one couldn't paint black but have to make black a dark blue to show shade on the canvass. That apparently is his success in achieving the realism.

He also went on to explain how many variations of dark blue form the shadows in one simple tree. This remark set my mind in motion. One cannot see black because black has no colour to show, but black is the colour most prevalent in the universe. One can see only by colour and since black is not a colour we should not see black, but we do.

If the darkness was the representation of "nothing", then that should be exactly what we must see, nothing but the stars. Taken from the top picture some stars and leaving the rest to nothing is what we see in the picture below. A blind person sees nothing but when we look at space, we see something that we think nothing of as we see as space. One cannot have the ability of sight and see nothing except by closing your eyelids and then you see nothing. But in that case you do not see "nothing" in contrast of "something" you see "nothing" without it contrasting to "something".

By the ability to see the darkness such action in it self renders the darkness a factor of forming something other than nothing and that changes the acquired value of the darkness from nothing to something. There is an eternal difference between something in infinity and nothing. That black stuff...the use of the word black by itself makes it a contentious issue. That is even before we are using logic by disregarding the nothing part which crosses over what borders the ridiculous throwing the whole argument into the mindless because it is ridiculous to think of anyone is able in seeing something that is made up of nothing. Yet I stand alone against the might of Mainstream science no matter how correct my views are and notwithstanding how far their senselessness are going into madness, I still stand alone in my thinking as I am again and again rejected by the Brainy Bunch.

The arguments introduced touches the most basic aspects of my work and by no means can such an introduction secure an opinion that I do realise. Yet, not once through all my long investigation in the past thirty or more years have I found any other person claiming such views that I have brought about even in this skimpy way. If you see it, what you se must be light because you see it.

We view an object over a distance and see what the light brings across. It reads as $k = a^3 / T^2$. I see the space that time holds by time developed as k the time factor of relevance. Time at present T^2 is holding the space a^3 in relation to how the cosmos developed the relevance k. If I move time back by bending time with a lens $T^2 = k / a^3$ then space will increase as the "distance" or relevant position of the space increase. I challenge all Newtonians to increase or decrease nothing. I can see with the aid of light. Light has to come to my eye to allow me to see. If the dot is to small it will find not enough representation in the overall picture of visible light coming to me. I still can see, however that which I interpret as light I can see and to my mind that which I cannot interpret I interpret as if I cannot see it. I can see it but I am unable to interpret it. If I have to acknowledge that I can't interpret what it is that I am seeing I am as stupid as a grazing cow not knowing what she is looking at when she stands staring at what it is she does not focus on while grazing. No mathematician that concerns him as smart as to accomplish the calculations God used when God invented the cosmos will ever think of himself or herself as so stupid he or she doesn't even know what they are seeing. They know exactly what they are seeing... they are seeing nothing and to top that they are to stupid too realise that they should realise it is impossible to see nothing!
 By increasing the one part the other part has to decrease. But that doesn't mean they have to go less blind or blinder because they are seeing what they are seeing. It is only a matter of finding a balance in what they do interpret and what they do not interpret.

Looking at light coming from the Sun there is a distinct interpretation that the light coming on is bright and it is hot and it is hot because it is bright. Finding myself in the darkness I find the darkness is cold and the light is dark. In that way I think of the light being bright as hot. That is because in relation to the cold of the light coming my way I am hot, In relation to the darkness and the cold in the darkness I am cold. With the bright light is my relation in the light that brings about that I am hot. I am the relative hot party because I feel the light touch me and where the light touches me it exaggerates the cold, which I interpret as a hot spot. One can't feel the light but I can feel my condition at the point I sense the light and at that

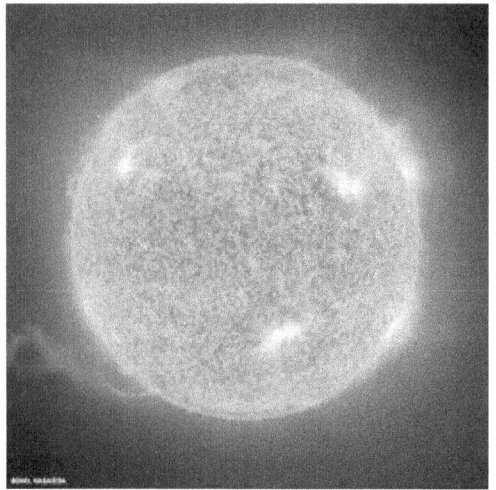

point I feel heat. Where I am within the darkness the darkness of outer space is not freezing me because I am freezing at the point I touch outer space. It is because I am so much colder than the hotness of outer space that I am freezing instantly. Remember I am freezing and not outer space. Lets take a look at the Sun . Where the Sun meets outer space the Sun is boiling and not freezing over. By the cold Sunlight touching me I am able to generate a lot of heat within me as the heat has a route to where it can conduct. By generating the heat, which I then am able to conduct, I feel heated and generated. On the other end where I touch outer the region close to outer space, there is an abundance of heat outside me and with me being so utterly cold in comparing to the hot outside of outer space, I cannot conduct a flow of heat outwards and that stops me from generating heat inside. That smothers my generating ability and life is generating motion in a quest to manipulate space-time. If the part holding outer space was cold, the Sun would freeze at that point with ice everywhere.

However, where the Sun is in contact with the outer space region, the Sun is boiling. That means the temperature the Sun holds must hot up and star cooking to meet the required heat at the point of contacting outer space. Outer space is raising the temperature of the Sun and not dropping it. I am in relevance to the heat, which makes me either the hot party where the coldness of the Sun touches me and I feel my body raising the temperature of the heat to bring it to an expectable level to my body heat. At the point where my body meets outer space I burn black because at the point outer space ingenerates my cold body with its heat.

At the point where the Sun touches the heat of outer space the liquid heat of the Sun starts to boil. It is not like the heart transmitting we are familiar with by an element heating the water from within. The contact is on the outside and the heat at that point releases from outer space as it flows to the Sun . The Sun has no ability to realise the heat into the Sun and the she heat turns the pebbles into small crystals we know as photons. It is heat particles like it was in the Sun , but at that point where it touches the atmosphere it becomes concentrated liquid particles with a gas envelope leaving it to be photons. Within the Sun , the liquid photons are like water with a much more solid binding. This is because the Sun is that much colder than that outer space is. The temperature of the Sun rises at that point where outer space comes into contact with the liquid of the Sun .

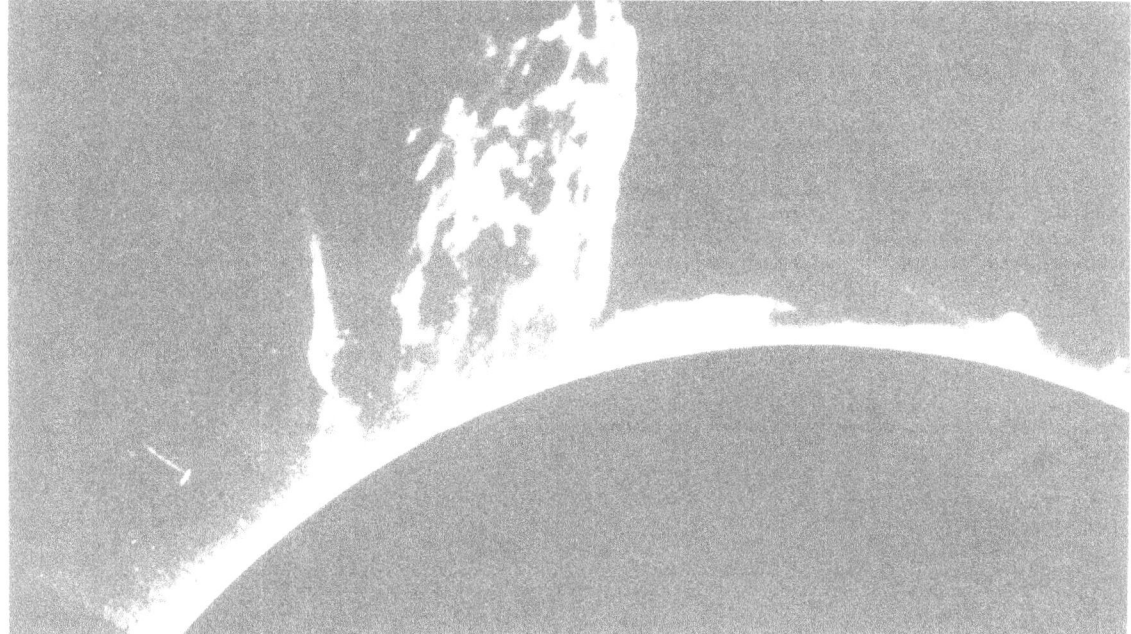

It is not the Sun that is loosing heat because then the Sun would freeze at that point. The Sun throws plumes into outer space because the heat entering from outer space distributes uneavenly and at certain points the heat rises to higher levels than at other places on the surface of the Sun . This can only be because of waves forming at that point just like the waves we see in the phenomena we call mirages. It is layers of different concentrations of heat. At those points it turns the cold substance within the Sun , the liquid in of the Sun into gas plumes. But even there the distribution is uneven and only where outer space truly touch the liquid can some liquid rise sufficiently to form a gas that will be absorbed by outer space. The rest that did not heat to a proper gas remain a liquid and drops back into the liquid Sun . If the plumes of heat forming the prominence carried the heat internally as the heat expanded outwards the plume would explode since all that lovely heat is suddenly exposed to cold and the plume would violently release the heat to the cold. It is not happening that way. The plume of liquid remains concentrated and only on the outer regions does it turn the liquid to a gas. That means the heat is not in the liquid but the heat is outside the liquid and it is turning the liquid to a gas from the outside inwards.

When we see dry ice boiling the boiling is in the fringes where the ice is in contact with the air. The ice does not heat from the centre because then the ice would explode from the centre outwards. The ice is stable on the inside but burns away at the fringes where there is contact with the much hotter atmosphere. The boiling would go on until the liquid ice has gone into a total form pf gas. But the process is on the wall of the ice and not from the inside and therefore the ice does not explode but turns to gas systematically. The Sun converts much more space to a frozen liquid than outer space can boil away the liquid interior of the Sun because with the exasperating movement of the Sun the motion freezes outer space by gravity, which is the freezing of time.

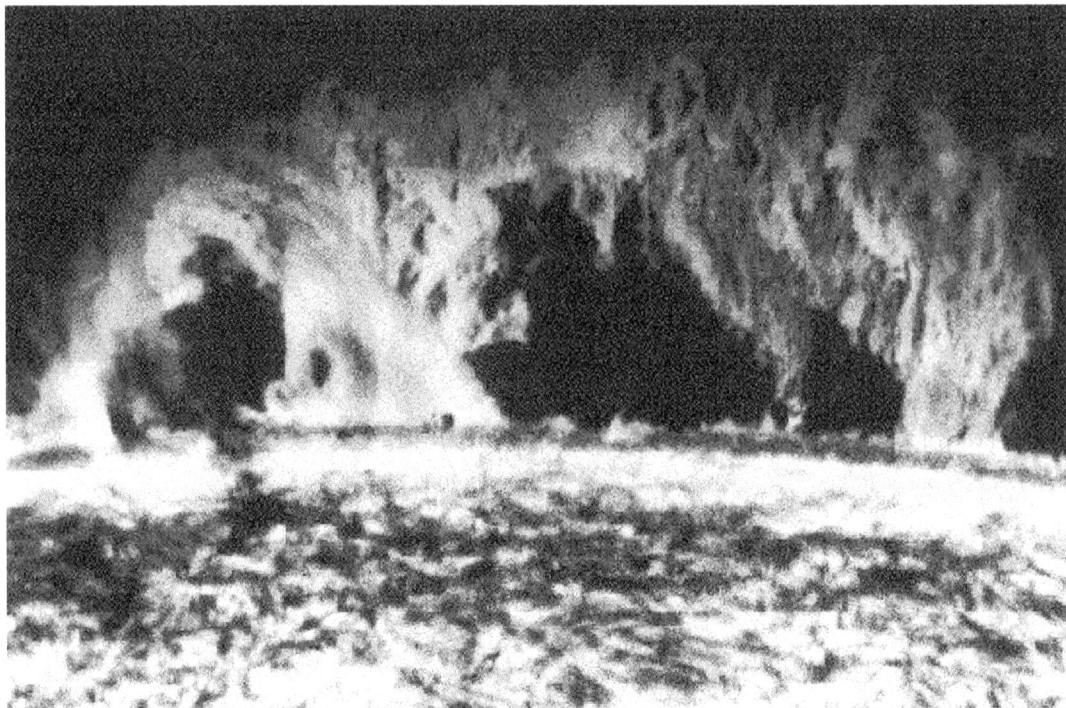

 Look at the flow of the prominence. If the heat was in the prominence the prominence would disintegrate the moment it release from Sun because the heat would expand into gas. But since the prominence is cold because the Sun is cold that makes the edges of the prominence hot. When the edges of the flowing liquid turns hot, the relevancy to the inside turns the prominence even colder since the edges of the province is hot. In that case the prominence does not expand with heat but contracts and the density the prominence has increases. The prominence then falls back into the Sun because it contracts and its density would not allow it to remain out of the Sun . If the heat were inside the prominence the prominence would boil away and not conserve the relative small amount of heat it has. Since the prominence plume is part of the contracting cold it falls back into the Sun , leaving the photon particles that became liquid gas prominence plumes to expand with the gas in outer space and allow the hot outer space to expand the light in all directions. The surface of the Sun is a pool of liquid. Where outer space touches the pool of liquid the heat wave formed by outer space breaks down the cold in the liquid if the Sun . The liquid breaks up into fragmented light particles and as the fragmented liquid light particles become a gas it flows through outer space and out. But by outer space also being a light with much more intensity than the liquid heat of the photon has, the photon puts the space it holds in relevancy 3^3 into motion of space $3\Pi^2$. And where the space is colder than the contact it makes the space forming the photon puts the heat difference between the space and the photon into motion exactly like a drop of water does when the drop lands on a red hot stove plate. The heat transforms to gravity and gravity is either expanding motion or contracting motion

Light is much more than the medium science takes it to be. Light connects the Universe in a way we cannot contemplate. Light being far apart originating from regions not in the same time or Universal space connects in a way that present us with a picture holding the Universe in an understandable content. From the point we stand and we watch the Universe the significance of what we see surpasses the sense of understanding of what we are experiencing. How can the few photons that our lenses catch coming from such an area as the night sky cover transmit the complete picture of what we see. Take a few seconds and study the picture of the night sky then rethink the picture applying the full content in the picture to what the size of you eyes is. Think how big the picture is that your eyes take in and translate that area to the size of your eyeball in an effort to determine a ratio. One will be forgiven if one thinks of the ratio as eternal to nothing. Yet a few pages back I showed that according to mathematics there couldn't be anything as nothing. Consider the path the light followed from the source connecting to light from all other sources where all particles of the other light may come from and bringing a full picture to the lens one use to look through. In your mind connect a line from every atom producing light and connect the lines to your eyeball and see how you can manage to fit all the lines, as small as the lines may be.

If I can bend something to increase the light by which I see, then it is light that I am bending. If I increase the light I use in magnitude in order to see, then I am increasing light. I am not decreasing the darkness by bending the light. I am bending the light because I am improving the focus I have in the light. By bending the darkness I am not squirting out more light, I am increasing a balance between that which I have a use for and that which I am unable to use. It doesn't mean that which I am unable to use is noting, unless I am a mindless animal that cannot realise that that what I cannot use, also exist in a way that I can use it if I had some intellect to do so. Newtonians are much better…they seem to realise that what they are unable to calculate cannot exist because the hell can be so stupid as to invent something that is outside the spectrum of their calculating abilities, after all, they are occupying the centre of the Universe from where they can see and calculate the lot.

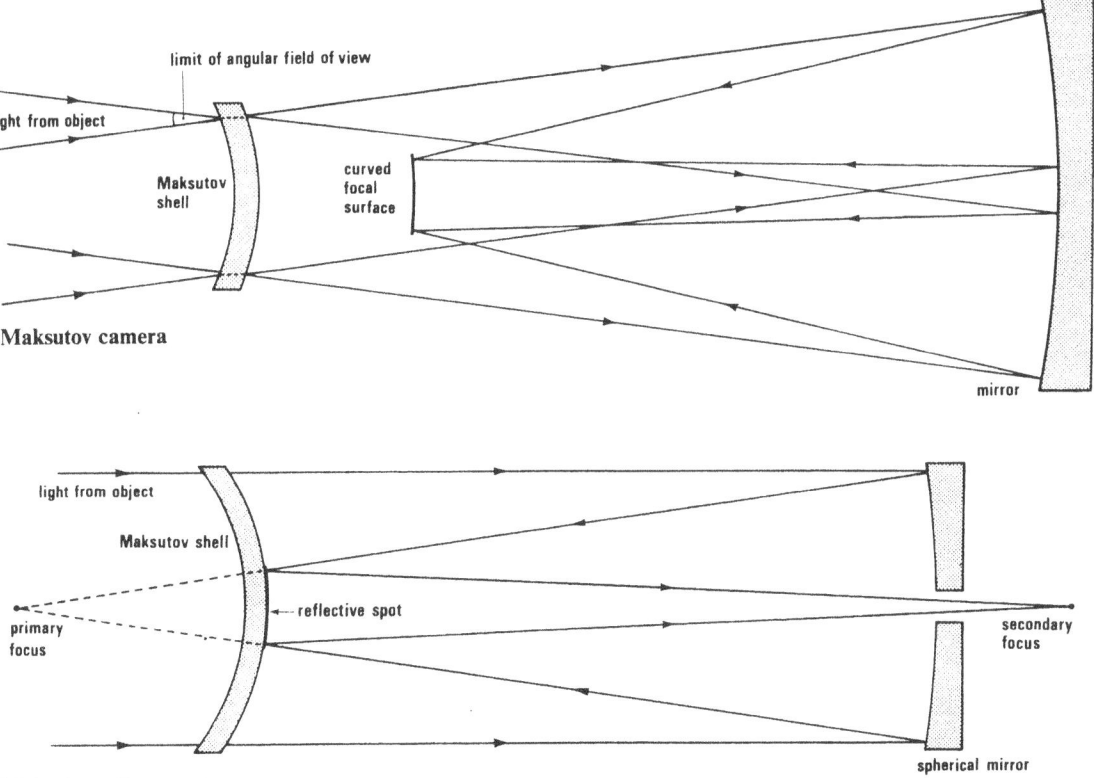

Maksutov camera

Maksutov—Cassegrain

It is so easy…if you can bend something more and afterwards see more in it must be a lens with a curve because we alter the curve. It you can read radio waves in the curve by bending the curve more with optic devises such as radio astronomy, then that which you bend is light. One can only reverse time as an optic elusion and bending something to gain an optic visibility improvement must be that there is an improving in the clarity of light. It is all part of the curvature of space-time since the curvature of space-time is the eternal sphere. We have to keep in mind that the first sphere was so small it was by today's standards not part of the cosmos because eternity parted from infinity. That came by measure of light or heat or whatever one wishes to call the substance the entire Universe is made of. By today's standards that, which was so small is so big the entire Universe fits into the parts that parted. Everything even what is so huge that we cannot see it still only is between the two points that was so eternally small.

If it is lenses that enable us to see what we can't see in outer space it also means we cannot see the light, which is outer space because we haven't got the lens to match the curb of outer space. Newtonians think of outer space as geodesic zero, with nothing in outer space but space. Geodesic zero means the light travels in a straight line from where it originates unhindered all across space to where the light connects the eye. Such an idea by itself is outrages because the stream of photons reduce in space to such a minute quantity that taken the area the photons travel and the space in vastness it covers, the chances of one photon coming across many hundreds of light years through billions upon trillions of cubic kilometres of space and selecting my eye to convey the electricity is less than infinite. Yet such conveying takes place every second of every minute.

The position of the location of the second singularity, which is the precise duplication of the first singularity but in a diminished capacity, is obvious to miss when one is not applying a detective mentality, as one should in scrutinizing the cosmos. Culture will have us believe that when one sees a colour shining from an object the colour is associated with the object. Logic tells a different storey. A yellow dot is all the colours in the spectrum but yellow because it is disassociating with the yellow. That goes for red blue and all other colours we may visualise. I think the norm accepts this as scientific fact with very little argument or substantiating proof about that required.

If light came as individual streams of photon flurries, then our visage would translate that as such shown in the fragmented picture above. It would be a picture unconnected bringing across some photons in the manner where every object stands apart not being related in any way and that will be what we see, if it is anything that we see. That we know is not the case but that means geodesic zero is as much rubbish as anything Newtonians regard with simplicity and with careless thought. Geodesic zero means nothing and how can I see nothing as darkness because "nothing" is not darkness, nothing is "nothing" and the darkness I see is darkness showing the darkness as something. What then about colours that are technically not colours as is the case with black and white? White is simple. By spinning all the colours in the spectrum the colour white shines through. Black is quite another matter. A friend of mine whom is one of the best painters I have ever come across told me that one couldn't paint black but have to make black a dark blue to show shade on the canvass. That apparently is his success in achieving the realism. He also went on to explain how many variations of dark blue form the shadows in one simple tree. This remark set my mind in motion. One cannot see black because black has no colour to show, but black is the colour most prevalent in the universe. One can see only by colour and since black is not a colour we should not see black, but we do.

It is quite obvious that any object sporting a specific colour is all the colours there are except the colour it is rejecting just because it is rejecting that colour If we see an object being yellow it is a human response to think the object is yellow Quite obviously it is not yellow because it rejects the yellow we associate the colour with. If that argument is true in respect to one set of circumstances it must be true through out. If the Sun rids it of heat it supposedly has the Sun must be cold because there is just no more heat left. If outer space absorbs all the heat on offer and shows no change that can be attributed to the collecting of heat, then outer space is as hot as it can get. If a yellow object is yellow on the outside it cannot also be yellow on the inside because where is all the yellow coming from?

If it was true about a yellow object not being yellow and a red object rejecting red and therefore not being red, the same must be true about dark and light. The bright object rejects all the light just because it is light from the outside. If it is light from the outside it has to be very dark inside. That too would count for what we believe it to be dark stars. Such dark stars must be most brilliantly lit because they keep all the light to their inside and well protected. They are dark because they keep the light on the inside where we can't see it and therefore out of our viewing range. The stars have gone so cold the stars have to conserve all heat to remain in gravity. With light being the highest concentrated form of heat, it stands to apparent reason that the light would be the energy of prime choice to contain. The same must then apply to outer space in that outer space is conserving all light and by keeping all light, outer space is brilliantly lit. We just are unable to witness the light because our position is such concentrated where as the light being dark is expanded to the full. We are able to see the galactica because the galactica represents highly concentrated light in one reduced area. The darkness contrasting the light we see as darkness because the light is expanded to the ultimate. The fact that we can see the darkness makes the darkness light, which we are unable to see. However with the space stretched to the maximum the lens we see light by has as far as our position goes, not even slightly curved because we are so small. Now you go and tell any mathematician in charge of theories this much and see how far you can get convincing him about your view. They wish to manufacture and design space whirls and not see reason.

The fact that we see light means that the dark next to the light cannot be "nothing", If the darkness was the representation of "nothing", then that should be exactly what we must see, nothing but the stars. Taken from the top picture some stars and leaving the rest to nothing is what we see in the picture below. A blind person sees nothing but when we look at space, we see something that we think nothing of as we see as space. One cannot have the ability of sight and see nothing. It is light that we see and it is light that we use, which enable us to see. That proves the darkness that we see in outer space is light that we see without recognising it as such. If the darkness was the representation of "nothing", then that should be exactly what we must see, nothing but the stars. Taken from the top picture some stars and leaving the rest to nothing is what we see in the picture below. A blind person sees nothing but when we look at space, we see something that we think nothing of as we see as space. One cannot have the ability of sight and see nothing. It is light that we see and it is light that we use, which enable us to see. That proves the darkness that we see in outer space is light that we see without recognising it as such.

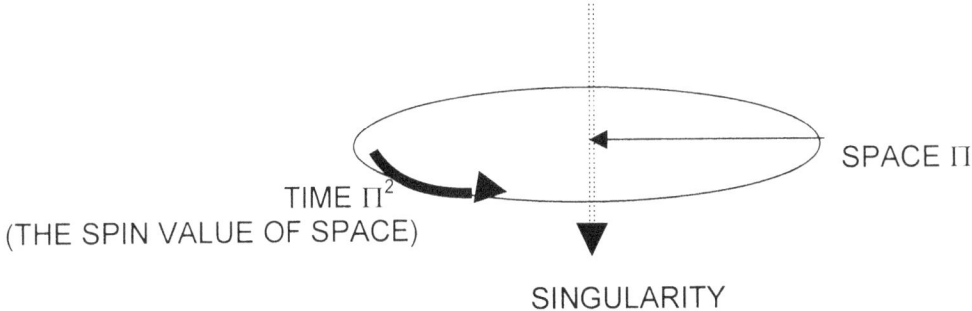

TIME Π^2
(THE SPIN VALUE OF SPACE)

SPACE Π

SINGULARITY

Gravity is the very same but it is the recalling of the space by creating motion in the space. Gravity is the retracting of heat by splitting matter as matter duplicate and reduce space by increasing space in expanding. As the space gets more and the time holding the space gets less per unit in time used the heat distribution is wider in less time and by such distributing the heat in relevance gets less because more gets distributed by a wider area in a shorter time frame. Gravity is the retracting of heat by cooling because of the expanding of heat increasing.

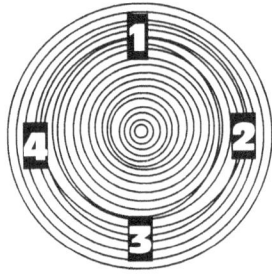

By moving from 1^0 to 1^1 and from $1^0\Pi^0$ to $1^1\Pi$ requires space. Yet such moving did not leave the realm or the domain of singularity. The motion was still within singularity because moving involved forming a relevancy between heat and cold between infinity and eternity, between space and time and most of all producing what will in the far future develop into a Universe that can even be a host for life albeit on a very small spot for a very short while in relation to the vastness space has and the duration cosmic time has.

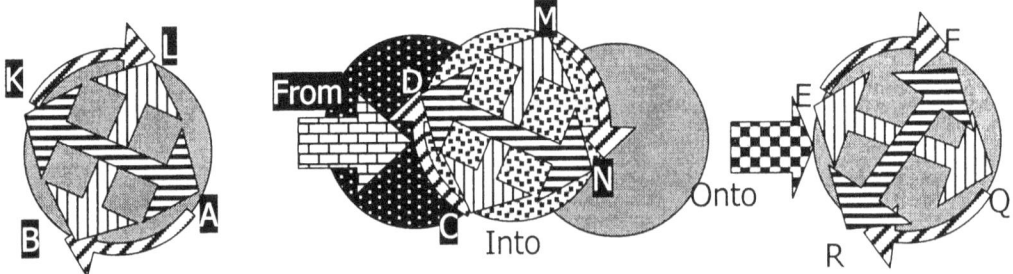

Relevancies came abut when the dot moves away from the spot but had no space to move. All that was possible was to charge singularity by relevance to comply in being activated into complying. Space-time is motion and movement are all the same things only separated by dimensions and dimensions are formed space, where the dimensions become space being

in motion and the space is motion by contraction or by expansion but because time is almost eternal at k^0 our perception of the universe we are in is a stable and steady eternal structure. Gravity is motion and motion creates space to the third by the third in the third that interacts with one but establishes ten.

The cosmos holds no constant and that is the only constant. Every aspect of the cosmos is relevancies where matter in different forms, form different relations to other matter also in various forms. As I indicated about time, where time is an ongoing repositioning of relevant matter locating relevancies in the position they hold to the time they apply. The cosmos cannot grow, as much as the cosmos cannot shrink. It is a never-ending flow of changing relevancies, where singularity meets singularity as much as space meets time. The point in singularity I named time, is the point where time started and where time ends as much as where time will finally fulfil its reducing of space. Heat made space renegade and time slowly contains space by reclaiming heat. That is the cosmos.

The atom forms the Universe as it shapes the Universe because the atom is the epitome of the Coanda principle. The atom spins and the spin provides an accumulative that drives stars, which in their turn give an accumulative spin that, drives galactica and in all that there is a centre governing singularity controlling every spin just as the simple top showed us.

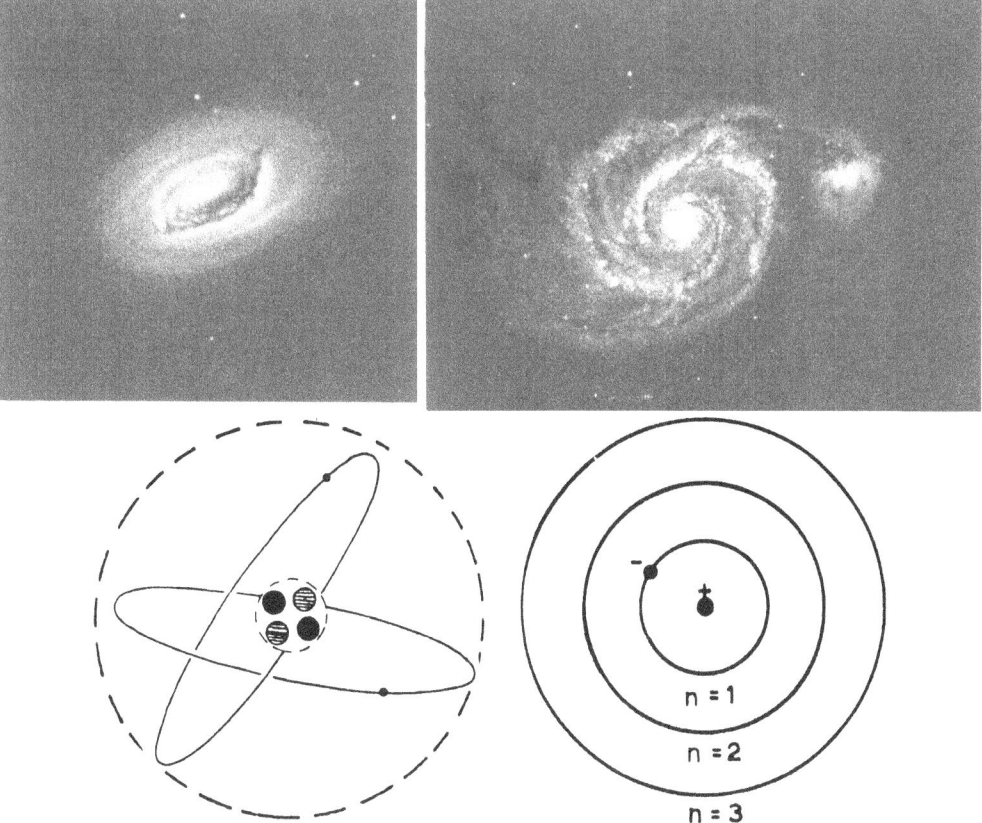

The atom is the template on which the entire Universe is formed. The atom is a galactica, it is a star, it is a bicycle, it is a supersonic aircraft flying in the atmosphere, it is a moon orbiting another containing cosmic structure, in fact only nothing does not use the atomic principle as a format in the Universe and nothing is the only fact not represented in the entire Universe. There is a solid centre, which is supported by a liquid outside, and the balance between the two determines the relevance applying. The atom therefore is the Coanda effect which is the Universe in its entirety.

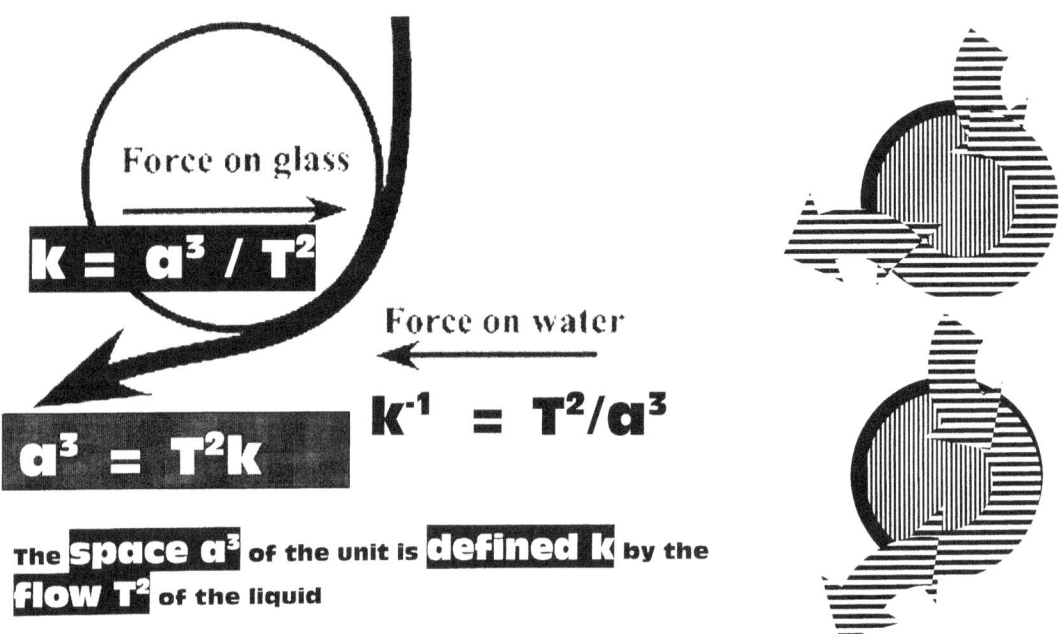

Force on glass

$$k = a^3 / T^2$$

Force on water

$$k^{-1} = T^2/a^3$$

$$a^3 = T^2k$$

The **space** a^3 of the unit is **defined k** by the **flow** T^2 of the liquid

That what I am about to explain may sound inconceivably simple but don't blame me for that. It is not my fault no one brought what I am about to say into the context of gravity and if I don't explain the most mundane in connection with gravity there is no one going to do it.

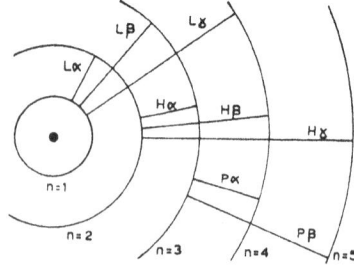

The control the atom has in regulating heat is plainly godly simple as it is Godly genius. The control the atom has on heat by expanding everything outside while accumulating heat, as a time retardant inside is that which creates the Universe just as much as it is that which produce the Universe.

By producing space in expanding when heating the atom not only control what belongs to the atom but also that what is in the control of the atom and what is not in the control of the atom since the entirety of the Universe is the world of the atom.

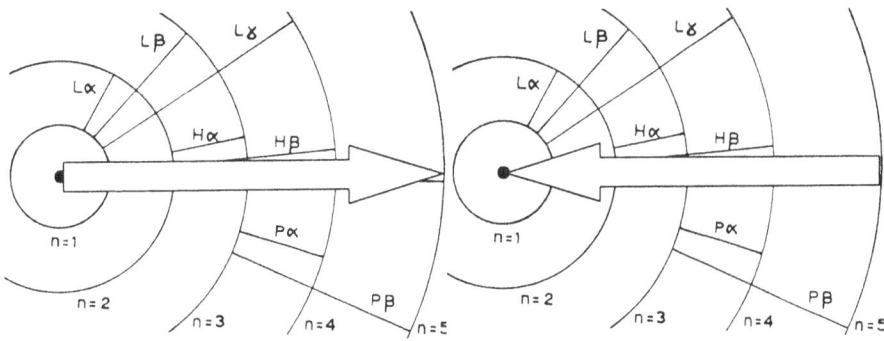

What we see as heat is relevancies because as the relevancy within the Sun changes the atom adapts to the changes. The atmosphere of the Sun becomes denser, which we see as being hotter and the containing becomes stronger. The atom has to reinvent it by adapting to the changes or different surroundings. In this manner the motion that the star provide which is so much more than what is the motion is we find in outer space that the hot / cold dynamic changes all together.

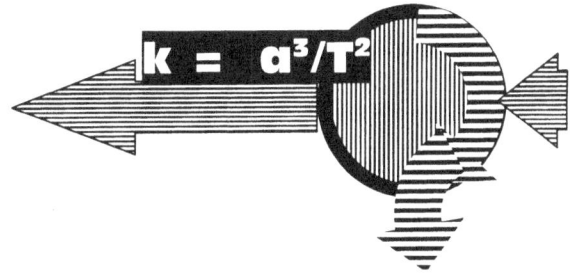

Depending on the flow of time, the atom will accumulate heat within the structure or on the outer side of the structure all depending on which at the moment holds the strongest relevancy between $k = a^3 / T^2$ and $k^{-1} = T^2 / a^3$. That is gravity committed by the flow of space-time and controlled by singularity $k^0 = a^3 / T^2 k$. **This was also what the Big Bang was all about.**

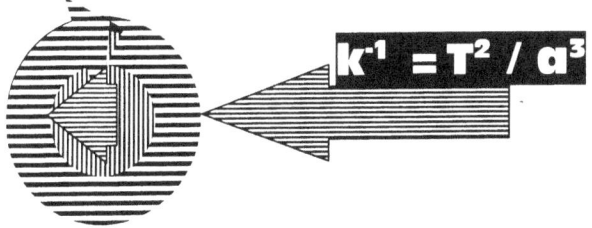

This became possible as the Coanda effect brought the speed of light as gravity (**k**), which brought the motion of the Universe in contraction to the speed of light T^2 that made the Universe be the speed of light a^3, which is exactly and specifically what the **GUT** theory proves.

The spin of the liquid proves the value of the relevancy. The stronger the motion is that the liquid generate, the higher would the contraction be and the lower the spin motion is that the liquid generates the higher would the expanding be. In that we find a definite favouring of either the factor of seven or the factor of ten depending on what the situation will dictate.

Depending on the balance there are the contraction will be totally dominating but never to a point where it annihilates expanding and on other occasions the circumstances would be that the expanding may dominate but also never to a point where it devastates contraction.

The motion of the liquid factor puts the aspect of time in eternity in relation with infinity where the ratio that develops gives infinity the chance to interrupt eternity. Infinity in the centre is immovable but has to associate with eternity since the tow parted by space. Therefore we have eternity having three positions that goes square since it develops an alternating stance with singularity and in that the three of eternity develops a square since the angle of developing cross 180^0 by the margin of 90 ^{0}.

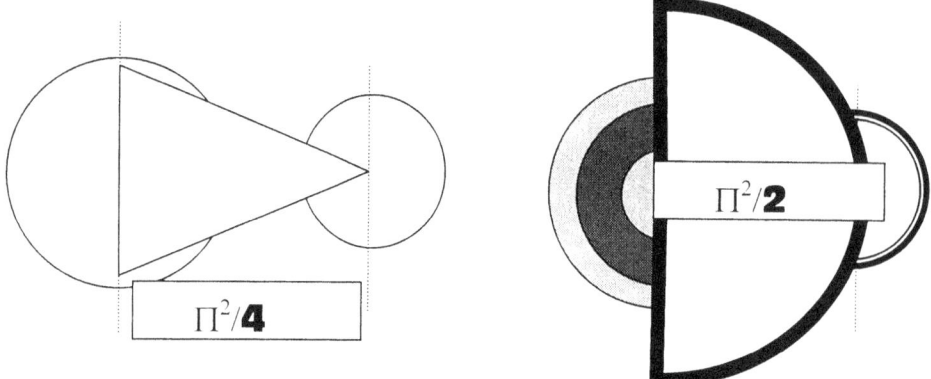

From the mathematical legacy we can see that time was forming 5 but with material not formed yet, that which later became material moved through time holding the value of 5, at a measure that was slightly less than half the value of time and still had the notion to move being $\Pi^2/4$, which was at slightly less half the value of 5.

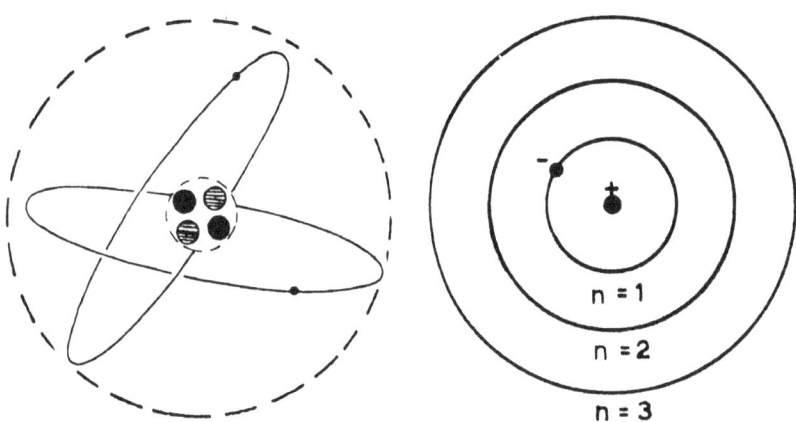

But as space grew into a stronger part of the Universe time came into a square having five on both sides of the divide and time gave space an integrate part of the motion that time provides. Then the motion became $\Pi^2/2$ and material found a better and a much more valued relation in time. This only happened while time grew less, which gave space more value in $a^3 = T^2k$. That produced the atom we now enjoy with three factors instead of only two, as was the case in the previous dispensation. There then was time, space to hold time and space lagging behind time. That part that was lagging

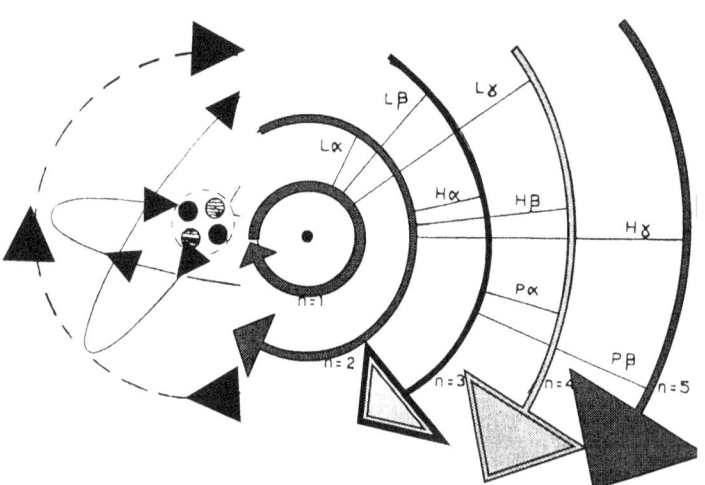

was excluded and that part formed a secluded unit that became the atom.

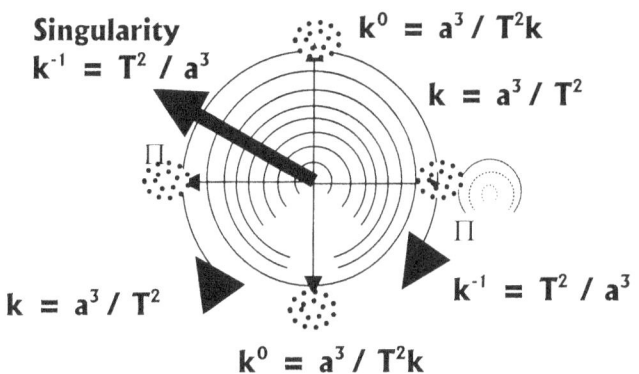

Every time the point rotating changes direction it crosses the divide and in that the entire Universe changes. What comes down then goes up and what went forward then goes back. The line is a cutting limit that divides one sector from the other like no other found anywhere. In that the divide produces changes that are beyond comparing to anything else and in that we find such conclusive distinction in gravity. That which cannot spin does spin by never actually spinning.

From investigating this we can presume with very little doubt as to how the universe came about.

The atom is the Universe that is true but that sounds like rhetoric. Let's break that down a bit by dissecting what this means. The atom is more than just the main ingredient because the atom is also a calendar of the Universe. The atom is a storage facility of the Universe and the atom is a building stone of the Universe. The atom is the final formation the one that will come last as much as it is the one that came first. The atom is the custodian of singularity and the guardian of maintaining the Universe.

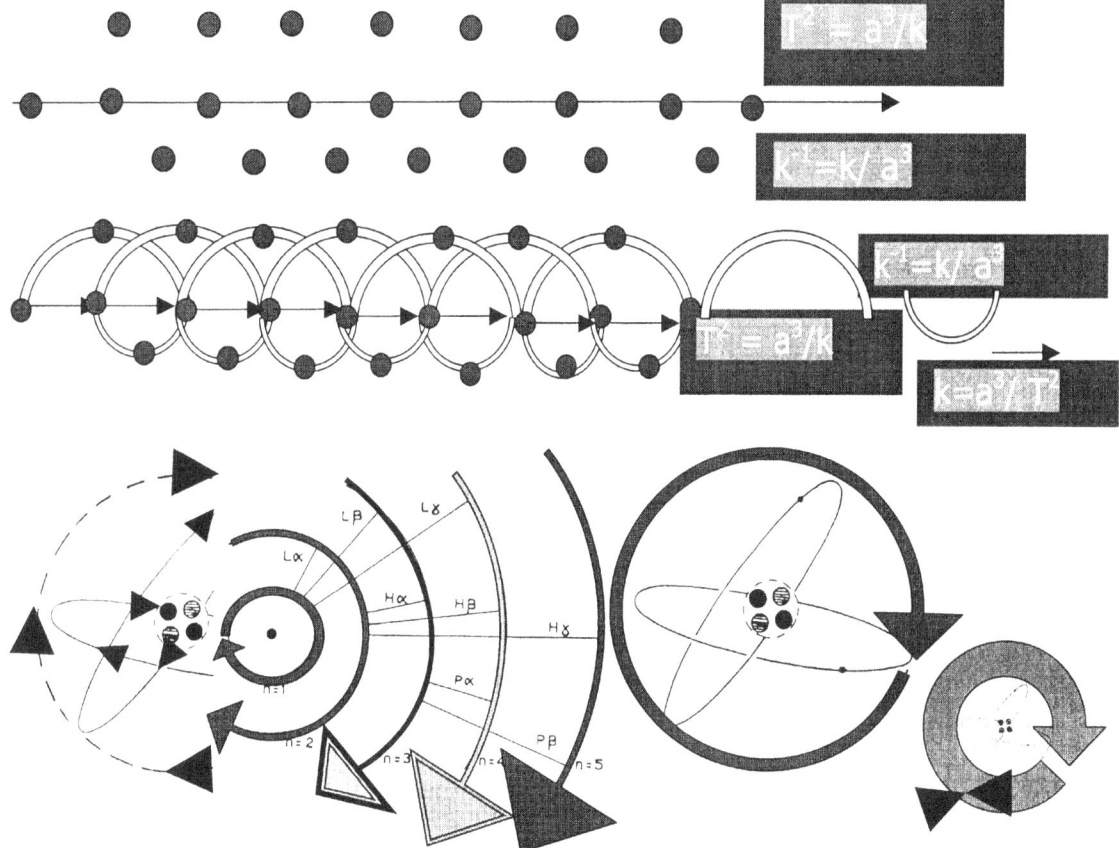

Going back to the start is going back to where space was a thought that divided eternity from infinity.

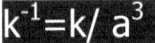   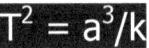

Time cannot move back but in relevancy time can respond to the nature of motion contradicting its repeat by the nature of the repeat. Time progressing $T^2 = a^3/k$ has a repeat of $k^{-1}=k/a^3$ which leaves a nett growth of $k=a^3/T^2$

When referring to how things develop I have this human tendency to put such

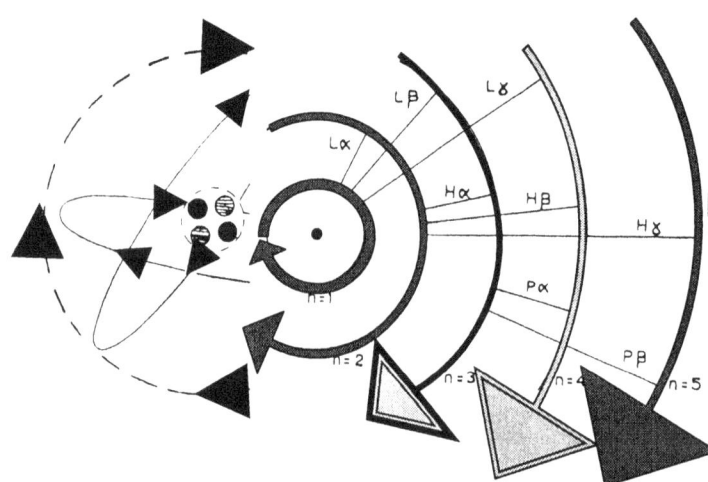

matter in relation to the past. In this matter I must beg your forgiveness but it is my human nature I sometimes find hard to control. That, which I refer to as if in the past is as current as you or I breathing, but since it does no match what our everyday experience would have us believe that it is occurring around us I differentiate in a human perspective commonly use. In this I use the past but please note that I state without excluding anything that in the cosmos there is no future or past in the sense we humans wish to observe time.

In science everyone has the opinion that material exist while the rest is nothing. It is the atom and around the atom is nothing. It is nothing keeping atoms or material apart. Such a view is extremely short sited and is evidence of a lack of thought. It is time we put true thinking power to task when dealing with matters in cosmic proportions. I do not put myself on a pedestal. To the contrary any one can see how shallow and little educated my work really is and yet, if someone with such a little educated background can see what I see, how much more is there to see when persons with true mind power star to truly think on cosmic issues. Three words that should be taken out of any cosmic context is nothing, maybe and suppose. From those three words I could find no evidence in my entire search.

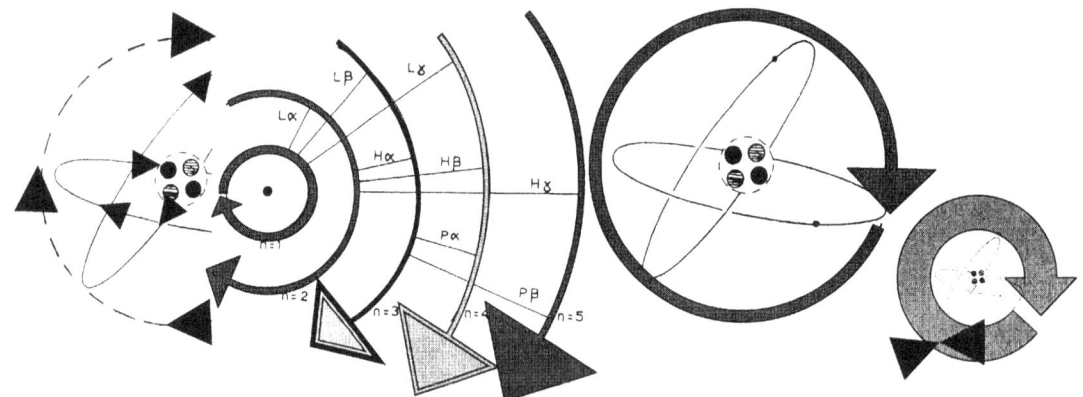

We find the atom spin and think of the atom as where we now are. Sure that is as correct as it can be but the atom spins because new alliances come about from previous conceptions. Being previous does not mean it has moved to the past, as did napoleon and his thirst for war, no it is just functioning holding prominence below the veneer of the current.

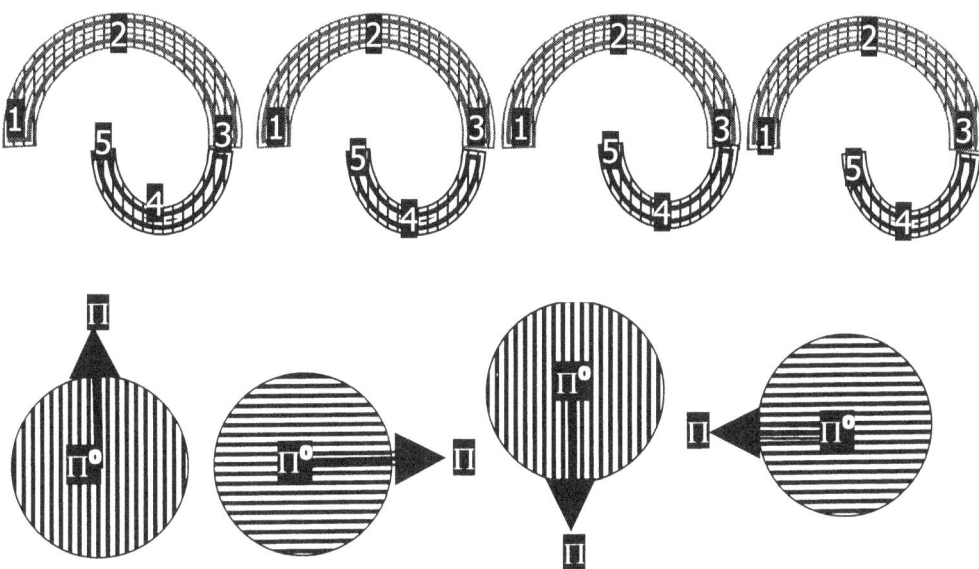

The motion of the atom still is the result of the motion of much smaller groupings of retarded heat that play a catch up game with time.

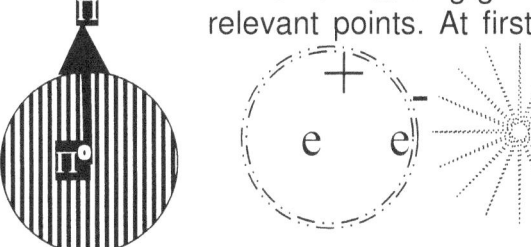

The time retarding grew as the spin involved a higher degree of relevant points. At first there were two points namely this side 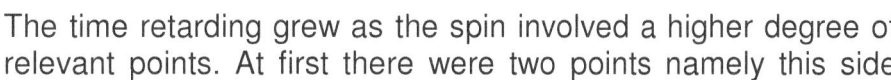 and the other side of the Universe. Then the four came into prominence. Later eight points came into affect and so it carried on. This was time developing as time always increased the back log it formed with the current situation prevailing at that moment.

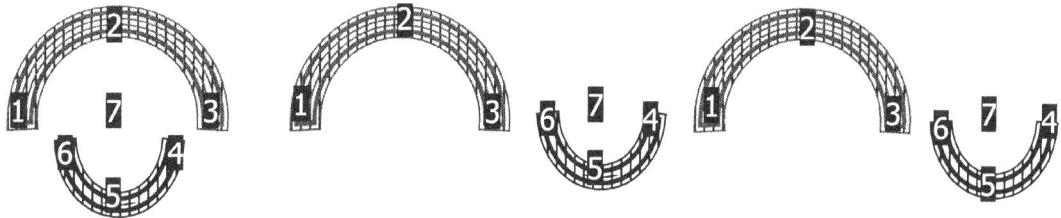

The flow of time is heat flowing past points that was there since eternity parted from infinity. It was there when the cosmos came about. What distances the pointers from each other is heat or light or whatever one would call the formless filling of space-time.

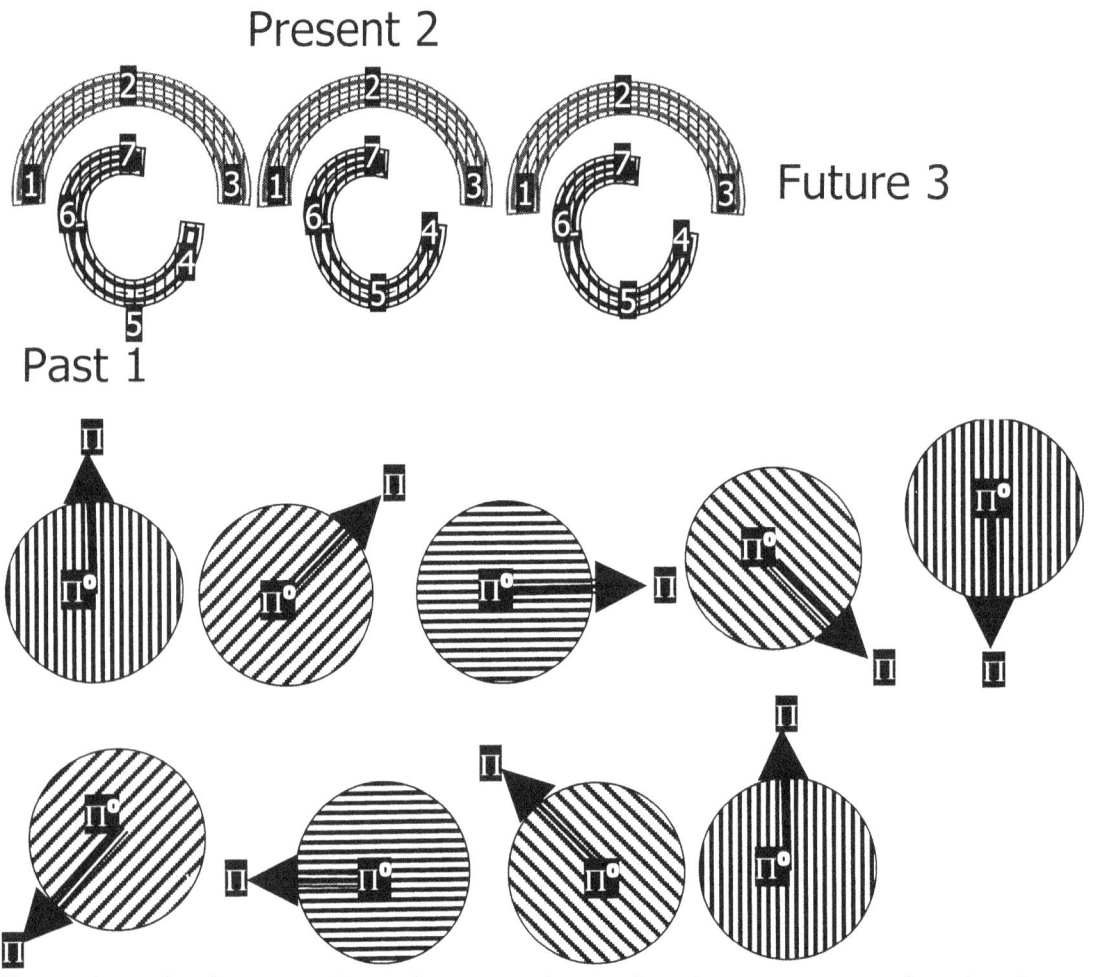

Every time the increase brought new relevancies that put new orders in placer to match the previous requirements. But also it brought about that duplication by progress of space increase was in advantage of the situation.

The relevance we find in the atom is a flow of heat. There is no such a thing as fixed particles. There is heat and there are points positioning the characteristics of the heat at such a location. The heat flow past and at that point the heat adhere to a positional interpretation lead on by points in reference to other points in singularity.

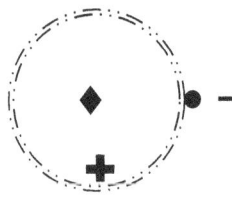

The Atom is a solid atom but it is space-time in reference to singularity pointers that mark those spots, which forms the relevancy within the markers where the markers perform the claiming as characteristics of the atom. Through the markers flow heat, which is time retardation and the pointers, are time in present. The time in present refer to the time retardation as allocated reference markers and in that allow the heat or time lapse to catch singularity. The atom is taking in heat and the heat in that instant inside the atom adheres to the form, which the pointers or markers prescribe. As retarded time moves on the heat will change form in relation to the allocated position and to what the position that holds the pointers dictate the form will be. It that sense the entirety of everything there ever can be inside our cosmos is liquid time and everything dictating what form the liquid will take on is singularity that is holding pointers which acts as the solids.

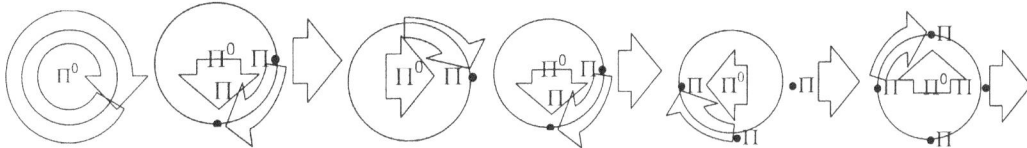

In the motion there are the four markers indicating the limit of the atom while what fills the atom is the motion that flows directional as time catching time by destroying the backlog of time.

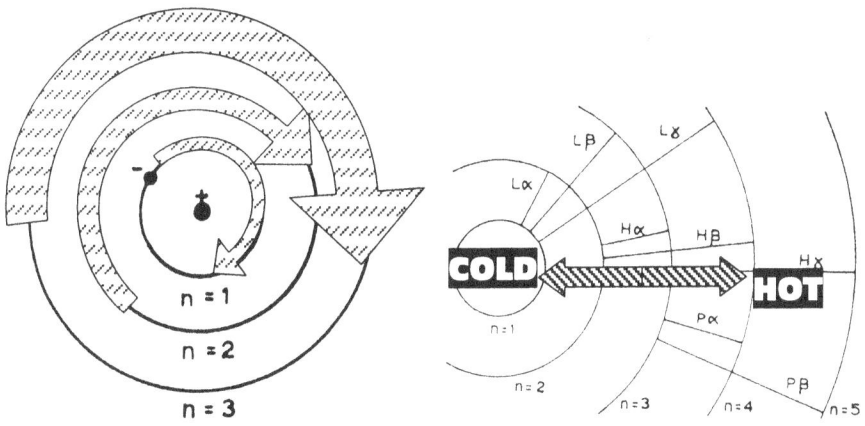

The liquid of time is part of what we see and the solid forming time is part of what we cannot see. The solid shape the liquid into forms and specifics while the liquid provides substance. The liquid is part of the Universe we materially hold and the solid is part of what we in spirit hold. The solid is not a part of the cosmos but is there in directing that which is apart of the cosmos. It is singularity that control heat and the flow thereof while it is heat that provide singularity that which is generated as time retarding.

It starts as follows:
ALL OF CREATION STARTED WITH Π = TIME AND AS TIME WAS ETERNAL,

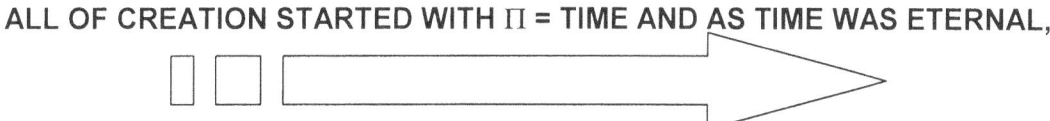

The spot grew into a dot by the value of 1^0 going onto 1^1. However this also was Π^0 going onto Π. However with no space yet available and the spot forming being so small (as we still can see when observing the centre of the spinning top because the line inside the line that is inside the line we cannot see is the line we are referring too.) The line is there as it was, but our inability to go that far into singularity places us at the disadvantage not to recognise the line as it is.

Because it is spinning it has to be Π and because it is spinning it also has to be a line and all of that is there to witness for all those disbelievers.

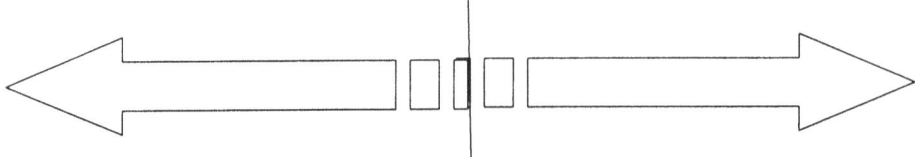

The line formed a spiralling line that was so small it was continuous and never broken due to the lack of space and those not believing me I advise you to inspect the spinning top. The line formed next to the running line and that too is in plane view of all but we cannot see. The line has to form a line on this side of the divide because it forms a divides as much as there has to be a line on that side of the divide for the very same reason

This places the single line that is half of the full line at a value of 180°.

That is also the value of the circle and the square both dimensional components of the cosmos. By reducing the one side of the square to zero (which it cannot be) the square will disappear. Therefore one has to reduce the one line of the square to infinity to produce a square holding a straight line with the line running in both directions from a point of infinity.

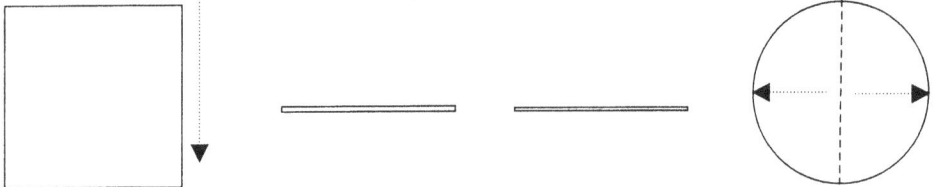

That same principle applies to the circle because the radius of the circle is one side to the round cube. That makes the square the circle and the straight line the very same thing of something totally different. The common denominator is the singularity of eternity finding infinity.

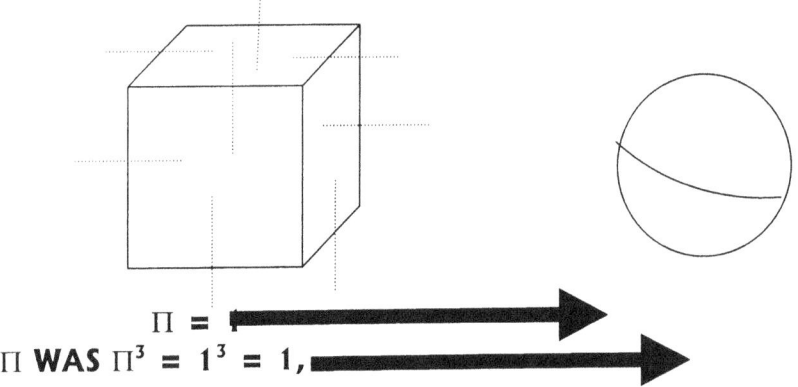

$\Pi =$

Π WAS $\Pi^3 = 1^3 = 1,$

Π WAS $\Pi\Pi\Pi\Pi = 1 \times 1 \times 1 = 1$ THEREFORE TIME WAS ETERNAL:

$E = \Pi = 1$

Let us have another look at the straight line

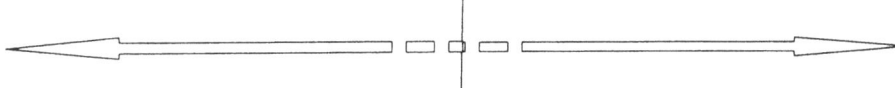

No line can start at zero because having a starting point of zero there is no line (0 X by what ever reduces whatever to zero). The starting point has to be infinity the shortest any line can be leading to eternity the longest any line can ever be. By having infinity there then has to be a VERTUALL ZERO (not zero) and from that point the rest of the line must start running the other way.

This establishes that no line can have a single direction and must have a continuance to both ends. Such a line has to have a value of 360° not 180° as believed.

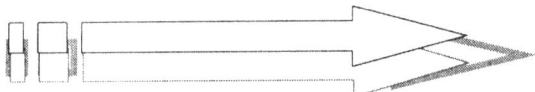

From infinity came the straight-line borders bordering the straight line on both sides of infinity

MATTER EVOLVED FROM TIME AS TIME = 7 ($\Pi^2 + \Pi^2$)
THEN TIME, BECAUSE IT WAS ONE, MOVED A DIMENSION UP AND DOWN, WITH BOTH DIMENSIONS BEING EQUAL, THE SAME IN EVERY WAY.
Let the lines be somewhat bigger than infinity where we can see the lines more clearly

MATTER RECEIVED THE VALUE OF Π^2 AS FOLLOWS:

THEN   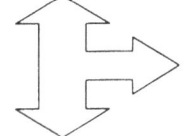 $\Pi^1 = 1\Pi$
$\Pi^0 = 1$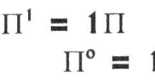

$\Pi = 1\Pi$

THEN TIME HAD A DIMENSIONAL VALUE OF Π = 1; Π =1 AND Π =1, BUT AT THE SAME Π DUE TO THE PROXIMITY OF SINGULARITY

POINT OF SIGULARITY

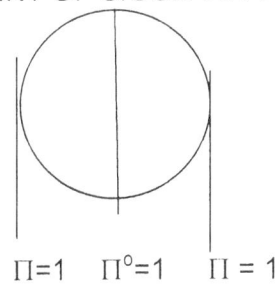

Π=1 Π^0=1 Π = 1

The Universe formed singularity in the straight line with two points of Π to both sides. There was no radius because the radius was infinite small. Then time formed as part of singularity in the value of Π going on to $\Pi\Pi\Pi$ going on to Π^3. The one Π remained in singularity Π while the other $\Pi\Pi\Pi$ became Π^2.

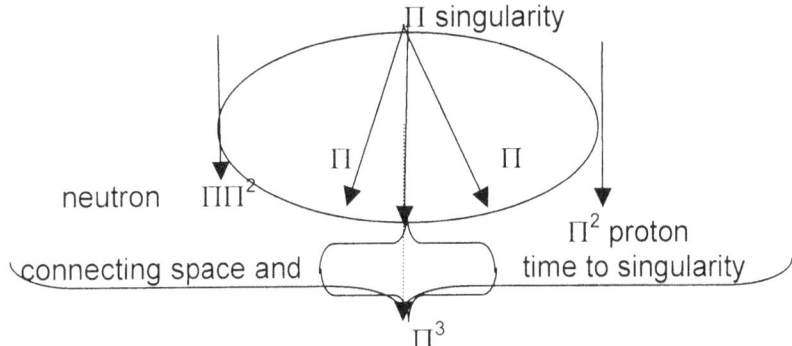

$$\Pi = \Pi\Pi\Pi = \textbf{SPACE} = 31 = 10\Pi$$
$$= \Pi\Pi\Pi\Pi = \textbf{MATTER} = 31 = \Pi\Pi^2$$
$$\Pi = \Pi\Pi\Pi\Pi = \textbf{TIME} = 31 = \Pi^3$$

TIME Π

While on the other side of the universal atom Π^3 formed space in time in the value of $\Pi\Pi^2$. Between the two universal atoms the line to singularity was the combination of Π^3 holding Π to Π^2 and forming space in matter to the value of $\Pi\Pi^2$.

In all it became 10 Π in all being in singularity

Whatever there is today started at a point we cannot trace. It started at a point we are able to envisage and locate but that is only with a mind that can accept what is not there to view. It is acceptable through intelligence because through intelligence we can detect it albeit outside the Universe we have. It is just like religion and worst is that it is generating what there is because what there is are not if not generated. There was Π^0, which was α^0 or if you would rather have it Ω^0 or it maybe was 1^0, but more correctly it was all the above and the beyond because multiplying what ever constitute the mentioned will bring about what is mentioned to a precise equality. It was a spot that was not. The spot is still there because the spot is still not there. It was a line that ran eternal but because it ran eternal and kept repeating exactly what was before to the precise what came afterwards, the line was there and was eternally running, while never changing in the least or growing by any measure. It was not one because before it could reach one, it returned to what was repeated and the process cycled back to before one was reached and even before one could be accomplished. It was such a continuing of the monotony, no change occurred and therefore never did the running produce progress because the progress was in the perfect repeat of what was before. The duplication brought contraction to the minutes detail. That is where our atheists get one hiccup. The repeat brought eternity and the repeat was so perfect that the repeat continued. The repeat still is with us as much as we are within the repeat. To bring change to the eternal repeating of the monotonous there had something beyond the Universe that institutes change. There was something that brought a difference and we are within that difference. That difference was time and that time is what we move through as much as what we see at night. Oh, how stupid and how thoughtless the minds of atheist and other atheistic animals are. Baboons do not recognize the light we are within because they cannot think and are therefore they are atheists. Spiders cannot think and therefore they are atheists, as they do not think where the line is that is not. Reptiles cannot think and without thought they are incapable to see what time is, how time that is not generate space that is in time that is not. Mammals cannot envisage what space is, what light is and what makes us see the darkness cannot be. All the animals I have mentioned are mindless atheists because they fail to see beyond the visible into the realms of the thinkable. All that I have mentioned passes them by including religion and the accepting of God because through being incapable of thought and reason they cannot envisage what only intellect can bring to mind. Because of the incapacity to think the animals are both mindless and they are atheists. Therefore atheists are mindless. The night sky is such a bright light our evolution protected our vision from the brightness in order to give as much better vision. Through evolution development our eyes are protected by how we remove the qualities from the light. However animals do use that light and not our light to see by.

You can shine a bright hunting spotlight onto an animal at night and the animal will not be able to see the light on it. The animal does not use the light to see better as the animal is totally unaware of the light. Then a prowling cat comes from the night and sees the antelope in the light the night provides. It does not use the light, which the spotlight casts and the light is not even traceable to either animal being the hunter or the hunted. From there we accept that during the day the animals must be using our light to see because the nightlight is inferior to see by. Who says they use the daylight much different from the nightlight because all evidence is there that they cannot recognize our light as light. It is very evident in the manner they go on hunting and grazing while being totally unaffected by our form of light.

That which you see at night because you cannot see darkness and you cannot see black is the light the Universe is painted in just like the Bible says. That is the light that started all because that is the light holding us away from the eternal darkness. This is not religion and it is not a sermon, it is hard-core and brutal basic science and it the most fundamental basic physics there is. It is the start of the mathematical Universe portraying the only physical way it could ever be. The light that came from the Command is the light allowing material to move.

My atheistic idiots, your mindless caught up with you!

Then came this light that the Bible refers to as the first of what ever was and what our stupidity tells us is darkness. This was moment-Alfa. The darkness was there and from the darkness heat came about. Only heat expands and it interrupted the true invisible darkness, the blackness of a Black Hole, the invisibility coming from within the Black Hole. Eternity tore from infinity. Darkness broke from light. Heat broke from cold. Relevancies parted by 1^0 going 1^1. There was one but also there was two too because one cannot be without two being there to ensure one is one. The marks are still with us but to see the marks requires a great deal of intellect.

$\Pi^0 \Rightarrow \Pi$. In this there was only space for one being one in the two forming one. It was $\Pi^0 \Rightarrow \Pi$ however there was no space to be $\Pi^0 \Rightarrow \Pi$ and there fore because of the lack of space to be which is the infinity of time braking the eternity of time the true measure was $\Pi^0 \Rightarrow \Pi$ but realized only 1^0 going 1^1. Π was to the future because of the motion of time involved and the space less ness of space at the time. By inclining to move the process crossed the Universe but also it took one eternity to accomplish the feat.

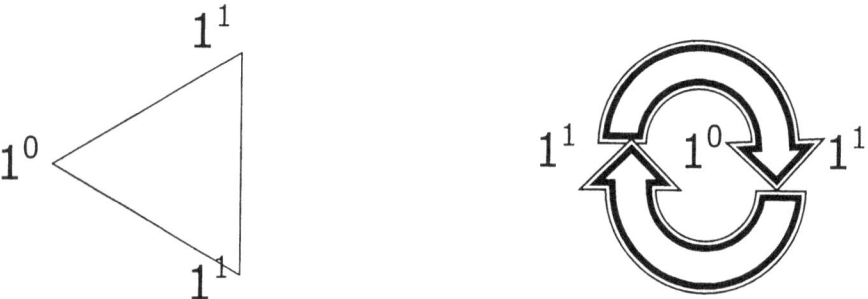

The fact that 1^0 going 1^1 brought movement can only become a reality as a result of light. Light is heat and the heat is expanding.

1^0 going 1^1 where $1^0 \Rightarrow 1^1$

1^0 going 1^1 1^0 ▶ 1^1 had to bring about 1^0 going 1^1 1^0 ▶ 1^1, because the eternal repeat of duplicating while contracting was not relieved from the Universe. Before the contracting was equal to the duplicating because by measure the heat was identical to the cold. It was eternity that was interrupted by one cycle of infinity and was in repeat of eternity. Once something is part of the Universe there is nowhere else to take it so it has to remain as a part of the Universe.

Then came three because motion was so limited that the least inclination to move threw what wished to move to the other side of the Universe, As it moves it also moved across singularity. It crossed the entire Universe as it moved because it moved and finding nowhere to move too. It crossed the entire Universe and it took one eternity less the measure of one period lasting infinity to achieve that. That brought to relevance three points where each was in measuring quantity exactly equal but also one Universe apart.

In the reality there was now two points holding singularity on both sides of the Universe because by crossing the divide that crossing set in place the two sides relevant of singularity governing. However infinity was bridges at two points holding infinity with which process eternity repeated the past into the future.

This then is the occasion where Pythagoras stepped in. Since it as a crossing of the divide the crossing involved a line that formed a half circle connecting a triangle. But the crossing was done in the space of half the Universe and since the Universe was 180^0 half of the Universe was 90^0. That involved Pythagoras as mathematics was born. Up to this point it was arithmetic with adding but now mathematics came into place. Remember we are a few eternities in side the development of the Universe.

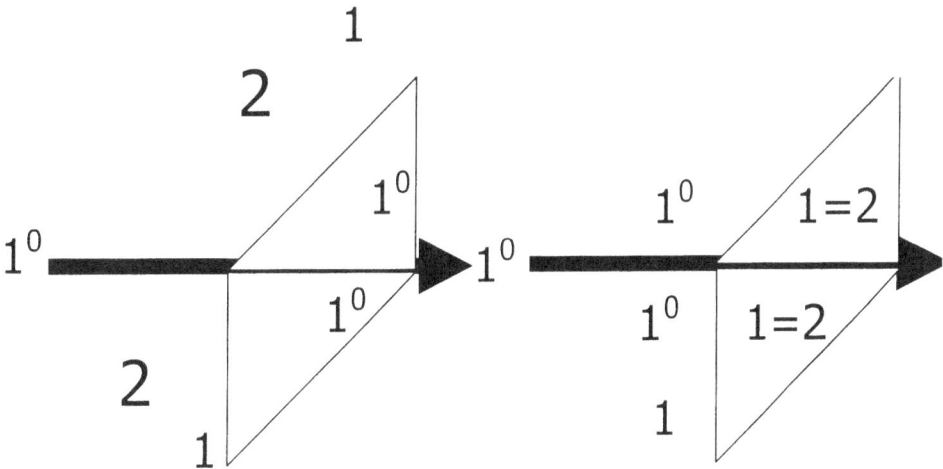

In the three came four that brought along five.

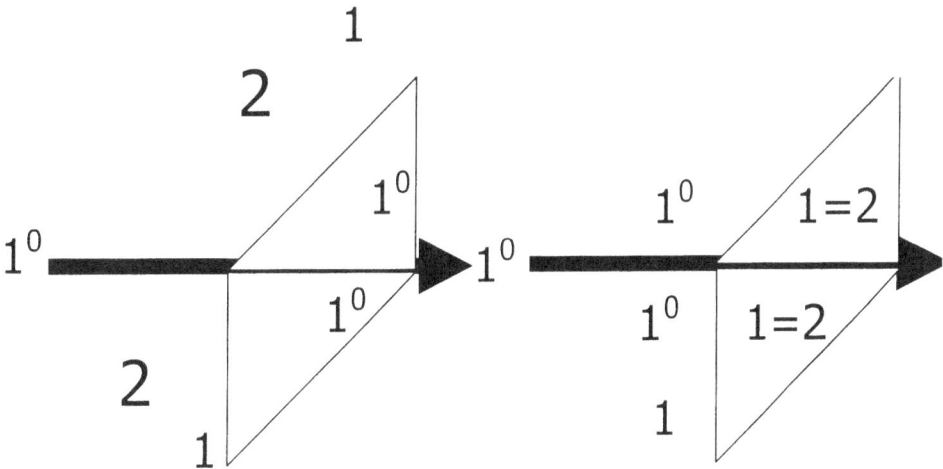

Then five filled the one half of the Universe that was able to contract and cool while the Universe divided the other half into sectors of what was (5) and what will be (5), which put material (7) in relation to the half of the Universe $\Pi^0 \Rightarrow \Pi$ in which the material was at that specific point (five relating to seven) in time.

$$1^0 \quad \blacktriangleright 5$$
$$1^0 \quad = 1 + 5 = 6$$

The motion consisted of Π moving to Π and thereby duplicating Π to relieve Π^0 of the burden of overheating. On the one side of The Universe there was Π^2 being relevant to Π which was forming on the other side of the Universe. The entire Universe had the combined value of Π^2 on the one side in addition of Π forming on the other side. I wish to remind the reader that any and all points formed by singularity was as much representing the Universe as it was the Universe at all times because $\Pi^0 = 1^0 = 1$. That made the entire Universe being any point affirming singularity by forming about singularity.

Reaching five is a benchmark because at that point half the Universe was finalised. From the five that formed the Universe continued and formed space-time. All progress balanced on the five that formed where two parts of equality parted eternally as liquid separated from solid, motion moved apart from the motionless, heat diverted from cold. The four in time grew away from the one point space formed being just outside the control of time. It was the start of material since from the point five formed just outside the four of time. There was a lagging behind of heat building and not being in the range of the immediate control of time. Time could generate a point in heat without bringing immediate demise to the point through cooling.

But that meant that the Universe was a total of $\Pi^2 + \Pi$ which when added was also $\Pi^2 + \Pi = 13.0$

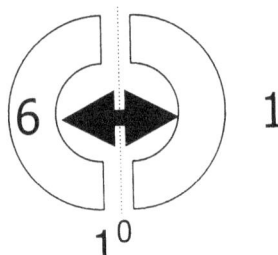

13.0 − Π^0 = 12 because singularity cannot be part of space-time developing as the space, which later was filled with the material that formed, filled this part.

12 / 2 = 6 Material formed at the point where six was located.

$$6 \;\; + \; 6 = 13 - (\Pi^0) = 12$$

Because singularity is a divide and is not part of space-time singularity as a factor removes from space-time. Why it adds with five to form six is because to the one side only singularity is in the other side of the divide. Only nothing can be in two places at the same time therefore on any one side was the half of twelve, which divided 12 in two parts. That then was 12 / 2 = 6

$$6 \;\; + \; 6 = 12 + (\Pi^0) = 13$$

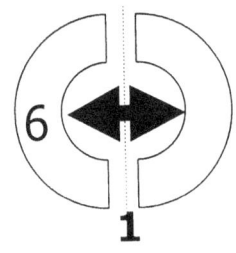

Developing six was an addition to the square as the line flowed and did not involve the crossing of the divide. Therefore Pythagoras was not involved by the forming of six.

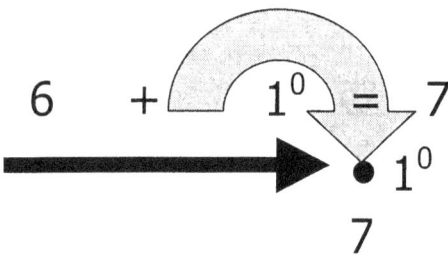

At the point where the space filled with heat meets the point in time representing singularity the end of material (6) confirmed the following spot (+1) at 7.

Forming seven very much involved singularity because it confirms appoint where space ends and space (8) begins.

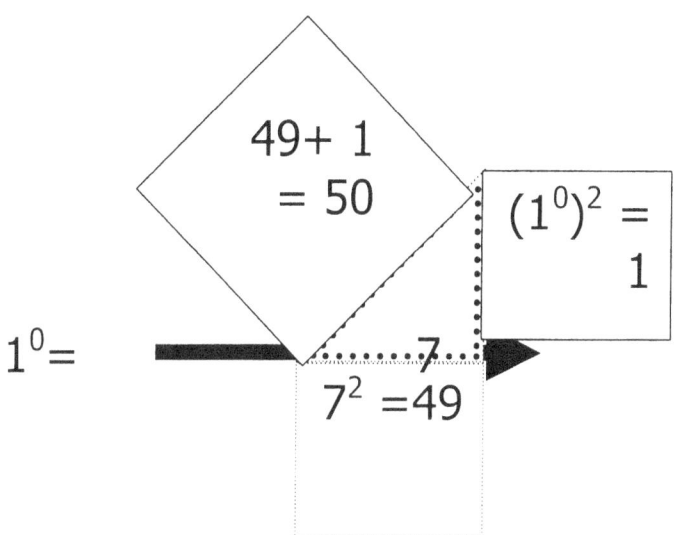

By taking singularity into Pythagoras and filling the Universe by halving the square of space seven completed the required circle within one half of the Universe in order to relate to half the time it takes material to fill time by duplicating. To find the necessary cooling required for control material has to use five points to be within because of the square involved. The there has to be another double five amounting to ten to fill the void from time in the past (position of five) and time in the future (another position of five).

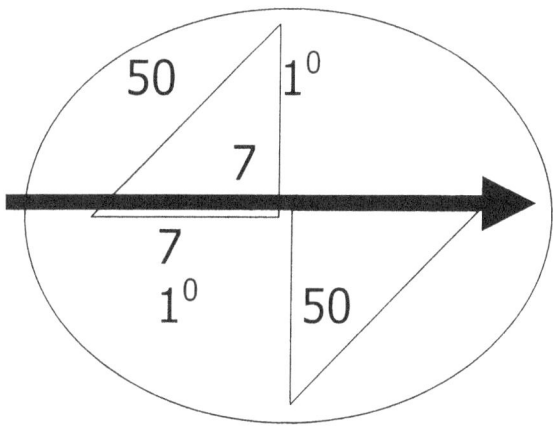

$$50 + 50 = 100; (100)^{1/2} = 10$$

$$(\Pi^0) = 5 \qquad 10$$

$$7$$

The Titius Bode require meant is seven holding relevance to ten and ten being relevant of seven while being in half the Universe $\Pi^0 = 5$

Then come eight causing a line of material to break.

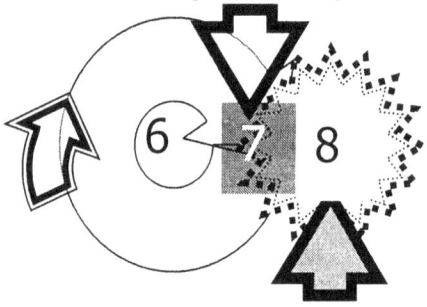

At seven the line completes at a point distinguishing material with in space from space without material

The circle of development has finalized a point. Seven has gone square 7^2 and realized with singularity half of the final of space in the absolute square.

It is this eight to ten science does not recognize and do not distinct as one other part of time. This in relation with the finality that came about at the point seven marked by using Pythagoras that another space, this time in time was developed to compromise for the lagging of time within space-time.

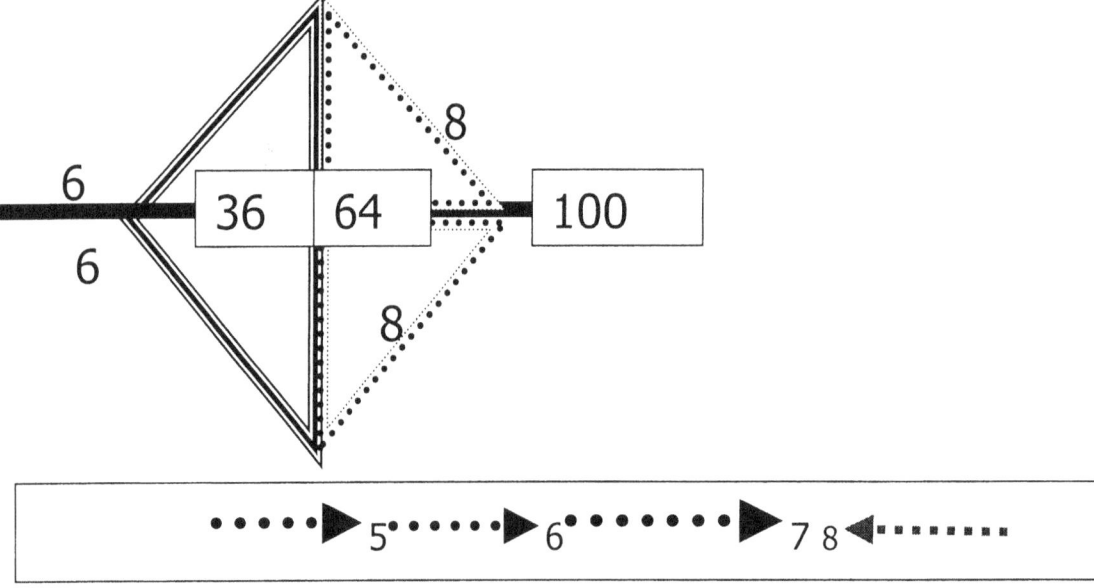

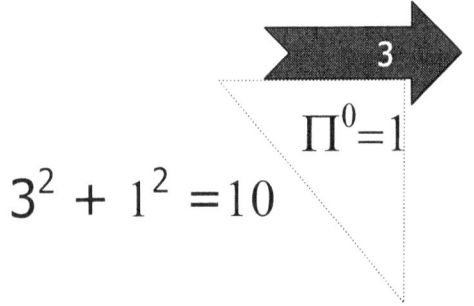

$$3^2 + 1^2 = 10$$

The cycle of eternity could then complete one more time by forming singularity once more

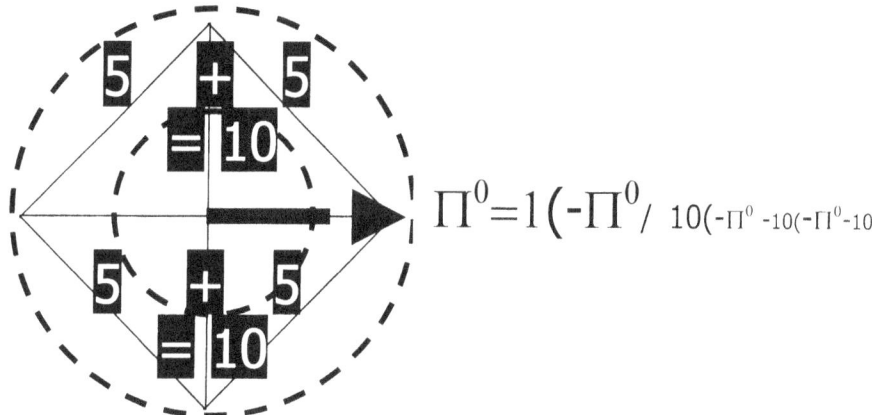

$$\Pi^0 = 1\left(-\Pi^0\big/\ 10(-\Pi^0 - 10(-\Pi^0 - 10\right.$$

With the Universe established at ten crossing the divide meant that Π^2 at four was a half and five was completing the one half.

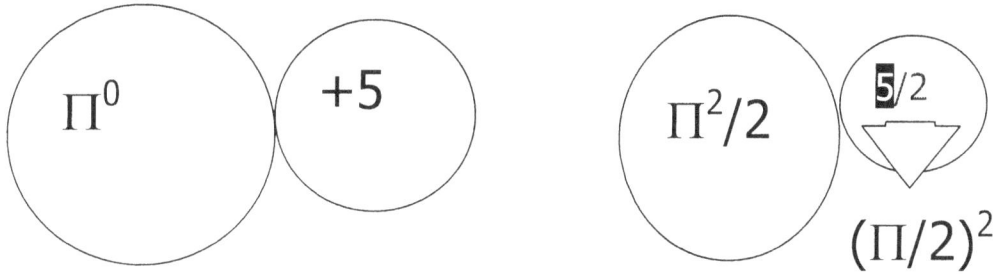

The Roche limit shows that singularity needs at least more than half the Universe (5/2) to share and…the Lagrangian system is at least half the Universe.

Therefore the circle extended to the point outside singularity where matter was holding a space value of 10 Π^3 and space holding matter was in 10 Π. Then there came the proton connecting space in time to singularity ON THE OTHER SIDE OF THE UNIVERSE.

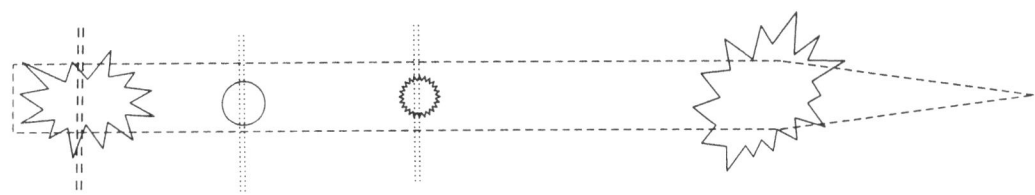

They say its Kepler's laws but that is where the inaccuracy starts because Kepler never said that. It is a derogative Newton established of what Kepler introduced as something very different.

The Law Newton gave as Kepler's law of orbits.
All planets move in elliptical orbits, with the Sun at the centre focus.

The figure shown below is set to show a planet with a mass of m, moving in such an orbit (blue dotted circle) around the Sun . The Sun holds a dominating mass of M. We assume that $M > m$ so the centre of the mass of the planet is virtually in the centre of the Sun . The orbit is described by giving its **semi major axis** a and its **eccentricity** e the latter defined so that ea is the distance from the centre of the ellipse to either focus F or F^1. *An eccentricity of zero corresponds to a circle,* in which the two foci merge to a single central point. The eccentricities are not large as the sketch would indicate but in reality they seem circular. In order to bring across the eccentricity the orbit discrepancy has to be exaggerated to find meaningful clarity. In the case of the earth the true eccentricity is only 0.0167.

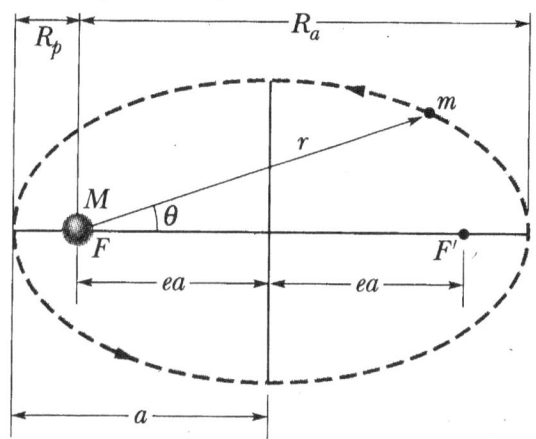

According to Newton and from what we can observe in the sketch the planet with mass m is at focus F of the ellipse. The other, or the *empty,* is focus F^1. Each focus is at a distance ea from the centre, e being the eccentricity of the ellipse. The semi major axis a, of the ellipse, the perihelion (nearest to the Sun) is at a distance R_p, and the aphelion (farthest from the Sun) is at a distance R_a are also present.

Notsofast.

The diverting from the main axis forming an angle θ is not a mathematical angle but is only time duration. In physics there might be a point in reference to another point but this is far from physics. It is cosmology. There is a point holding eternity 1^1 by three positions $1^1 \times 3$ in relation to infinity 1^0 and the three positions is time. There has to be a past to achieve the present and to achieve the present there has to be a future. The atom has to have been there if it is to be where it is in order to go too it will be next. There cannot be a top or a bottom or a front or a back because there is an infinite centre one relating to six other position which numbers eternal positions all sharing one unit. Those being strong in the field of mathematics, can take this challenge and find the possible number of 1^1 in relation to the centre with no sides 1^0 and calculate how many possibilities there might be on the edge of any sphere that will correspond by six to the total number of possibilities not there in the centre singularity 1^0. In the Universe there are no fixed point. On Earth there is the all - dominating centre which grants physics relevant formation, but in the cosmos there are no such points anywhere. There is in the relation one point holding a singularity concerned with expanding, which has the position of 1^1 and there is a singularity concerned with contraction 1^0. The motion depends on getting al the atoms from one location to the next location and this involves the distributing of all the factors concerning 1^1 in relation to every one of the factors relating individually and as a group to 1^0. This reference of F relating to $F1$ is corresponding not to the planet but it is corresponding to every atom individually where the group forms a centre, which corresponds to a centre that forms with the aid of all the atoms forming motion within the Sun . The relation that this group has, forms a partnership with the motion within the entire Earth and from that a centre of governing gravity is selected by the motion of every independent atom that responds to the motion of every subatomic particle notwithstanding how incredibly minute they may seem. The Sun turns with the turning of every atom within the Sun which generates a combined motion that generates a selected centre through the all inclusive effort of every proton that provides the expanding and contracting balance we find in the Sun . By the motion there is also a conflicting motion that supports the motion where the one part is lateral and the other is rotating. The space $\mathbf{a}^3$ moves partly in a circle $\mathbf{T}^2$ and partly in a straight-line $\mathbf{k}$. That is what Kepler said when Kepler said $\mathbf{a}^3 = \mathbf{T}^2\,\mathbf{k}$.

An eccentricity of zero corresponds to a circle, **in which the two foci merge to a single central point.**

If that was true then the circle in which the planet orbits the Sun should be dead canter because with an eccentricity that is obvious there has to be favouring one side and then favouring the other side. The zero correspondence is not there since there is a dual eccentricity detected in all the planets in orbit around the Sun . There is much more a swapping of prominence between the two points in foci.

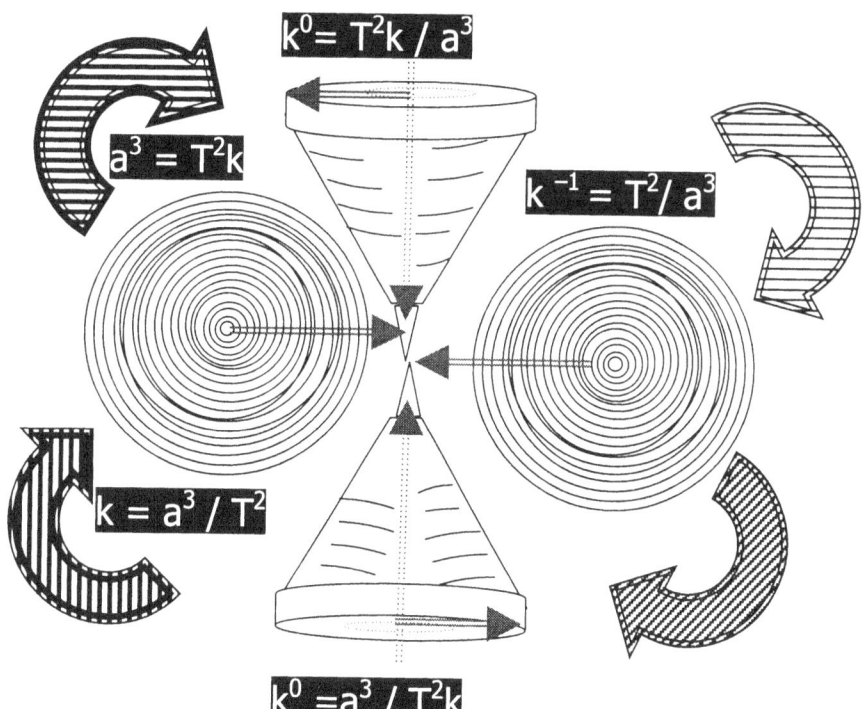

$$k^0 = T^2k\,/\,a^3$$

$$a^3 = T^2k$$

$$k^{-1} = T^2/\,a^3$$

$$k = a^3/\,T^2$$

$$k^0 = a^3/\,T^2k$$

There is a divide that separates one side of the orbit from the other side of the orbit and the one is opposing the other side of the same orbit. This has to do with the changing of the motion in relation to the orbit where the changing of relevancies produce gravity, but not by mass because mass is something in Newton and in Newtonian imagination. Mass is a hoax, a lie and fraud to such an extent as the world has never witnessed.

The Law Newton gave as Kepler's law of areas.

A line that connects a planet to the Sun sweeps out equal areas at equal times. Qualitatively, this second law tells us that the planet will move most slowly when it is farthest from the Sun and move most rapidly when it is nearest to the Sun . As it turns out, Kepler's second law is totally equivalent to the law of conservation of angular momentum. Let us prove it.

In time Δt, the line r connecting the planet to the Sun (of mass M) sweeps through an angle Δθ, sweeping out an area ΔA. The linear momentum p of the planet and its components.

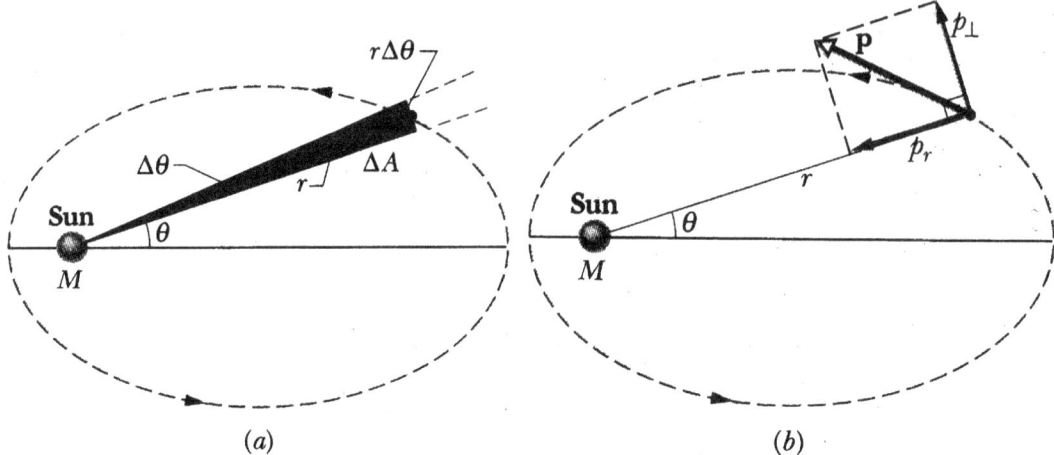

(a)   (b)

The area of the shaded wedge closely approximates the area swept out in time Δt by a line connecting the Sun and the planet, which are separated by a distance r. The area ΔA of the wedge is approximately the area of a triangle with base r Δθ and height r. Thus $\Delta A \approx \frac{1}{2}r^2\,\Delta\theta$. This expression for ΔA becomes more exact as Δt (hence Δθ) approaches zero.   The instantaneous rate at which area is being swept out is then

$$\frac{dA}{dt} = \frac{r^2}{2}\frac{d\theta}{dt} = \frac{r^2\omega}{2}$$

in which ω is the angular speed of the rotating line connecting Sun and planet.

The figure shows the linear momentum p of the planet, along with its components. From, the magnitude of the angular momentum L of the planet about the Sun is given by the product of r and the component of p perpendicular to r, or

$$L = rp_{\perp} = (r)(mv_{\perp}) = (r)(m\omega r)$$

$$= mr^2\omega$$

where we have replaced $v_{\perp}$ with its equivalent ω r . Eliminating $r^2\omega$ leads to

$$\frac{dA}{dt} = \frac{L}{2m}$$

If dA / dt is constant, as Kepler said it is, then that means that L must also be constant – angular momentum is conserved. So Kepler's second Law is indeed equivalent to the law of conservation of angular momentum.

Newtonians share a dubious opinion that the second law is totally equivalent to the law of conservation of angular momentum. If so someone forgot to tell the person holding this opinion that there is a "small" side and there is a "larger" side and this does not stroke with the mass pulling mass idea. If this was correct then it would be that mass in motion neutralise mass in centre

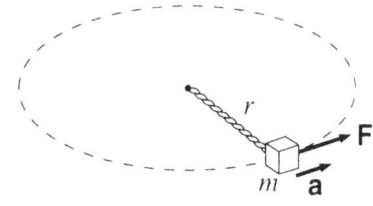

and not mass pulling mass by reducing radius to the square. On earth the swinging object will hold a perfect radius and if not the spinning object will destroy the centre object as the imbalance will cause the centre to shift because the thrust varies. I still do not agree with the Newtonian view on the physics part but in that I have no mission.

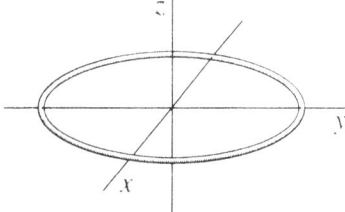

A ring is placed on the x y with the centre point spacing the circle evenly. That represents space-time. I shall get to Newton's mass in a short while… The motion we get in rotation honours the straight line because it admits to the circle being present and

combined they pulsate between the triangle representing one another in relation to the other side of the Universe and each other. Lets study the process of the motion in the most incredibly small minute context in order to exclude any and all time delays forming the material that is filling the space. We go to where there still are no

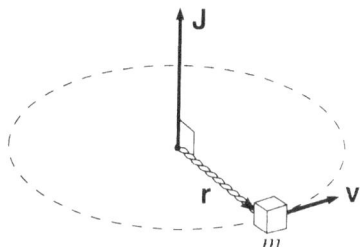

mass present even by Newton's imagination. Lets make the Force F the overheating and the mass m the overheated while area a, is the expanding.

The first acknowledgement we have to make is singularity is charging motion and that would be 1^0 that are charging Π^0. We are now where $6\Pi^0$ 1^0 is going to form Π. In the forming of Π time continues to run in a line. The line is forming Π because Π is forming on the inside of time being on the outside. That which is in the process of

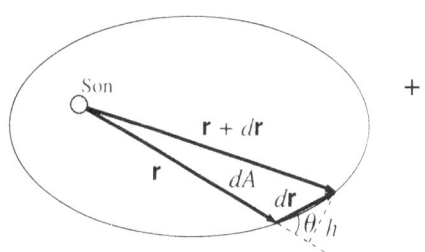

forming is to the outside of time because time is eternally with no end and no outside on the outer limits of Π.

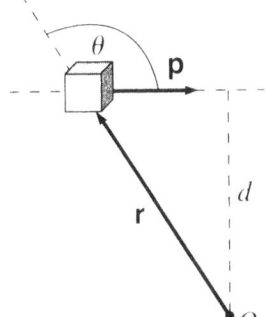

On the one side the motion formed a straight line to the value of 180^0 and at that moment on the other side of the Universe (the Universe being 1^0) a half circle also to the value of 180^0 that in relation forms a triangle also to the value of 180^0 and that is making what there is in duplicate in both directions (**k**⁻¹ and **k**) formed as the dot 1^1 coming from the spot 1^0 went around the Universe because the spot1^0 could turn just as much as it still cannot turn. That is the reason 1^0 has to charge 1^1 to commit motion as Π^0 and dorm a line that represents a half circle on the other side of the Universe. We can see that Θ is equal to p while d is the motion representing p that is acting as r but since there are no possibility of motion O is actually going over to the other side of the Universe and from (**k forming k⁻¹**)

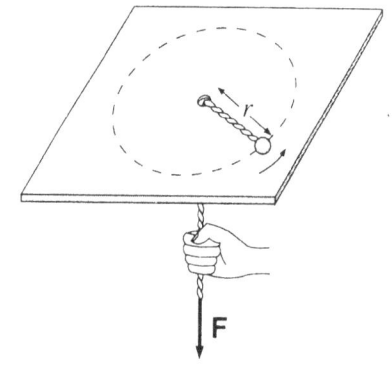

In order to understand what truly is happening is that eternity cannot expand to the outside. It has no outside. Eternity has to reduce in order to expand and by reducing it is releasing infinity which has no star by giving infinity a start to the outside where eternity that has no end finds an end in infinity with no start point where infinity starts. I am incapable of putting it more practical and Newtonians with not the mental capacity to understand must at this point find a motor mechanic to explain to them what is going on.

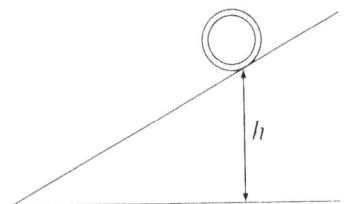

The motion coming about is eternity squeezing the daylights (because it is heat or light) out of infinity that already has no space where in it can be squeezed.

The circle finds the drive it will have from the outside where the outside is in contact with eternity and eternity finds a line to guide (not drive) the rolling circle. It is powered by the discrepancy in the motion of time on the rim in relation to the singularity centre that still is unable of motion. The drive is in the space-time that the motion charges intro action. The fact that the top stand erect alone is most significant. The drive- line that charges the motion that drives the top has its significance from a line spun around the outside of the top. Life with its manipulating qualities wrapped a rope on the outside and with a release of the top while still controlling the lien the top found locomotion. The biggest energy issue at this is admitting the energy origination. The energy is life induced. It is not cosmic.

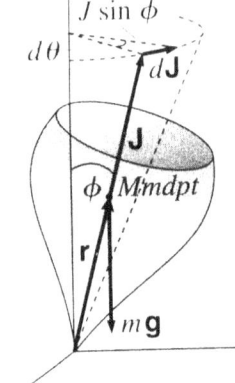

The energy cannot have an equal in "mass" meaning size or "mass" meaning duplicating tempo since the atomic or particle presence within the top has no strong enough motion accumulated between the lot to displace the singularity relevance to a centre where from such a centre a governing singularity can be charged and maintained.

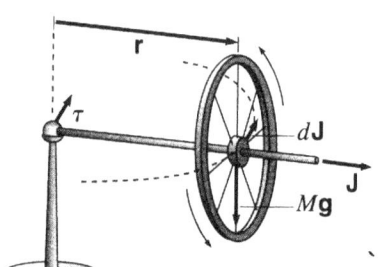

The whole motion locomotion is life inspired and life has validity on one small spot in one tiny solar system that is maintaining a most insignificant solar object of very little cosmic substance. The gyroscope finds the drive coming from the outside because the very inside is unable of motion. The motion drives by means of both dJ and Mg where Mg has no mass reference. The required increase in driving effort comes from the larger ratio the outside of the wheel hold to the significance of the inside that still is too small to move notwithstanding the increase in ratio between eternity and infinity.

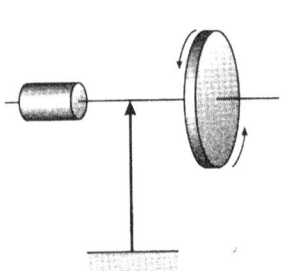

The significance of a drive line is placing the generating there is between that which moves eternally and that which eternally can't move in relation to another but connected Universe where that which eternally cannot move has a conflict in ratio to that which moves eternally.

Notsofast.

From the planets view the planet is moving away from the Sun by a measure of k = a^3 / T^2. Therefore according to the planet it is moving straight ahead way from the Sun without even glancing back.

From the view the Sun is enjoying the situation the planet is coming back and the planet has no time to look back into the open sky as it is on its return by the measure of $k^{-1} = T^2 / a^3$.

In reality from an observer's view going by the name of Johannes Kepler space a^3 is in a relation with time by the straight - line k as well as the partial circle T^2. The planet is returning as fast as the planet is escaping and every factor finds a position of victory while equilibrium is the only victor.

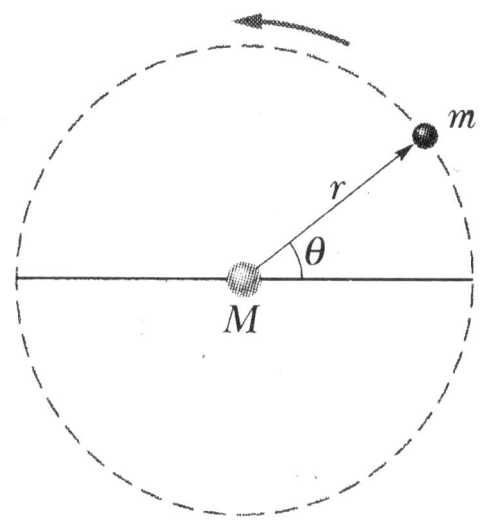

A planet of mass m moving around the Sun in a circular orbit of radius r.

The Law Newton gave as Kepler's law of periods.

The square of the period of any planet is proportional to the cube of the semi major axis of its orbit.

To see this, consider a circular orbit with radius r (the radius is equivalent to the semi major axis of an ellipse).

Applying Newton's second law, F=ma, to the orbiting planet yields

$$\frac{GMm}{r^2} = (m)(\omega^2 r)$$

Here we have substituted from for the force F and used to substitute $\omega^2 r$ for the centripetal acceleration. If we replace ω with $2\pi/T$, where T is the period of the motion, we obtain Kepler's third law:

$$T^2 = \left(\frac{4\pi^2}{GM}\right) r^3 \qquad \text{(law of periods)}.$$

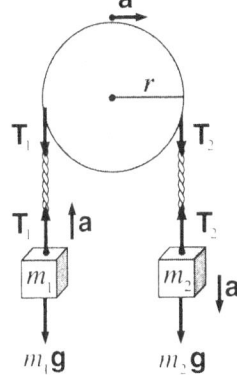

The quantity in parentheses is a constant, its value depending only on the mass of the central body.
The Equation also for elliptical orbits, provided we replace r with a, the semi major axis of the ellipse. This law pinned on poor old Kepler who was rather innocent is the biggest swindle mankind was ever hoaxed by.

Newton suggested that the two structures were working in a relation as pulleys would. To do that a point has to be in place on which the pulley may pivot. Then the one mass will pull(ey) the other mass and bring the one mass closer to the other mass by diminishing the radius to the square because it is done from both sides. He caught the world hook line and sinker for three hundred and fifty years and every one was caught because no one would admit they were to stupid to understand what the most brilliant brain of all time saw.

Notsofast.
This is the biggest scam the world was ever caught by. Nothing can even compare to the stunt Newton pulled off. This made Newton the Master swindler who is matchless by miles. Some sell the Eiffel tower and other tricksters walk through the China wall, but let Mr. Houdini repeat this one. Newton never (not once) had to prove mass because mass was never a cosmic factor and never being there it was never there to be proven so everyone accepted it as proven. Mass is Earth born and it is Earth related.

There has to be a resistance stopping the motion or resisting the gravity to bring about mass because the mass comes about when duplication by motion is prevented from moving while the moving still manage to preserve the independent nature of the structure holding the considered mass factor. In outer space everything must move therefore the only restriction is the conserving nature the particular elements reserve.

For three and a half centuries no one had the idea to return and see why mass is what mass is when mass is and when mass is not. Good God (that a prayer for my fellow humans), is everyone that stupid, that blind and that naïve. The most ridiculous part of all is the joke they pin on me. When they are cornered and their explaining has to go beyond the ridiculous, they say it is I that cannot understand Newton. It then becomes me being uneducated and at every occasion they pin to my jacket the label that I am mentally underdeveloped.

Then they counter the blame on me arguing that I am so retarded that I cannot understand Newton because I am uneducated. Suddenly my being a motor mechanic convince them most of all because of all brainless things any one can be a motor mechanic is the worst there is. Then it dawns on them that it was I that never even for one semester was at any University to be educated in the science of Newton. By all these standards I am ridiculed and labelled because I can't understand Newton! I thank God I can't understand Newton! Lets look at what is so obvious even I can see it and yet that all the educated never question or sow any doubt.

This law predicts that the ratio T^2/a^3 has essentially the same value for every planetary orbit around a given massive body. The table above shows how well it holds for the orbits of the planets of the solar system.

If the Newtonians could just once for one instant show me ho mass does effect the behaviour of planets orbiting even in the least, then I too can boast about me finally being able to understand and appreciate Newton. I have tried for so long on so many occasions and in so many respects to find grounds for Newton's persistence on blaming mass in relation to motion created. I have committed so many arguments that I could not take in any direction as to why Newtonians would follow the untested and meaningless arguments so sheepishly just to follow their Master. There are slight anomalies but one can see there is a persistence of three in relation to the time space holds in all cases.

PLANET	SEMI MAJOR AXIS a (10^{10}m)	PERIOD T (y)	T^2/a^3 (10^{-34} y^2/m^3)
Mercury	5.79	0.241	2.99
Venus	10.8	0.615	3.00
Earth	15.0	1.00	2.96
Mars	22.8	1.88	2.98
Jupiter	77.8	11.9	3.01
Saturn	143	29.5	2.98
Uranus	287	84.0	2.98
Neptune	450	165	2.99
Pluto	590	248	2.99

KEPLER'S LAW OF PERIODS FOR THE SOLAR SYSTEM

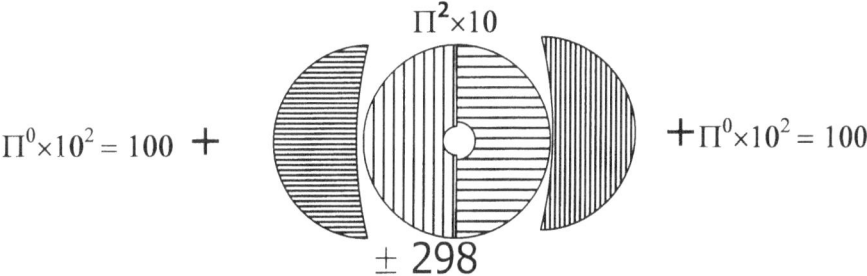

$$\Pi^0 \times 10^2 = 100 \quad + \qquad \qquad + \Pi^0 \times 10^2 = 100$$

$$\Pi^2 \times 10$$

$$\pm\,298$$

Time in spot 1
$\Pi^0 X\ 10^2\ +$
Time in spot 2
$\Pi^2 X\ 10\ +$
Time in spot 3
$\Pi^0 X\ 10^2 = $ 3D 3 Dimensions in total $\pm$ 298 or then two
positions (spot 1 and spot 3) and one allocated in (Time in spot 2 $\Pi^2 X\ 10$) gravity-motion.
Notwithstanding the planet arrangement time relates to space by $k^{-1} = T^2 / a^3$.

Spot 1 Spot 2 Spot 3

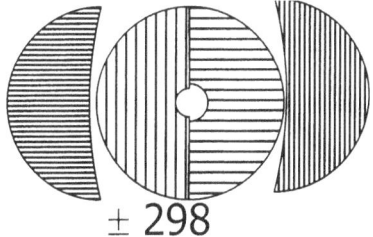

$$\pm\,298$$

From the above table one can clearly see the three positions in time that space holds. By placing time in the relevance to space we also can clearly see that the space is flowing where the relevance is reducing time $k^{-1} = T^2 / a^3$ or $\Pi^0 = T^2 / 3a^3$. The space is flowing towards a dominating centre and that we know is the centre of the Sun .

$k^{-1} = T^2 / a^3$ which then is $3 = T^2 / a^3$
$T^2 = a^3 / k$ which then is $T^2 = a^3 / 3$
$a^3 = T^2 k$ which then is $a^3 = 3\,T^2$
$k = a^3 / T^2$ which then is $k = 3\,a^3 / T^2$

$k^0 = a^3/ T^2$ which then is $k^0 = 3 a^3/ T^2$

There is no mention of mass or any discrepancies in any form or shape or thought. All the planets serve the Sun on equal terms notwithstanding whatever notion humans try to add for what ever disguised pleasure the human may find in such an argument.

Here is a typical Newtonian response to a very formidable question about the masses of the planets don't matter when determining their orbits. What counts on this situation is the equilibrium between the gravitational attraction and the centripetal force. Look at these equations:

Fc = (mv^2)/r   and   Fg = (GMm)/r^2

*Where Fc is the centripetal force, Fg the gravitation, m is planet's mass, M is Sun 's mass, v is the orbital velocity of the planet (which is really not constant, but we can consider it this way) and r is the orbit's radius. G is the gravitation constant. Its value is 6.67E-11 m^3/(s^2 * kg).*

The orbits are stable, otherwise we wouldn't be here to discuss this problem! So, the attractive force (gravity) has to be equal to the centripetal force, which tends to make the planet escape from its orbit.

Fc = Fg

(mv^2)/r = (GMm)/r^2   and with some simple manipulation:
r = (GM)/v^2

As you can see, the radius of the orbit doesn't depend on the planet's mass. It depends only on the Sun 's mass, and inversely on square velocity. What determines the position of the planets in any solar system is the system's initial conditions, which aren't well known... yet! The initial angular momentum, the mass distribution discontinuities in the dust cloud that originated our solar system (or any other), and some other factors, were the conditions that led the planets to be arranged this way. Yes, they were "built" about 5 billion years ago on almost the same orbit they have today!
*Fc is the centripetal force, Fg the gravitation, m is planet's mass, M is Sun 's mass, v is the orbital velocity of the planet (which is really not constant, but we can consider it this way) and r is the orbit's radius. G is the gravitation constant. Its value is 6.67E-11 m^3/(s^2 * kg).*

There is no indication whatsoever in any language ever proposed that mass has any indicative role or influence on any planet in any way. Not the motion. Not the orbit. Not the rotation. Not the velocity. That which is supposed to be is in our minds and not on paper. That is how we wish to interpret cosmology. The larger person is the heavier person holding more momentum. The larger planet is the heavier planet forcing more gravity and is therefore the strongest. That is rubbish. That is the mindless gargle of the brainless. I am sorry but I cannot be less explicit about this matter. It is out of touch with reality it could be part another fairy tale. It seems no one has learned! Go on and calculate which planet is generating most gravity per cubic what ever. Find out in simple terms that the big structures are generating almost nil when compared in direct relation to the smaller planets. It has nothing to do with their being gas and everything to do with the inverse square law.

Again I reiterate ands repeat my question: If the gravitational constant had the means to eliminate the mass factor in the orbiting of the why is there a need to use it in any form of calculations? Why persist with a meaningless proposal.

Jupiter has a MASS of 18955.872 x 10 ^{24} / GRAVITY of 24.89778 = 761.58
 That means every 761.58 parts holding space generates one Nm of gravity

Saturn has a MASS of 5686.76 x 10 ^{24} / GRAVITY 10.556 = 538.7
That means every 538.7 parts holding space generates one Nm of gravity

Neptune has a MASS of 1027.872 x 10 ^{24} / GRAVITY 11.2815 = 91.1
That means every 91.1   parts of space held generates one Nm of gravity

Uranus has a MASS of 866.52 x 10 ^{24} / GRAVITY 8.96634 = 96.64
That means every 96.64 parts holding space generates one Nm of gravity

Earth has a MASS of 59.76 x 10 ^{23} / GRAVITY 9.81 = 6.091
That means every 6.091 parts holding space generates one Nm of gravity

Venus has a MASS of 48.7044 x 10 ^{24} / GRAVITY 8.87805 = 5.485
That means every 538.7 parts holding space generates one Nm of gravity

Mars has a MASS of 6.418224 x 10 ^{24} / GRAVITY 9.81 = 0.654
That means every 0.654 parts holding space generates one Nm of gravity

Mercury has a MASS of 3.3029352 x 10 ^{24} / GRAVITY 3.64932 = 0.905
That means every 0.905 parts holding space generates one Nm of gravity

Pluto has a MASS of 0.11952 x 10 ^{24} / GRAVITY 0.3924 = 0.3045
That means every 0.3045 parts holding space generates one Nm of gravity

PLANET	MASS	GRAVITY	GRAVITY/MASS	%
Mercury	0,05527:1	0,372 :1	6,821	682,1 %
Venus	0,815 :1	0,905 :1	1,11	111 %
Earth	1 :1	1 :1	1	100 %
Mars	0,1074 :1	0,38 :1	3,538	353,8 %
Jupiter	317,8 :1	2,538 :1	0,007986	0,80 %
Saturn	95,16 :1	1.075 :1	0,0113	1,13 %
Uranus	14,5 :1	0,914 :1	0,063	6,306 %
Neptune	17,2 :1	1,14 :1	0,663	6,63 %

Because of the invert square law the existing of such a law puts Newton's theory on mass in doubt. The more the size is the less the force of the mass will be in proportion to space occupied. The bigger is not the more powerful or the strongest because of the invert square law. If there were no invert square decrease of space to power ratio, then yes it would make sense. But the Sun has the least gravity of all in the solar system just because of the size it holds. Can any Newtonian out there please explain what mass has to do with the rotating of planets around the Sun and what do you accomplish by calculating the mass as a factor. That has no practical implication even in the least in relation to the orbit. It implies a massive degree of senseless ness shown by the educated to prove his education to be bizarre and null - in void. It is like blaming the ice-cream sales in Alaska for the persistent drought in the Namib Desert and then invents non-relevant factors to prove a link. If mass was a factor then Jupiter must either propel faster or slower than the rest or Jupiter should be farther or closer or have some indication that places the size relevancy in another category than the others. To say the gravitational constant even things out means to say the gravitational constant is nullifying the mass and then that proves by point, the mass had nothing to do with the price of eggs in the first place. Every one is bullshitting one another by using invalid factors to calculate non-existing quantities in no relevant and banal calculations. The mass does not make a planet go faster or slower or influence the structure in motion. Then why the hell use it in calculations. The mass does not influence the allocated orbit in any way, then why the hell use it. The mass does not pull the planet closer to the Sun so why the hell bring mass in as a factor. Mass does not indicate any difference even in the smallest indication between cosmic structures so why the hell use it. It is used to make the mathematician feel worthy and superior even if it also shows his absolute mindless ability in rational arguing. If a factor has no influence on whet is proposed what their calculations wish to measure except to prioritise the ego of the mathematician then using it in calculations is outrageous. Using mass has the same implications in the calculation than it has to bring in India's net pepper production and substitute that with any factor so why not use that as a factor in the calculations. This is how the layout of planet orbits are in reality.

Mercury	Venus	Earth	Mars	Jupiter	Saturn	Uranus	Neptune	Pluto
0.055	0.86	1.0	0.11	318	95	14.5	17.2	0.002

This proves distinctly that there is no proof that big and small has any consideration in the cosmos but the idea of differentiating between big and small is a human concept that holds no cosmos merit.

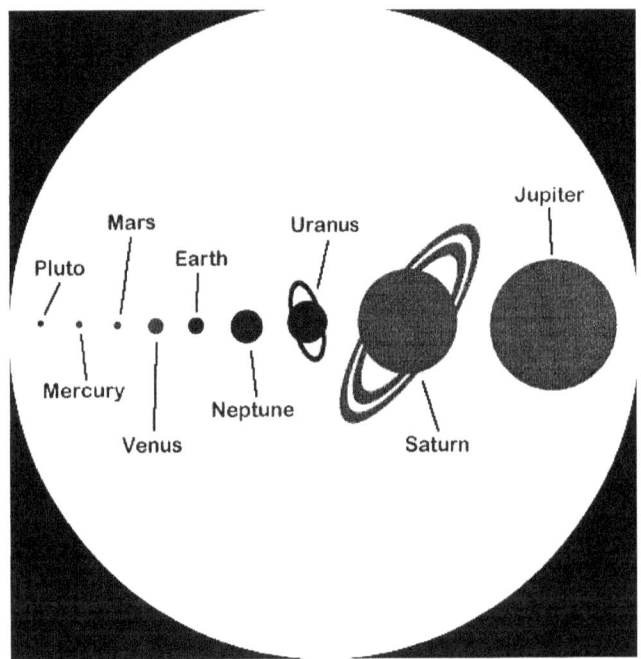

The big question to ask is why would mass not show as a mathematical indicator when orbit velocities of different planets are concerned. Why would time show a flow that turns to be negative in relation to space when time is put in division of space? These are the mathematical answers that hold truth and not the criminally motivated mismanagement of mathematical principles and mathematical rules that Newton advocated. If mass does not show as a factor one cannot go and grant mass a premium but one has to surge for a reason why mass does not show it presence as a factor. If mass had any influence on the orbit as the calculation formulas would suggest then the orbit arrangement should be as the illustration suggest. If mass had anything to do with gravity we would find either the largest in the inner orbit since it has the largest mass that will erode the radius the fastest or we will find the least in the inner orbit because the least mass will produce the least motion. Then the planet orbit layout would be as follows:

Jupiter has a MASS of 18955.872 x 10 ^{24}

Saturn has a MASS of 5686.76 x 10 ^{24}

Neptune has a MASS of 1027.872 x 10 ^{24}

Uranus has a MASS of 866.52 x 10 ^{24}

Earth has a MASS of 59.76 x 10 ^{24}

Venus has a MASS of 48.7044 x 10 ^{24}

Mars has a MASS of 6.418224 x 10 ^{24}

Mercury has a MASS of 3.3029352 x 10 ^{24}

Pluto has a MASS of 0.11952 x 10 ^{24} /

This is the manifestation of the fairytale where it shows the naked reality when the sublime goes mad in the practical implication. It is equal to the fairytale that tells about the King that paraded around naked because only the stupid would realise he was without clothes and since only the mindless would not be able to see his magic clothes. Then he including his subjects pretended to marvel at the beauty because they would rather bullshit their minds than to admit that they were to stupid to see the clothes the King was wearing. So everyone pretended not to be stupid, who showed exactly how stupid he or she were. In this case it was the intellectually Superior that was privileged to be stupid just to show the rest who was the most stupid of them all and never to ask what the hell mass has to do with the whole affair. So every one who thought he was superior showed how big an idiotic fool he would be in order to prove the point. That made him not more a fool than the fool he or she was trying to impress and both idiots impressed other idiots by accepting that mass plays a major part while mass has no reason to be part of any of the facts relevant. In this case it is the using of mass that everyone echoed because by not realising the use of mass would show the any individual that needed to become part of the Brainy Bunch how mindless he or she was not to see the advantages of the use of mass. So the mindless then accepted there is some lunacy. That is the fairytale but in reality the dead is criminal. If I go about and suggest facts, which is untrue and while I know them to be untrue I still support the facts and support the system declaring the facts as to be true, the system that includes me is a fraud. No wonder the whole lot is devious and mindless atheists. A question never asked and a thought never put to words is that if there is a Gravitational constant guiding all cosmic matter by the force of gravity where is it going and where is the centre where it is taking all the material.

That statement is criminal with dubious intent of misleading the public and is a conspiracy outweighing crime syndicates such as drug lords and the mafia.

Again from the lips of the Newtonian into your face and then you tell me who is the criminal misconduct. If any person goes into a contract with such devious preconditioned misleading an reliable evidence with the intent to mislead and defraud by giving unreliable and untrue facts deliberately while full well knowing what is suggested has no truth to bear, you are a swindler, a cheat, a criminal and you belong in a safe place away from society where you and your malice may not defraud others any more. If you do not like my saying this prove me wrong. Prove how your swindling behaviour to cover Newtonian defrauding is not criminally intended!

This might seem harsh criticizing but it is reality. Producing fictitious facts in order to mislead and present untruths is criminal. Any academic wherever can either sue me for wrongful slander but doing so that academic must prove without doubt how mass plays any part in initiating or contribute to motion of whatever major or minor influence on whatever scale. It is criminal to go around and falsify information in order to protect a corrupt complot and counterfeit facts with the intent to deceive and spread deception. I say astro physics is a hoax from the start up to the present and up to now the academics scandalously avoided me by ignoring me in order not to face up to their deceit. I challenge who ever to charge me with slander and prove me wrong. Prove in what way does mass bring about any implication to the orbits of any cosmic structure. No individual has the right to claim the privelidge to atribute any factor where there is no clear evidence that such a factor has a validity in its presence. One cannot tell Jupiter it has mass of what ever measure we humans dedicate to Jupiter justr because we see a flow of more light coming through a lense that Pluto show bwcause Pluto show less light relflecitng. If theey all orbit equal they are all alike, and it is a human duty not to tell the cosmos what Newton says the cosmos should be but to take guidence from Kepler and allow the cosmos to inform us about matters that prevail in the cosmos.

	Distance (AU)	Radius (Earth's)	Mass (Earth's)	Rotation (Earth's)	# Moons	Orbital Inclination	Orbital Eccentricity	Obliquity	Density (g/cm^3)
Sun	0	109	332,800	25-36*	9	--	--	--	1.410
Mercury	0.39	0.38	0.05	58.8	0	7	0.2056	0.1°	5.43
Venus	0.72	0.95	0.89	244	0	3.394	0.0068	177.4°	5.25
Earth	1.0	1.00	1.00	1.00	1	0.000	0.0167	23.45°	5.52
Mars	1.5	0.53	0.11	1.029	2	1.850	0.0934	25.19°	3.95
Jupiter	5.2	11	318	0.411	16	1.308	0.0483	3.12°	1.33
Saturn	9.5	9	95	0.428	18	2.488	0.0560	26.73°	0.69
Uranus	19.2	4	17	0.748	15	0.774	0.0461	97.86°	1.29
Neptune	30.1	4	17	0.802	8	1.774	0.0097	29.56°	1.64
Pluto	39.5	0.18	0.002	0.267	1	17.15	0.2482	119.6°	2.03

Mass has precious little influence on gravity because gravity is the motion, whereby there is interaction between time serving as a liquid flowing by contraction to the centre of the Sun and the planet by duplication is part of

the flow. Since there is a flow and duplication set in balance the duplication a^3 is in the range of the motion T^2 while the flow of three is in the straight-line k that serves time.

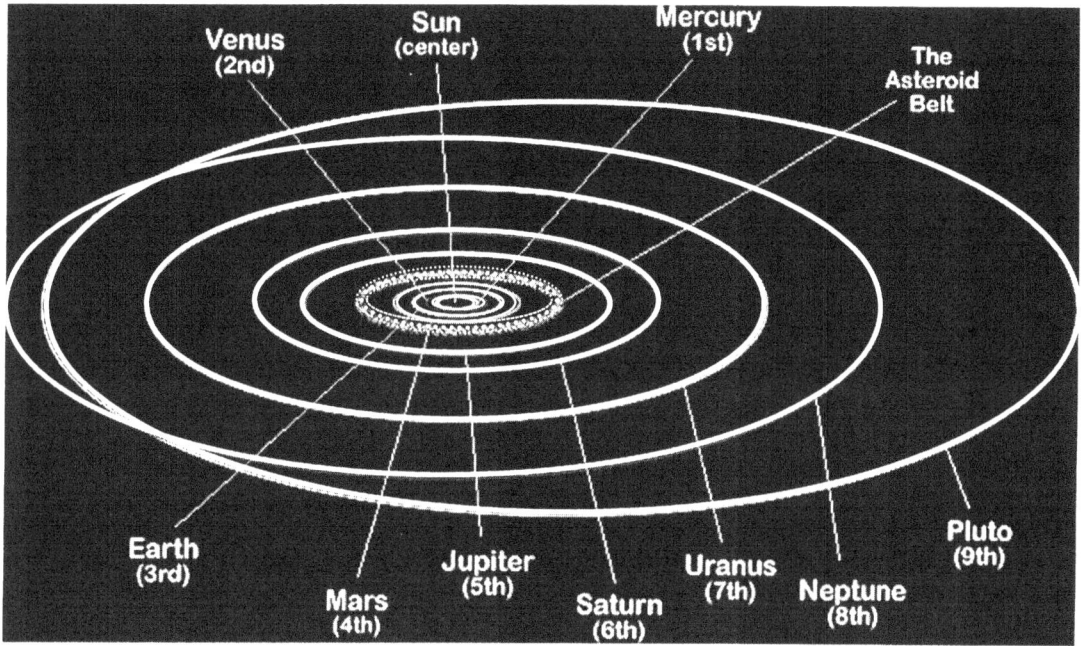

Just as improbable and therefore impropriate is the manner that science suggests that "by the magic of gravity" stars accumulate dust to form stars. The suggestion alone that dust by the force of gravity can accumulate into solid matter is fraud and then to wilfully with deceit intoned to use tax money to investigate such dubious ludicrous nonsense is over the top. It is fraudulent to use tax money on something that is obviously a hoax. If any banker or politician did the same criminal offence he or she would have been branded as a swindler and paid a penalty in a penal institution for a very long time.

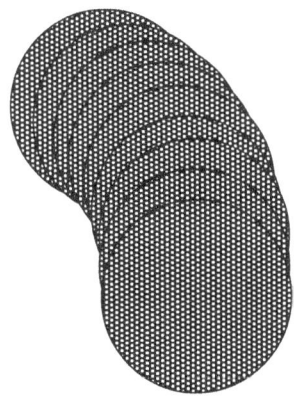

Let us for once and for all accept that Newton's mass activated gravity is inspired by his imagination. The process is when liquid in the form of outer space lines up against a solid such as a palate or the Earth. There is motion in the one department that acts as if it is a solid but is in fact the partner that holds the motion. Then there is the liquid, which by being the stationary acting the part of the solid but is a liquid all the same.

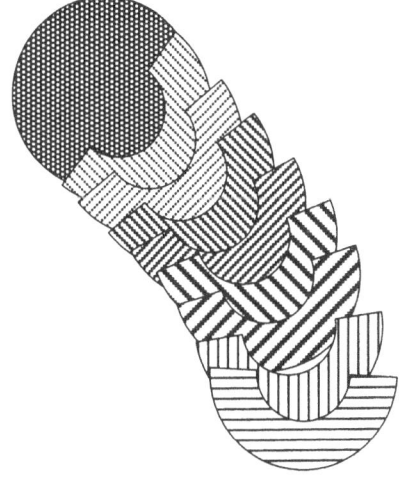

The body delivering the motion is a solid that forms a unit as space. The part that serves as a cosmic liquid is the partner that is also stationary and serves as an immobile liquid. This has things rather confused in the manner that gravity in the cosmos operates. Remember the planets is not a normal set up and even stars with micro stars to attend to is holy unnatural. It is very seldom in combination and when it is things get as confusing as we find it to be on Earth. The norm in the cosmos is a lone star that spins on an axis while in motion around a galactica. The galactica presents the same layout as a star and the working process in the galactica that apply in a similar mode, as stars with layers would have.

The body delivering the motion is a solid that forms a unit as space. The part that serves as a cosmic liquid is the partner that is also stationary and serves as an immobile liquid. This has things rather confused in the manner that gravity in the cosmos operates. Remember the planets is not a normal set up and even stars with micro stars to attend to is holy unnatural. It is very seldom in combination and when it is things get as confusing as we find it to be on Earth. The norm in the cosmos is a lone star that spins on an axis while in motion around a galactica. The galactica presents the same layout as a star and the working process in the galactica that apply in a similar mode, as stars with layers would have.

Do not look for the pumping going on where one can see time that meets space.

The pumping action is going on where the proton pumps time into singularity by expanding and then contracting in the very heart of the atom nucleus. There the duplication present the expanding and the contracting which feeds the star with the motion either in duplication or in contraction that the star requires to comply with the demand space-time insist on as gravity. The reducing of heat by motion is presented as cooling since motion reduces space and by reducing space it is cooling. To establish that rapid cold the proton moves 1836 times faster in order to restrain the heat from the

value the heat had when the the electron relevance. At the heat was already at the speed therefore the atom removes all freezing the heat into the every atom is a black hole. All as a general reducing of space governing singularity in charge. in the star is a pump that to cold and transfers singularity singularity 1^0 to regain what during   moment-Alfa.   The the star is not nearly the gravity the planets.

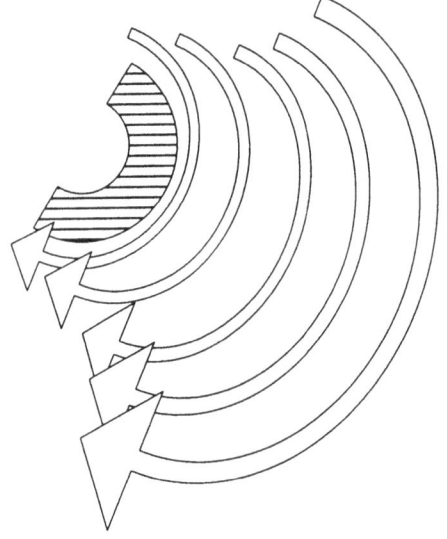

heat was at electron the of light and heat   by oblivious this adds up by   the Every atom coverts heat 1^1 to was   lost gravity   in going about

In the case of the planets there motion that puts liquids in

is an orbit relation to

solids without the much pumping being the dominant factor. The liquid allows the solids space within to move. As the solid pushes against the liquid the liquid bears down on the solid and some liquid give way but the inner liquid increases the density at the point and just above where it touches the solid moving structure. The liquid pushes down the solid while in accordance with the Coanda principle the space expands to a point directly relevant

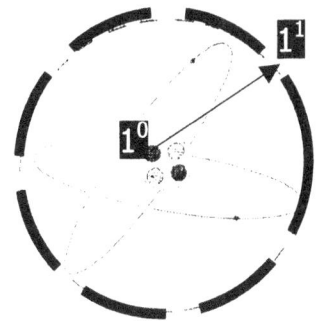

to the motion that the solid provides. That which is without motion is secured by the liquid to be part of the Earth while that which is liquid is secured onto the Earth as an extension of space.

There is an allocated line designated by the extending of the solid that includes the liquid to gather that liquid into the unit forming the solid. We gave that so many names ending with sphere even the thought of all these sphere makes ones head spin. How Newtonians fit the sphere as in stratosphere and atmosphere and what not into gravity is still a puzzle, which is eluding me in the manner that Newton's vision on mass was eluding me. In the end of all this there is a line that is the friction point and it is at that line where liquid tear from solid while the solid is actually intensified liquid.

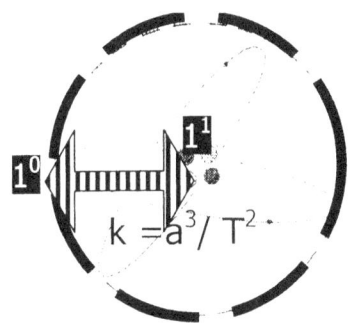

In the case of comets the Sun is the solid that forms stability while the comet is the solidity that moves and outer space is the liquid that does not move. In the case of comets the cosmic law is transgressed. The Sun is an atom. The Sun consists of a unit forming an atom where the Sun is the atom in compiled group but also where the group serves the unit. Every layer in the Sun is a liquid to the top where the bottom serves as a solid to the top layer forming the liquid. The proton puts time at motion where time puts space in demise. Time devour space as eternity meets infinity. The atom is Black Hole with matter in between infinity and eternity and this fills the black Hole with substance that is forming space - time. The final conclusion that any star can arrive at is when it takes

The proton serves as 1^0 to the neutron being 1^1 where the neutron serves as 10 to the electron being 1^1. The atom forms a Universe that hold both eternity and infinity apart by allowing motion to separate time. The atom concludes the Universe because the atom is what concludes the Universe as much as it started the Universe. In the end all star will be one atom in the hydrogen atom but that sis the final conclusion where the last era arrives. The atom is the Universe.

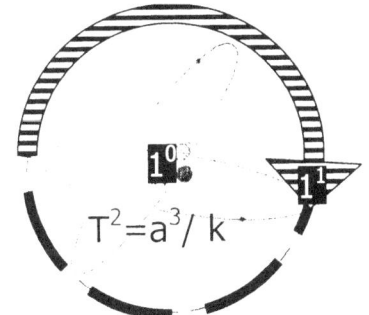

The atom maintains relevancies where the core within the atom serves as 1^0 and the orbit serves as 1^1. The core is the solid and the electron is the liquid. The electron provides the motion because in relevancy at the point where the electron is located it is the electron that is in motion while the core within the atom is a solid that does not move. The atom serves as movement because singularity generated by all atoms forming the unit provides the motion. All atoms forming the star are allocated the value of motion being 1^1 while singularity charged with governing the star is 1^0.

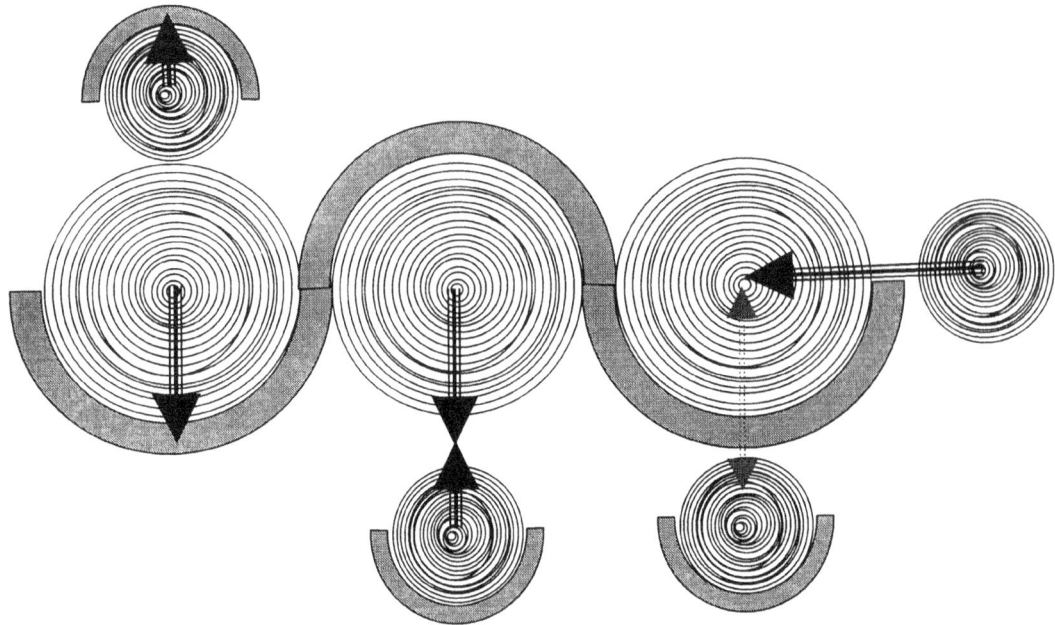

However the only constant in the Universe is that there is no constant applying. Everything is in cyclic shifting as the relevance relocate and alternate positions. In order to get a flow of space - time 1^0 and 1^1 must be forever alternating. The fact of constants are that constants are as Newtonian as mass can ever be and constants are as much a fact that does not apply as mass where then mass has the same position. The planet forms an electron to the Sun becoming the solid and the Sun allow the planet to spin while the planet receives it alternating which forms motion from the Sun that provide the governing singularity not only to the Sun but also everything orbiting the Sun as an electron Because the planet is just an electron the planet will rotate about the Sun as any good electron would do.

When the planet is on one side of the Universe where the centre of the Sun forms the Universe the planet resist the flow of time. When the planet is on the other side of the Universe the planet. The Sun is 10 but the planet alternate 11 and 11 because the planet land on one side of the Universe and then on the other side of the Universe.

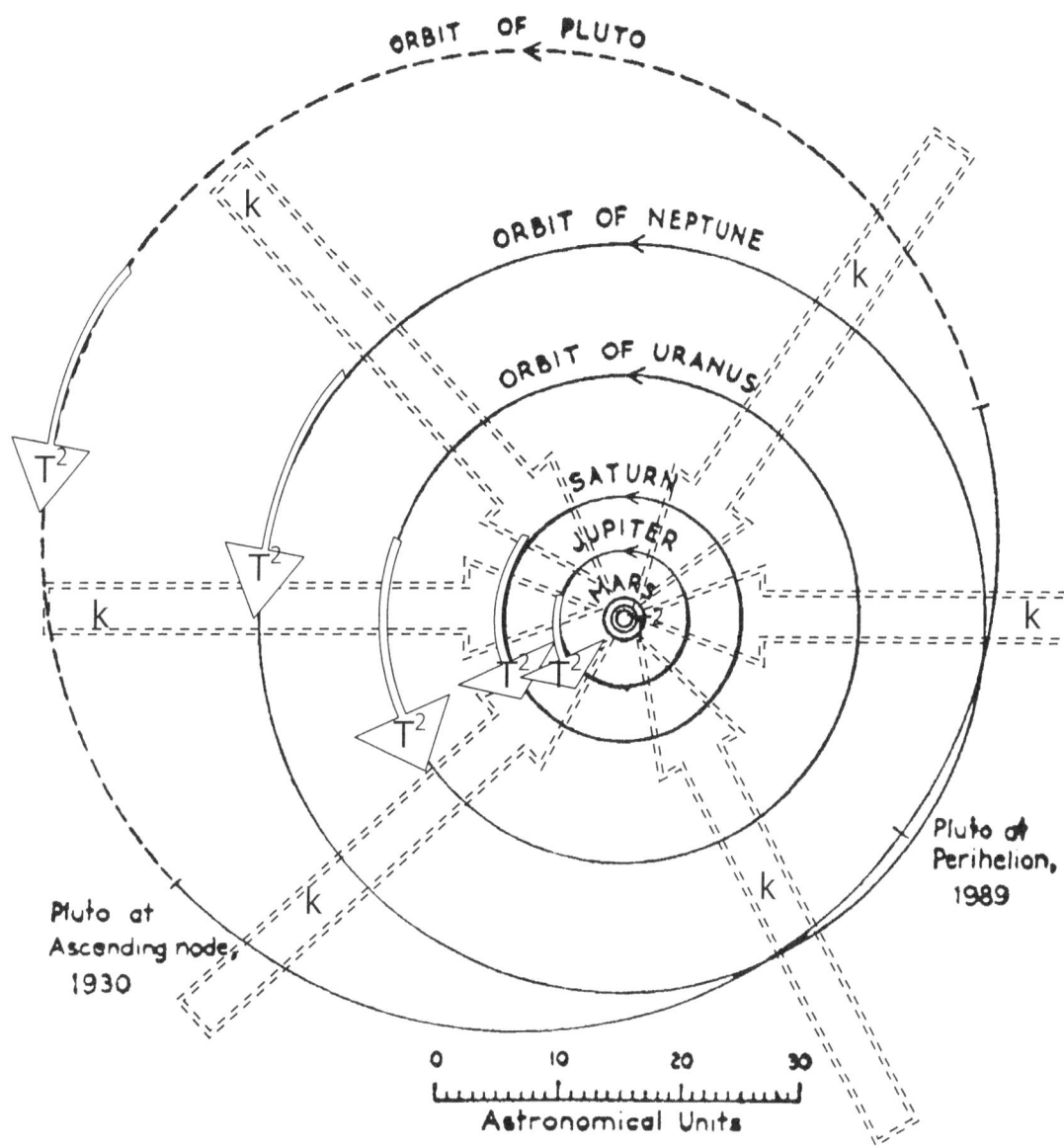

Kepler's formula insists on the flow of space-time. Gravity us never mentioned because gravity is a name coupled to an idea. If the idea is incorrect the coupling is unjustified. The space is floating in the moving time. $a^3 = T^2k$.

The Sun is contracting time towards the centre and by doing that it is implicating the process we gave the name of the Coanda effect. The space is duplicating the space it holds in order to flow in the time being contracted. By duplicating the space the space is generating what it was to where it is to the location it is going to be. That is space-time and that is gravity and that is motion. Restricting this process leads to mass applying as a restraining because no object can share space just as much as no object can be in two places at one time.

There is a flow of space-time running towards the Sun at a duplication tempo where three portions of space are in relation to one unit of time. Material stand related to space by measure of the past position, the present position and the future position and that the cosmos told Kepler by a language of mathematical equations. It is not my say so. It is not the say so of Kepler. It is not the hearsay of

Newton. It is not the interpretation of the visions of Hubble or the guesswork of Hawkins or the calculations of Einstein's perceiving. This comes from the horse's mouth. This has nothing lost in translation. The cosmos told this to Kepler without ceremony or private preferences colouring and tainting preconditioned favouring of prejudice. That the cosmos spoke in a language all can appreciate because no dialect my sound some incorrect abbreviations...and Newton still managed to bane all the information to impress with preconceived ideas.

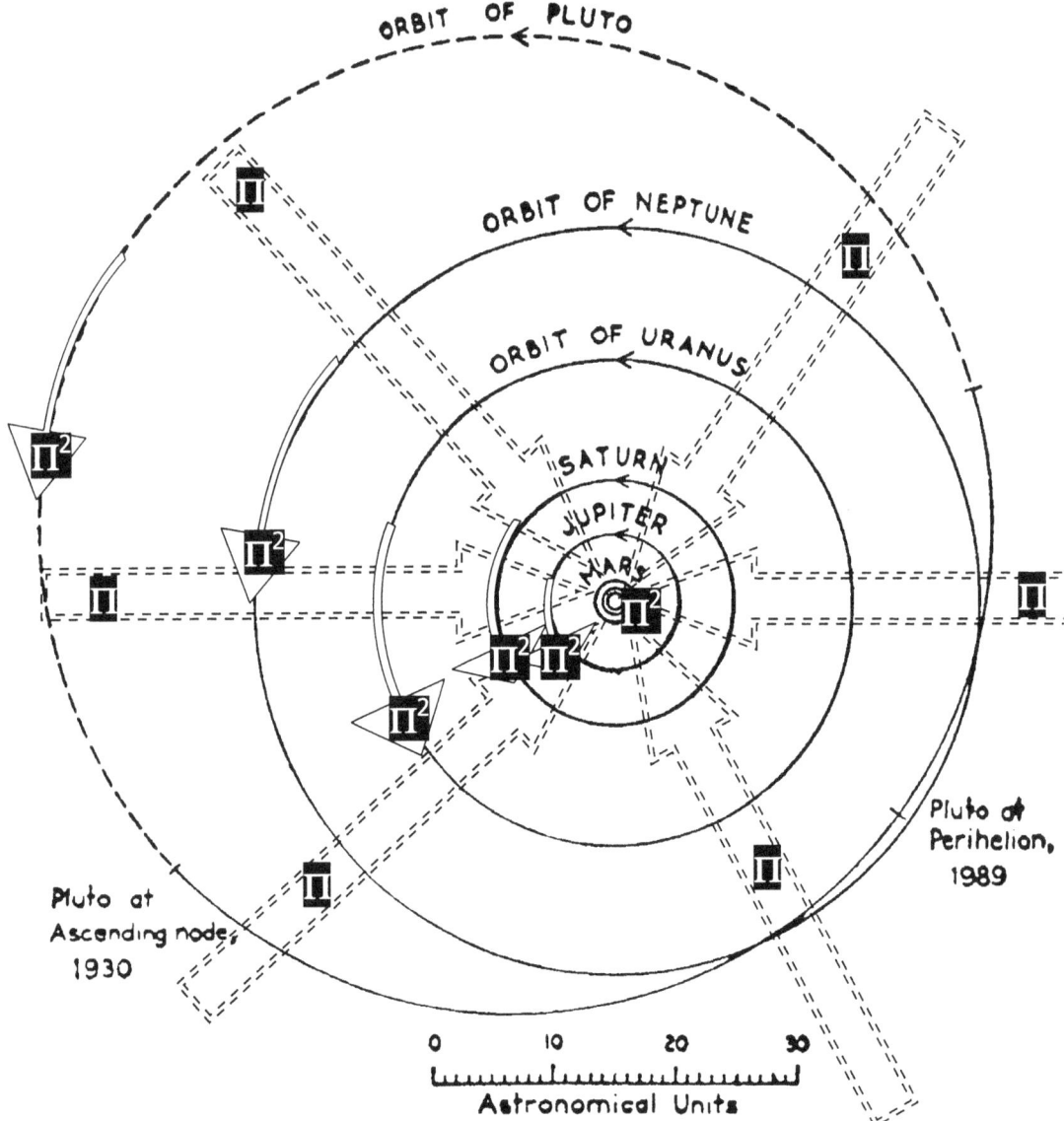

It is the atom that is in charge of the Universe because it is the atom that charges motion. The rotation of the object produces the proton duplication while the rotation around the contracting centre is gravity or motion unblemished and the time flowing towards the centre places the orbiting structure in the 3 dimensional space in time $(10)^2$ square. In the final analysis it still is the atom that produces the atom, which serves as a star. It is the proton at $(\Pi^2+\Pi^2)(\Pi^2\Pi)3 = 1836$ that forms the Universe in more ways than any human can appreciate.

It applies simply because in time there is a ratio whereby space duplicate in expanding as well as contracting and this ratio is serving what ever the cosmos might be. The cosmos said it is $a^3 = T^2 k$ therefore it is little surprising that Newton 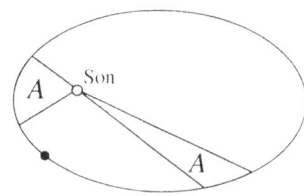 did detect a ratio between space-time and time flowing about space. It is more surprising that he missed the rest. Much more surprising is that every one of his dedicated followers and mathematical geniuses missed the rest. Then I better put my mouth where my proof should be and explain what the rest is that all missed.

The cosmos is not expanding because it is shrinking. When it started that which cannot be bigger parted from that which cannot be smaller. It started where infinity was eternity. Then infinity parted from eternity leaving eternity bigger than infinity but because of the size of infinity, eternity was not much smaller than eternity. The star is the atom within the star by the multiplied motion of all the atoms forming a unity that charges the motion in the star. The star is a culmination of the efforts of the star.

Eternity • Unified with infinity
When the first moment came the two factors being eternity and infinity was the same. Then they parted company putting a reference between them and not much more that just a reference of division

The ratio is increasing because it is decreasing the Universe. Let's put it this way:

Eternity •• parting with infinity   → the Universe cannot expand because that what was at first is eternally big with no possible end and in that is the reason why there is not possible expanding of the Universe.

The Hubble constant is the measure whereby the Universe is reducing since there is no possible room for any expanding. The part being infinity is the part that cannot reduce since it is as small is infinity can ever be smaller. It is so small it has no sides but all points share on spot. Any further reducing will bring about an increase in size and by reducing further it is increasing what never can further reduce. With infinity being there without having a possibility to reduce it is increasing what cannot reduce and by increasing that which cannot reduce it is reducing the part that cannot increase. While neither of the two is capable of changing in the direction they represent because they represent the entirety there is to represent, the increasing of that which cannot reduce is reducing that which cannot increase because without ever changing the two are growing apart. By eternity never changing while growing apart from infinity that aspect too never can

growing • 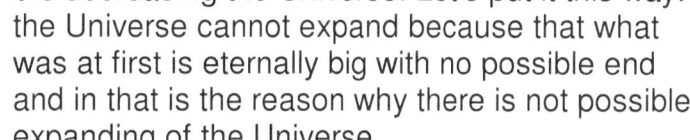 **Eternity away from infinity**   change, and while never changing that part that cannot increase is increasing the part that cannot reduce while the part that cannot reduce is reducing the part that is incapable of increasing. Those, the ones that cannot change is doing what it can do best to the other part that cannot change and in changing its relation with the other part it is remaining the same while the other side in reference then changes the reference.

What Hubble saw was a Universe that was shrinking away into the oblivious because that part that cannot reduce is reducing the part that cannot increase and since the part that cannot increase holds eternity, it can shrink the part that it has in the smallest side by shrinking its reference to that into the oblivious.

It all ends with the Black Hole (not quite but to go into detail about that requires another half a book of explaining and proving as the Black Hole has two more steps to involve mathematically). Let's put the Black Hole as the biggest there is while correcting this error at the same time. The Black Hole is so huge it fits a Universe inside but that means when the Black Hole was huge it was so big it parted eternity from infinity. That came when eternity was just bigger than infinity because the two then parted their shared unity no that long ago. At that point when the Black Hole was the liquid star sloshing away and shining as bright as it could the Universe was separating eternally big from infinitely small by a margin of Π. Eternity parted 1^0 from 1^1 by a margin of Π^0 that increased to Π by becoming Π^2. The increase came as Π move to Π forming Π^2 that came to a total of Π^3. Still it was at a time when infinity was just smaller than eternity and eternity was just bigger than infinity.

The difference at that at the start when 1^0 was going onto 1^1 which was going onto Π^0 that was forming $7\Pi^0$ and was combining time as $(10 + 10 + 1.9991 = 21.99991 / 7\Pi^0 = \Pi)$ this was forming Π^3 but there was little else to show on the eternity side as well as the infinity side and little split the difference. That was just about the environment the Black Hole encountered (okay there was more...but not that much more) and in that the Black Hole was what split infinity and eternity at the time. This we may deduce on the grounds of the facts we now see the Black Hole represents. At present the Black Hole is so small it can reduce the entirety of eternity into infinity while it remains so large that it can absorb the entire eternity into infinity without needing any matter to produce a time delay. It still continues to have the power to carry on what it started with. It started a process putting a bridge between eternity and infinity and counteracted when it shrunk eternity into infinity while it is expanding infinity onto eternity. It took this job when the Universe was wasting away.

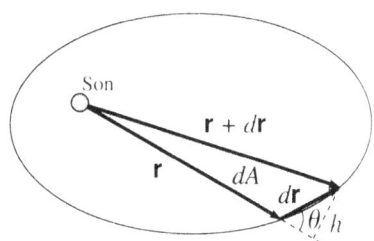

The wasting involved that when what cannot expand reduced and as it reduced it pushed apart that which cannot separate by decreasing that which cannot decrease as that which cannot expand did expand. The split was about that which cannot decrease to part from farther from that which

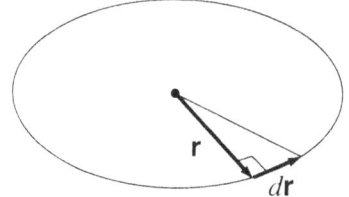

cannot increase while the difference brought about an increasing of the reducing Universe. After the long storey about mass and a centre point and a shared centre point with one point favouring both, it boils down to see how material move. Elsewhere I explain what material is. Material is seven point that has no space but is generating space claimed by that which pretends to heat and the retarded heat circle compacted as matter around a centre forming a sphere as small one are not able to imagine. It has no name because no name giving fame seeking Newtonian can get to it. Only the heat in retarding spin is a part of the Universe but that substantiating and controlling the heat confirms the allocation of the heat by swerving singularity Π^0 which is maintaining $7\Pi^0$ to become the smallest sphere there may be.

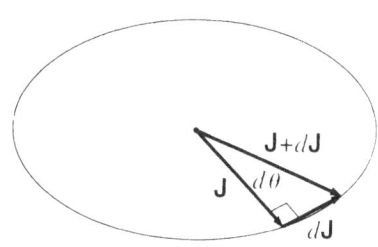

In the centre of the smallest matter runs a line that actually is just an expanded point and the retarded heat spins around this point where the rotating is holds matter as matter with the rotating. It is the fact of rotating that is producing the retarding of time and the more of this 1^0 to $7\Pi^0$ there is the more retarding of time going backwards there is. That is a part of the "nothing" Newtonians put into outer space as outer space. That line is where all possible points serving the line lands on one point that line is a dot with a spot in the centre that has no sides and all point in this line falls on the very same point.

The value of this point is 1^0. Because this line focuses all the possible sides on one point and this spot including all the sides that fit into it then still has no sides and fits still with the adding fits all into one place while it serves the rotating centre of retarded heat that finally combine as the atom, one finds that all the lines serving all the retarded heat fits into the next spot that form a centre line that also has no sides in one spot. Eventually the lot forms a combining line that includes all the material within the atom. This eventually then finally combines as $(\Pi^2+\Pi^2)(\Pi^2\Pi)3 = 1836$ and we find it serving our Universe in the capacity as an atom.

The atom is the atom because all the lines that centre the rotation of all the retarded heat combine in one line forming a centre to the atom. This is possible because all the lines has no sides just like the atomic governing singularity centre line also still has no sides. From there all the atomic governing centre lines fit into a centre line that can hold all the atomic lines because it holds all the atomic lines on one spot as one spot. The line finally forming as the governing singularity driving the motion of the star which is at that point representing all the retarded heat centres which project to the governing centre singularity of the star.

All the positions in eternity relate to every position in infinity and the line holding infinity is the result of the accumulation of all the points serving infinity in the retarded heat. All points that hold 1^0 projects to one point holding 1^0 because all points have no sides therefore all points fit into one point that form a centre line. All the possible lines by all the possible atoms is projected to one centre line since that one line holds all the possible points there can ever be, on the only one point in a point there possibly can be. This of course also applies in the case of and well as also to the orbiting satellite. Since that point cannot be smaller it can all fit into as well as fit all into that one point that cannot ever be smaller. Since all the points are therefore exactly equal to the extent they all are being the same point, all points everywhere are then the same point concentrated into one point while spread out as far and wide as the point may reach.

The point is singularity referring to singularity. Mathematically this point is expressed as 1^0, which by all mathematical rules are equal to 1^1, which is equal to $\Pi^0 = 1$. In that number 1^0 the entire Universe units as one unit that incidentally also never can be because everything fits into one spot that is not. To infinity eternity is 1^1 on the condition that infinity then is 1^0 and just the reverse is applies in the relation that eternity holds infinity because the reversing holds infinity at 1^1

as long as eternity can be 1^0. Since both are 1^0 while holding the other as 1^1, the roles inherently have to change when crossing over to the other side of the Universe. The one side of the Universe will gauge infinity as 1^0 while eternity is 1^1 and at the same moment on the other side of the Universe eternity will be gauged as 1^0 while viewing infinity as 1^1. The end to the Universe is not near as the start is far away (eternally further than the idiotic 13.5×10^9 years Newtonians give the Universe. The end will arrive when the difference between infinity and eternity will be so large infinity will again join eternity while eternity at that point will be so overburden extreme that the incorporation will go unnoticed by all that it happened. The planet orbit the Sun since time placed material at that location and allocated a time delay as to the position the Sun holds and the Planet holds. This same argument also serves the same way as it applies in the Sun and the orbiting satellite holding their relation secure.

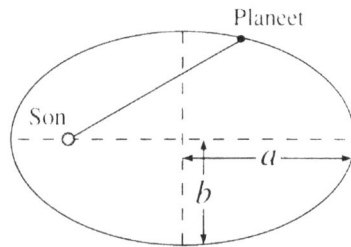

In regard to all these mentioned facts we can deduct that all the matter that the satellite holds one line charged that has no sides. Since this line has no side and takes up no space it fits into the governing singularity that has no sides and all possible lines of all possible matter fit into that into that one charged line that charges the next line where all the lines also fit into. This charging and combining of lines that has no space claims and therefore holds all points on one allocated position eventually form the atom that form the layer that form the star that form the motion in duplication and conserving the duplication by contraction.

Since that one line is exactly in equality to the next line running to the compiling centre line in the Sun, the centre line in the Sun therefore is also the centre line in the slightest piece of independent time delayed matter of the satellite and the satellite then becomes 1^1 to the Sun being 1^0. From the particle there is a reverse reality that the Sun forms 1^1 to the satellite particle line that holds 1^0, and all this represents eternity in reality departing from infinity forming reality and to the one, the other is 1^1 while that one holding the reference is 1^0.

To the Sun it is the satellite that stands between eternity 1^0 and infinity 1^1 while to the satellite it is the Sun parting infinity 1^1 from eternity 1^0 where eternity drives it in motion. This is the result because the same eternity drives both the Sun and the satellite as individual location where the other to the point that one has becomes part of the cause why the eternity parted from infinity. While the one is playing a blaming game on the other by taking president in the relation since it holds eternity and the other forms a factor of infinity, the one holds 1^0 to the position that one then allocates to the other as 1^1. In all the talking and all the explaining matter is in reality not even reality because matter is three points that lagged in time behind three points that in time serves as time to follow four points spinning in a centre where the four points all share the same spot on the fifth centre spot.

When the spot Π^0 became functional and established all relevancies possible, heat parted from cold as eternity parted from infinity. The expansion was not clear motion but more a parting of relevancies where a centre formed a relevancy because the centre could not provide motion. Without being capable of motion, the centre established four points, which also served singularity. From the inverse square law we know that the centre doubled by producing the four points holding singularity.

When the cosmos came to motion, motion was not yet defined. When the cosmos brought about motion, the first motion was relevancies. Cold parted from hot. Eternity parted from infinity. Motion parted from motion absence. Infinity broke the laboriousness of eternity for the duration of infinity. The spot became and grew into the dot. From what the spot was to what the dot now is might be just a mathematical implication of going from 1^0 to 1^1 but in reality

The Spot becoming the Dot

When space brought division between eternity and infinity

that first motion was the creating of and establishing of an entire Universe with all possibilities now in it. Never again can that much growth become a reality, although to us the growth is beyond what we ever can notice. But it is because the growth is so massive and we are so small that we are unable to notice such almighty growth.

If it is that simple then why is it complicated.

BEST WISHES,

PETRUS. (PEET) S. J. SCHUTTE

www.ingramcontent.com/pod-product-compliance
Lightning Source LLC
Chambersburg PA
CBHW080616190526
45169CB00009B/3199